TEXTILE COMPOSITES AND INFLATABLE STRUCTURES

Computational Methods in Applied Sciences

Volume 3

Series Editor

E. Oñate

Textile Composites and Inflatable Structures

Edited by

EUGENIO OÑATE

Centro Internacional de Métodos Numéricos en Ingeniería (CIMNE),
Universidad Politécnica de Cataluña, Barcelona, Spain

and

BERN KRÖPLIN

Institut für Statik und Dynamik der Luft- und Raumfahrtkonstruktionen,
Stuttgart, Germany

Springer

A C.I.P. Catalogue record for this book is available from the Library of Congress.

ISBN-10 1-4020-3316-8 (HB) Springer Dordrecht, Berlin, Heidelberg, New York
ISBN-10 1-4020-3317-6 (e-book) Springer Dordrecht, Berlin, Heidelberg, New York
ISBN-13 978-1-4020-3316-2 (HB) Springer Dordrecht, Berlin, Heidelberg, New York
ISBN-13 978-1-4020-3317-9 (e-book) Springer Dordrecht, Berlin, Heidelberg, New York

---

Published by Springer,
P.O. Box 17, 3300 AA Dordrecht, The Netherlands.

Printed on acid-free paper

Table of Contents

PREFACE

The objective of this book is to collect state-of-the-art research and technology for design, analysis, construction and maintenance of textile and inflatable structures.

Textile composites and inflatable structures have become increasingly popular for a variety of applications in - among many other fields - civil engineering, architecture and aerospace engineering. Typical examples include membrane roofs and covers, sails, inflatable buildings and pavilions, airships, inflatable furniture, airspace structures etc.

The ability to provide numerical simulations for increasingly complex membrane and inflatable structures is advancing rapidly due to both remarkable strides in computer hardware development and the improved maturity of computational procedures for nonlinear structural systems. Significant progress has been made in the formulation of finite elements methods for static and dynamic problems, complex constitutive material behaviour, coupled aero-elastic analysis etc.

The book contains 18 invited contributions written by distinguished authors who participated in the International Conference on Textile Computer and Inflated Structures held in Barcelona from June 30th to July 2nd 2003. The meeting was one of the Thematic Conferences of the European Community on Computational Methods in Applied Sciences (ECCOMAS, www.eccomas.org).

The different chapters discuss recent progress and future research directions in new textile composites for applications in membrane and inflatable structures. Approximately half of the book focuses in describing innovative numerical methods for structural analysis, such as new non linear membrane and shell finite elements. The rest of the chapters present advances in design, construction and maintenance procedures.

The content of the different chapters was sent directly by the authors and the editors cannot accept responsibility for any inaccuracies, comments and opinions contained in the text.

The editors would also like to take this opportunity to thank all authors for submitting their contributions.

Eugenio Oñate
Universitat Politècnica de Catalunya
Barcelona, Spain

Bernard Kröplin
University of Stuttgart
Stuttgart, Germany

On the Design Process of Tensile Structures

Rosemarie Wagner[1]

Fachhochschule München
Fachbereich Architektur
Karlstrasse 6, D-80333 München, Germany
R.Wagner@fhm.edu
Web page: http://www.lrz-muenchen.de/~architektur

Summary. *The influence of the development of computer programmes and automatic generation of cable nets and membrane structures will be shown in some examples. The main interest is laying on new evaluation methods of cable nets and membrane structure and the design process of membrane structures, integrating the material behaviour of coated fabric.*

Key words: Design process, cable nets, membrane structures, inflated structures

1 Introduction

The design process of pretensioned structures such as cable nets and membrane structures is influenced by the development of computational methods. While the first methods of form finding had been physical modelling with fabric, wire nets or soap films, today several numerical methods of form finding are developed based on the force density method [1,2], the principle of minimal surfaces [3,4] using dynamic relaxation [5,6] or other approaches in fulfilling the three-dimensional equilibrium. Further process has been carried out in the form finding with an anisotropic stress distribution [7]. All methods have in common that no material laws are necessary finding an equilibrium of the three dimensional shape for given stress distributions, boundary conditions and supports. These shapes of equilibrium should ensure in the built structure a homogeneous distribution of the tension stresses. In reality the material behaviour, process of cutting patterns, manufacturing and pretensioning on site influencing the stress distribution, wrinkles and regions of over stress are obvious, can be seen and measured.

The design process of pretensioned structures needed to be extended taking into account evaluation methods for shapes of equilibrium in relation the material behaviour and process of prestensioning. More realistic modelling of membrane structures is necessary including the strips in width, orientation of the fabric and seams for analysing the load charring behaviour. The process of reassembling flatten strips had already been proposed for a rotational symmetric hat type tent [8]. The process of form finding can be embedded in a design process including cutting pattern und structural behaviour under external loads. The load bearing behaviour

E. Oñate and B. Kröplin (eds.), Textile Composites and Inflatable Structures, 1–16.

can be evaluated by redundancy, flexibility or a stiffness value in relation to the curvature and the elastic strain of the materials.

2 State of the Art

The design process of tension structures such as double curved cable nets or membrane structures such as tents, air support halls or airships can be divided into form finding, static analysis and cutting pattern. The result of the form finding is a shape of equilibrium for a certain stress distribution and boundary conditions. The shape of equilibrium ensures the geometry of the double curved surface which has only tension and avoids compression in the surface. From this geometry the structural behaviour is exanimated and the cutting pattern is made of. The flattening of the double curved surface is a geometrical process without considering the stress distribution and the material behaviour. In the recent development of cutting pattern methods the stress distribution is taking into account [9,10]. The analysis of the structural behaviour is carried out without the influence of the width of the strips, the seams, the orientation of the fabric and the process of pretensioning. The separation of the structural behaviour and the cutting pattern leads in built structures to highly inhomogeneous stress distributions which can be seen in wrinkles and measured in stresses which are two times higher than required. The difference in the stress distribution and geometry between the numerical found shape of equilibrium and the real structure causes in the non consistent design procedure see Fig. 1.

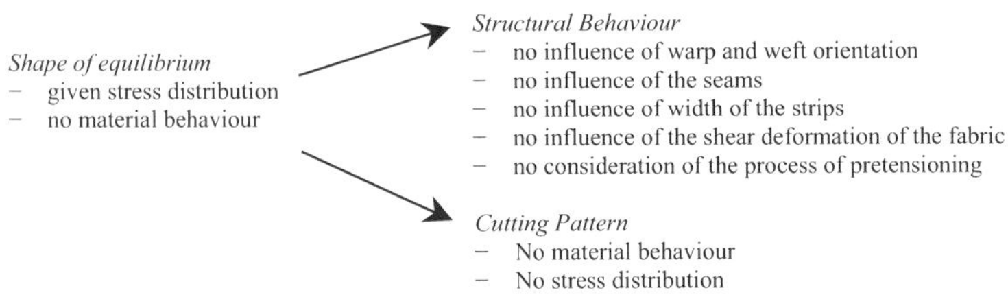

Fig. 1. Common design process of membrane structures

3 Enhanced Design Process of Tension Structures

An enhanced design concept will be based on 5 design steps defining the shape of equilibrium, generating the cutting pattern, reassembling and pretensioning the cutting pattern, the structural analysis of the reassembled structure and the evaluation of the structural behaviour. The material behaviour is considered in the last three steps: flattening the shape of equilibrium, reassembling and load bearing behaviour. The length and the width of the strips has an influence to the shear deformation of the coated fabric. The orthotropic behaviour of coated fabric influences the process of pretension and the stress distribution in the reassembled structure. The numerical process allows after evaluation modifications to reach better results in the reassemble structures considering stress distribution and deformations.

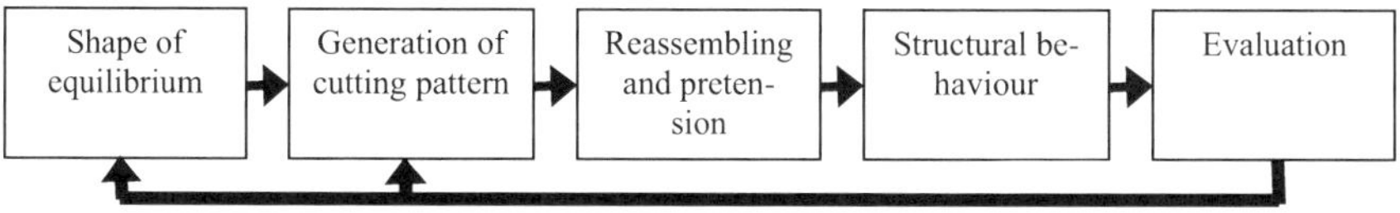

Fig. 2. Enhanced design process of membrane structures [11]

3.1 Shape of Equilibrium

The development of Computer Aided Geometric Design (CAGD) marked the start of changes in geometry endorsing new and free forms. This generation of double curved 3-dimensional surfaces is restricted by few limitations. Theoretically there are an unlimited number of forms to be numerical generated and represented. However, the manufacture and realization of such double curved surfaces are subject to numerous boundary conditions and restrictions. Using cables and membranes for the load transfer only tension forces can be carried, the cables and membranes can not withstand bending moments and compression forces in a global point of view. The structures have to be pretensioned activating the geometric stiffness or to be able carrying compression forces by reducing the pretension. The shape of equilibrium defines a pretensioned geometry of a doubled curved surface for a cable net or a membrane structure. The relation between the tension stress, geometry and equilibrium allows three possibilities to introduce the tension into the membranes and influences the shape of equilibrium, see Fig. 3.

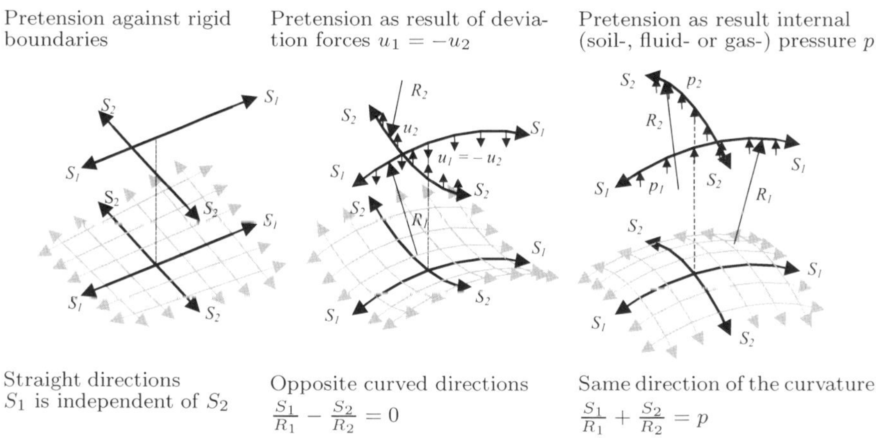

Fig. 3. Relation between tension forces and curvature

The pretension against rigid boundaries enables plain tension structures, tension structures with single curvatures and double curved tension structures if the direction of the cables or yarns is along the evolution line of a hyper parabola. The tension forces are independent from each other in this case.

The tension forces in surfaces with negative Gaussian curvature result of the deviation forces at the nodes and this leads to a relation between tension forces and curvature. Fulfilling the equilibrium at each node the tension forces are related to the radius of curvature in the both directions. Equal forces in both directions require the same curvature of the cables. The ratio of tension forces and radius of curvature is constant by meaning the higher the forces the lower the curvature to ensure equilibrium.

Stabilising the membranes with internal pressure leads to surfaces with positive Gaussian curvature and a dependency between the internal pressure, the tension forces and the curvature. The tension forces are directly related to the internal pressure and the lower the curvature the higher the forces.

Cable nets with square meshes are cinematic systems, the thin membrane without bending stiffness is statically determined. In both cases the double curved surface is a result of the three dimensional equilibrium at each node for given tension forces in a cable net or at each point for a given stress distribution in a membrane. The equilibrium is fulfilled without taking into account the material behaviour and is influenced by the boundary conditions such as high points, boundary cables or rigid boundaries.

For cable nets the numerical solution is based on the constant ratio of cable force and length in the first step. The ratio of cable force and length is described as force density [12] and the shape of equilibrium is calculated from a plane net with a square gird by moving the nodes in the third direction, forced by the fixed points and boundaries which don't lie in the same plane as the cable net. Depending on the change in length from the cable links in the plane into the three dimensional surface the cable forces changes, the longer the cables the higher the forces. The result is a doubled curved surface with a steady change in the forces along each cable related to the change of curvature of the surface. This method can also be exceed to cable nets which are statically indeterminate such as nets with triangle meshes because of the constant ratio of force/length. Both forces and length of the links are free parameter searching for the three dimensional equilibrium.

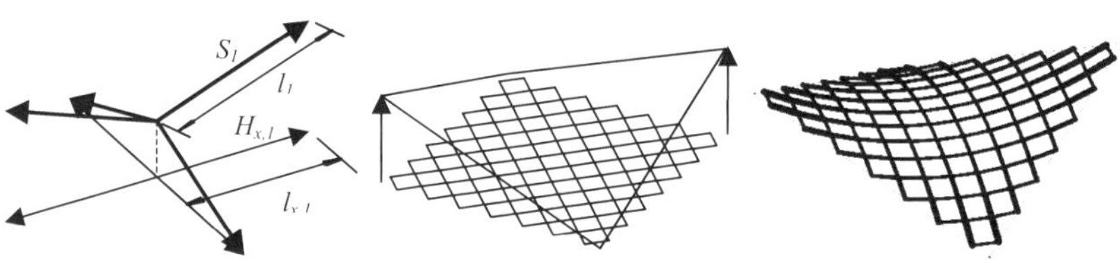

$$\frac{S}{l} = \frac{H_x}{l_x} = \frac{V}{l_z} = const.$$

Plane net with square meshes and constant forces in each link

Shape of equilibrium change in forces

Fig. 4. Shape of equilibrium fulfilling vertical equilibrium

Fulfilling the equilibrium in the tangential plane at each knot allows adjusting the link length and leads to constant forces in each cable. The cables are oriented along geodesic lines onto the surface and the angles are not constant at the nodes between crossing cables.

In membranes the equilibrium has to be fulfilled at each point of the surface. Plane state of stress assumed shapes of equilibrium are also the result of a given stress

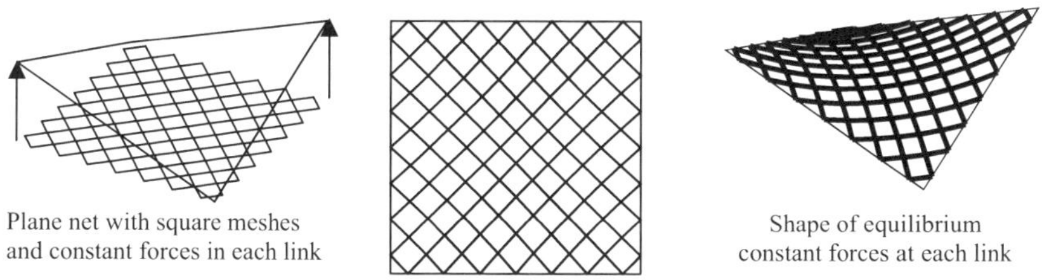

Fig. 5. Shape of equilibrium fulfilling vertical and tangential equilibrium

distribution. In general and in covariant description [13] the equilibrium normal to the surface is written as:

$$\sigma^{\alpha\beta} b_{\alpha\beta} = 0 \quad , \quad \text{with } b_{\alpha\beta} \text{ as tensor of curvature}$$

Related to main axis

$$\sigma^{11} b_{11} + \sigma^{12} b_{12} + \sigma^{21} b_{21} + \sigma^{22} b_{22} = 0$$

The orientation of the coordinate system in direction of the principle stresses ($\sigma^{12} = \sigma^{21} = 0$) or principle curvature ($b_{12} = b_{21} = 0$)

$$\sigma^{11} b_{11} + \sigma^{22} b_{22} = 0$$

Tension stresses in both principle directions $\sigma^{11} \geq 0$ und $\sigma^{22} \geq 0$ requires a negative Gaussian curvature, with $b_{11} = \frac{1}{R_1}$ and $b_{22} = -\frac{1}{R_2}$ the equilibrium normal to the surface results in

$$\frac{\sigma^{11}}{R_1} - \frac{\sigma^{22}}{R_2} = 0$$

The equilibrium in the tangential plane of the point can be written in covariant description as

$$\sigma^{\alpha\beta}_{|\beta} = 0$$

Assuming the stress is constant at a certain point leads to

$$\sigma^{\alpha\beta} = \sigma g^{\alpha\beta} \quad \text{with } g^{\alpha\beta} \text{ as metric tensor}$$

Substituted

$$(\sigma g^{\alpha\beta})_{|\beta} = 0 \Rightarrow \sigma_{|\beta} g^{\alpha\beta} + \underbrace{\sigma g^{\alpha\beta}_{|\beta}}_{\Rightarrow 0} = 0$$

and results in

$$\sigma_{|\beta} g^{\alpha\beta} = 0$$

The metric tensor has a certain value at each point in a double curved surface; this means the deviation of the stress has to be zero. This requires a constant stress distribution also to neighboured points and describes the hydrostatic state of stress. Therefore has to be $\sigma_{11} = \sigma_{22} = constant$ and with

$$\frac{\sigma_{11}}{R_1} - \frac{\sigma^{22}}{R_2} = 0 \Rightarrow \left(\frac{R_2 - R_1}{R_1 \cdot R_2} \right) = 0 \quad \text{and } R_1 = R_2$$

The tension stress in the surface is isotropic and homogeneous by meaning the stresses are at each point and in each direction constant and this is named as hydrostatic state of stress. The stress can be set as a constant value and reduces the description of shapes of equilibrium to the geometrical problem searching for the minimal surface by given boundary conditions. Physical models of minimal surfaces are soap films, in earlier times one of the few methods describing double curved surfaces which are at each point under tension.

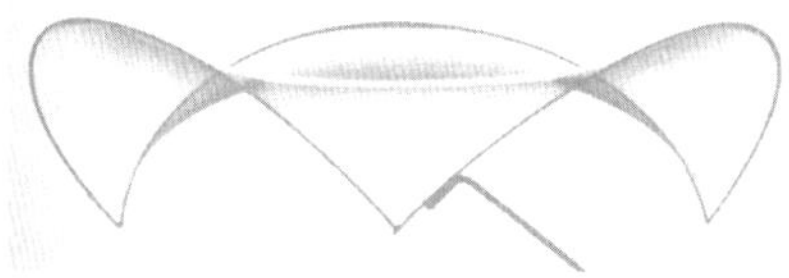

Soap film [14]

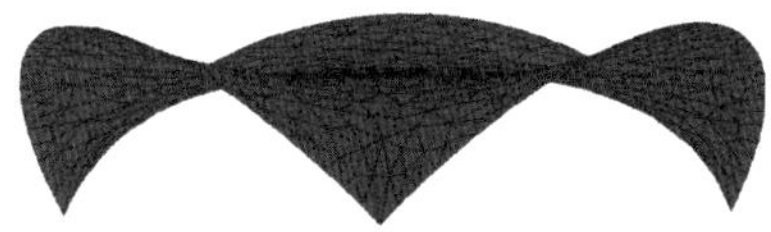

Numerical solution of the minimal surface [15]

Fig. 6. Minimal surfaces as soap film and the numerical solution

3.2 Cutting Pattern

The shapes of equilibrium are characterized by no material behaviour or by the material behaviour of soap films without shear resistance. The real shape of the tensioned structures is influenced by the material behaviour and the difference between the shape of equilibrium and the materialized, pretensioned shape resulting in the non existing shear stiffness of a cable net, the orthotropic behaviour of coated fabric or the relatively high shear stiffness of foils. Known from the globe is the fact that double curved surfaces cannot be flattened without distortion. Furthermore the fabric is manufactured in width up to max. 5 m and this requires the assembling of the whole cover with patches or strips of a certain length and width. The common way of generating the cutting pattern from the shape of equilibrium is described in four steps. The shape of equilibrium is cut into strips mostly using geodesic lines for the cutting lines. The whole structure is then divided into double curved strips. These strips are flattened with different methods such a paper strip method or minimizing the strain energy while flattening the strips. The compensation as final step is necessary to introduce the tension forces by elongation of the fabric. All strips have to be decreased in width and length in relation to the stress and strain behaviour of the fabric in the built structure.

Differences in geometry and stresses between the shape of equilibrium and the built structure are caused by the orientation of the fabric, the shear deformation of the fabric, the stiffness of the seams und the process of pretension. Reducing the mistakes in the cutting pattern which can be seen in wrinkles and can be measured in local stress peaks is possible by taking into account the jamming condition of the coated fabric. The load carrying compounds in a fabric are the yarns which are protected by the coating. In a woven fabric warp and fill will kept in place if the tension stress acts in direction of the yarns. Shear forces lead to a rotation of warp and fill against each other up to an angle when the yarns touch each other. The

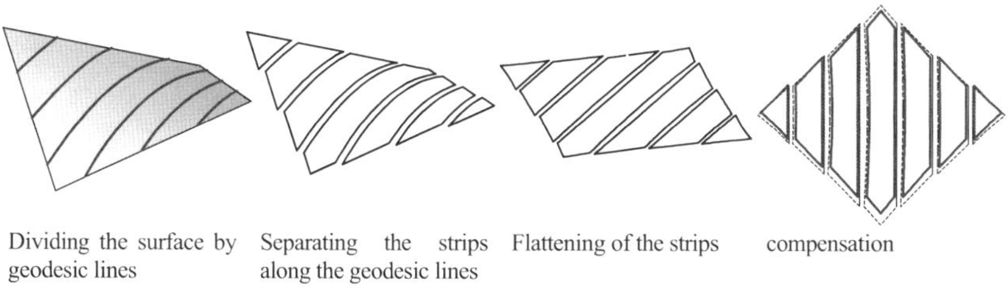

Fig. 7. Generation of cutting pattern [16]

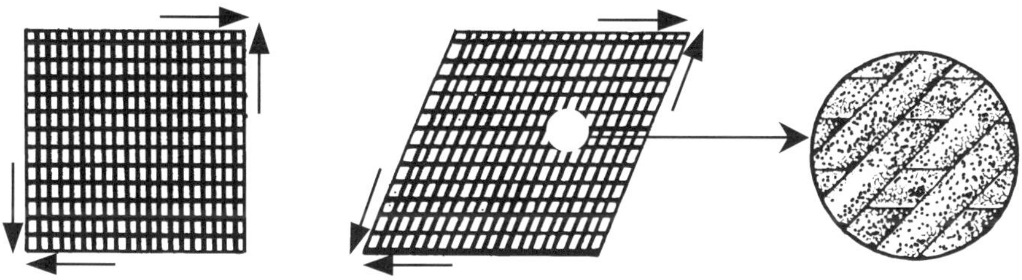

Fig. 8. Shear deformation of woven fabric [17]

maximum shear deformation is depended by the thickness of the yarns, the distance of the yarns and the flexibility of the coating. If the rotation of the yarns is larger than the required distortion to flatten the doubled curved strips the flattening is only a process of strainless deformation.

If the process is invert and still definite needs further examination because the manufacturing of membrane structures is from the flat and assembled strips into the double curved and pretensioned structure. Already known is the shear deformation which is used to build double curved surfaces with cable nets. The cable net can be put onto the doubled curved surface just by changing the angles between the cables; the distance between the nodes is kept as constant. The rotation of the two layers of cables against each other is related to the curvatures of the surface.

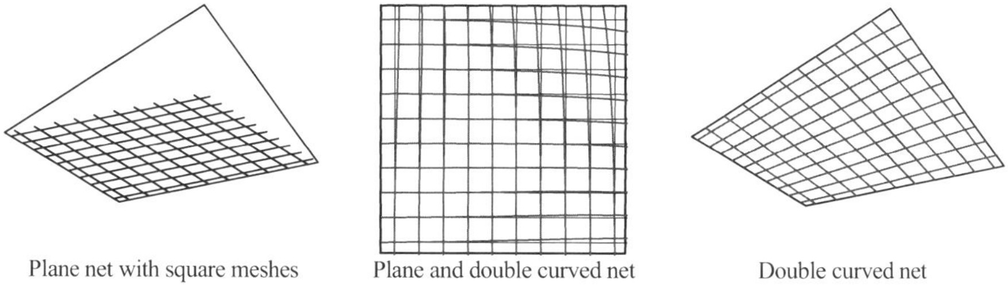

Fig. 9. Shear deformation from the plane into the double curved net

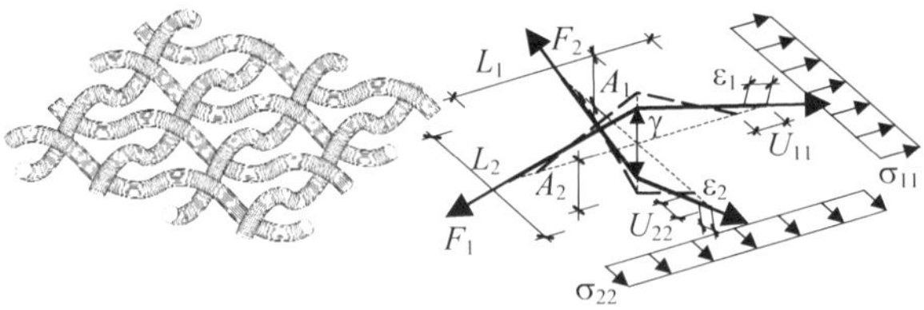

Fig. 10. Model describing the behaviour of a woven fabric

3.3 Reassembling and Pretensioning

The tension forces can only be introduced into cable nets or membranes by elastic strain of the cables and coated fabric. The numerical process of reassembly requires the description of the material behaviour in which both the change of the geometry and the elastic strain is considered. The change in geometry is for cable nets mostly the in plane shear deformation reaching the double curved surface. The change of geometry in woven fabric is related to the elongation of the yarns. The simple model is useful enough describing the behaviour of a woven fabric, developed in 1978 [18], refined and tested in 1987 [19] and finally numerical transferred in 2003 [20].

Neglecting the influence of the coating the behaviour of a woven fabric can be described by the

- Geometry of the fabric such as thickness and distance of the yarns (warp A_1, L_1 and inclination $m_1 = A_1/L_1$, fill A_2, L_2 and inclination $m_2 = A_2/L_2$)
- Stress-strain-behaviour of each yarn (warp F_1, ε_1, fill F_2, ε_2)
- The change in the thickness of the fabric (γ) and
- The equilibrium of the deviation forces at each knot

The ratio of unstrained to strained length is described by:

$$\mu_1 = 1 + U_{11} \text{ and } \mu_2 = 1 + U_{22}$$

With the ratio of undeformed and deformed inclination of

$$k_1 = A_1/\mu_1 \text{ and } k_2 = A_2/\mu_2$$

is the elastic strain of the yarns

$$\varepsilon_1 - \mu_1 \frac{\sqrt{1 + k_1^2 m_1^2}}{\sqrt{1 + m_1^2}} + 1 = 0 \quad \text{and } \varepsilon_2 - \mu_2 \frac{\sqrt{1 + k_2^2 m_2^2}}{\sqrt{1 + m_2^2}} + 1 = 0$$

The constrain of the distance between the yarns at the knots is

$$k_1 \mu_1 A_1 + k_2 \mu_2 A_2 - A_1 - A_2 - \gamma F_1 \frac{k_1 m_1}{\sqrt{1 + k_1^2 m_1^2}} = 0$$

Equilibrium of the yarn

$$F_2 \frac{k_2 m_2}{\sqrt{1 + k_2^2 m_2^2}} - F_1 \frac{k_1 m_1}{\sqrt{1 + k_1^2 m_1^2}} = 0$$

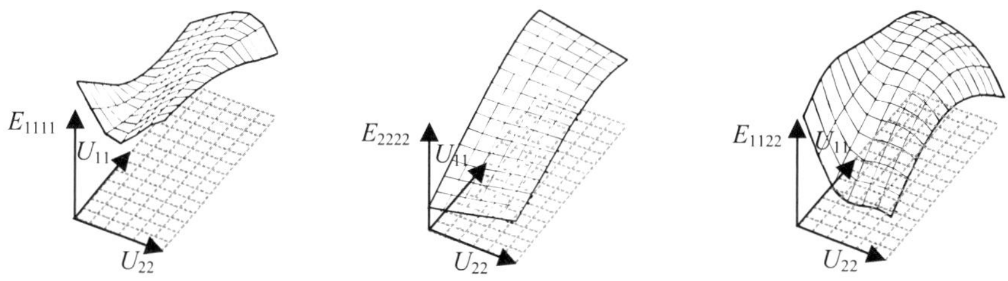

Fig. 11. Young's Moduli and Poisson ratio as function of the fabric strain, PVC coated fabric [19]

This set of 4 equations serves a non linear system of equations for the four unknown values $\varepsilon_1, \varepsilon_2, k_1, k_2$. After solving the equations the stresses of the fabric can be defined directly by

$$\sigma_{11} = \frac{1}{L_2}\left(\frac{F_1}{\sqrt{1+k_1^2m_1^2}}\right) \quad \text{and} \quad \sigma_{22} = \frac{1}{L_1}\left(\frac{F_2}{\sqrt{1+k_2^2m_2^2}}\right)$$

The calculated strains and stresses enable to define the stiffness E_{1111}, E_{2222} and E_{1122}. The elastic stiffness are non linear and closely related to the strain ratio in warp and fill direction. Even the Poisson ratio E_{1122} is non linear and depending to the strain ratio of the yarns.

The numerical process of reassembling enables taking into account the behaviour of the fabric, the influence of the seams and the distribution of the tension stress through the whole surface. The plane strips have to be remeshed, sewed together and pretensioned by moving the sewed structure into defined boundaries, moving support points into their position after reassembling or putting internal pressure onto the system. The stress distribution and geometry of the sewed and pretensioned structure is different from the assumed stress distribution of the shape of equilibrium. The differences are depending on the curvature of the surface, the orientation of the strips in relation to the main curvature, the torsion of the strips, the distortion of the load transfer along the seams, the stiffness of the seams, the assumed compensation of the flatten strips, the width of the strips, the of the surface, the shear deformation of yarns and in the shown example of the load transfer between the boundary cables and fabric, see Fig. 12.

In the shown example the stress distributions varies in a single strip and changes from strip to strip. Relatively low tension stress in the middle strip can been see as result of less compensation. The influence of the stiffness of the seams can be shown in the difference between deformation in vertical direction comparing the geometry of the shape of equilibrium and reassembled and pretensioned structure. For the shown example the difference is app. 20% of the span. The antimetric deformation is caused by the inhomogeneous stress distribution in the cross section along the high points. The tension stress perpendicular are unsteady, low stress leads to high vertical deformations and high stress kept the fabric down which can clearly seen in the up and down of the differences.

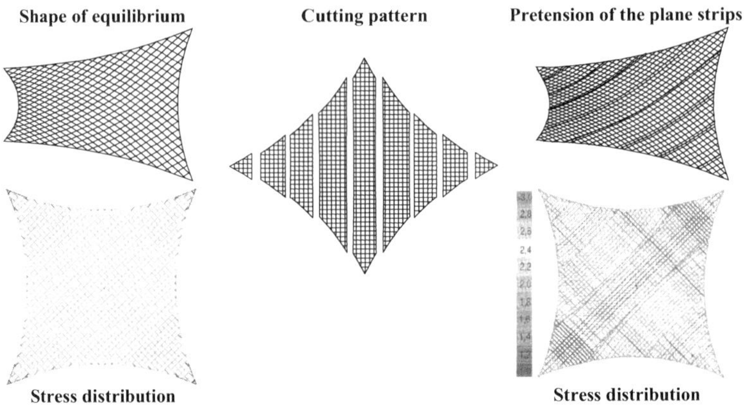

Fig. 12. Influence of the cutting pattern to the stress distribution of a membrane [21]

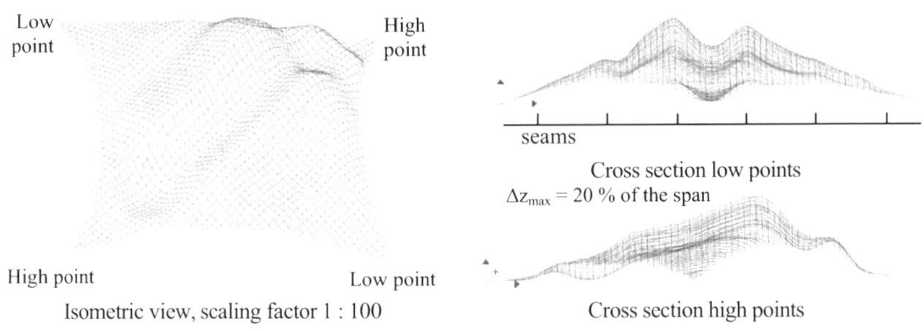

Fig. 13. Difference in z–direction between the reassemble shape and the shape of equilibrium [21]

3.4 Load Bearing Behaviour

The load bearing behaviour is in general depended on:

- The flexibility of the structures including masts, bending elements such as arches, stay cables, anchorages and foundations
- The height of the pretension related to external loads
- The orientation of the cables or yarns related to the main curvature of the surface
- The curvature of the surface and
- The stress – strain – behaviour of the material

The stability of the cable net or membrane structures is depending on the pretension. In structures with negative Gaussian curvature the pretension is only to reduce the deformation. The slag of the spanning direction causes a change in the system but no instability.

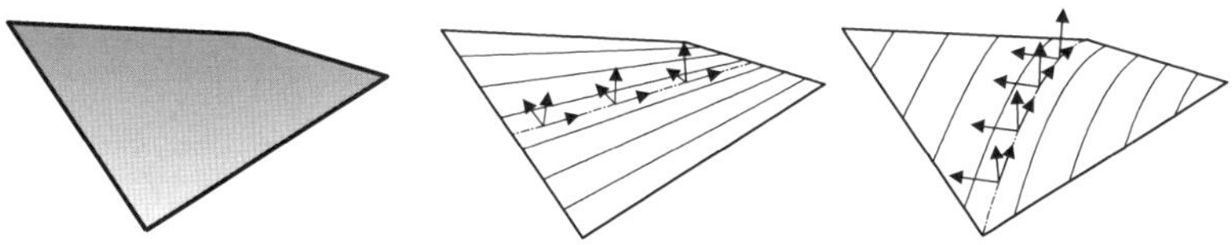

Fig. 14. Orientation of the strips in relation to the curvature of the surface

The isotropic and homogeneous stress in minimal surfaces allows at first any kind of orientation of the cables or yarns on the surface but the curvature of the cable layers or yarns, the elastic strain under external loads, the strainless deformations and the shear deformation is depending on the orientation. In the shown example, see Fig. 14, the yarns are straight if the strips are parallel to the boundaries and the torsion of each strip is high. Compared to the orientation parallel to the main curvature of the surface the torsion of the strips is zero if the centre line of the strip is equal to a line of main curvature.

Although both membranes have the same shape of equilibrium the load carrying behaviour is totally different. The membrane with straight yarns has large deformations under a constant and uniform distributed load, the stresses in warp and fill will increase. The membrane with the yarns oriented to the main curvature carries load be increasing the stresses in the sag directions and decreasing the forces in the span direction. The deformation is compared to the membrane with straight yarns very low.

Curvature and elastic stiffness have different influences to the load bearing behaviour of the membranes. Membranes with large curvature react under external loads highly linear, increasing and decreasing of the stresses are nearly independent of the elastic stiffness if the spans and the materials have been chooses for usual membranes and loads in Europe. Only if the elastic stiffness is relatively low the behaviour is starting to get non linear because of the large elastic strain. The vertical deformations are large because of the total length of the yarns spanning between the supports or boundary cables. The opposite behaviour occurs for membranes with low curvature by meaning the change of the stresses under external load is non linear and the deformation are decreasing nearly linear by increasing the elastic stiffness. Membranes with less curvature are more sensitive to changes in the elastic stiffness considering the change in stress and deformation than membranes with a higher curvature.

3.5 Evaluation

In relation to the influence of the curvature and the elastic stiffness to the load bearing behaviour a stiffness value can be defined for any doubled curved surface, developed by R. Blum [17].

For a constant and uniformly distributed load the equilibrium normal to the surface is written in covariant description as following:

$$\sigma^{\alpha\beta} b_{\alpha\beta} = p^3 \tag{1}$$

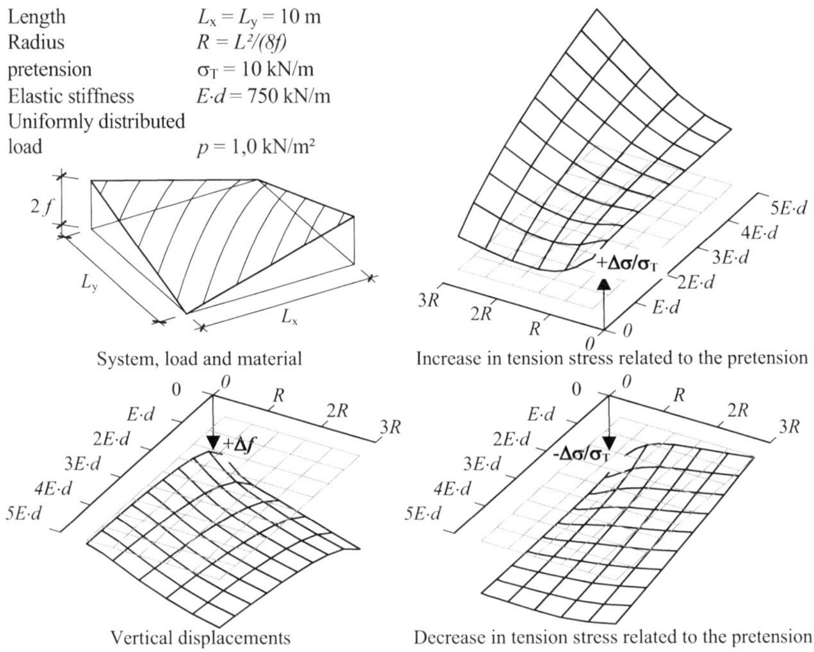

Fig. 15. Influence of curvature and elastic stiffness to the stresses and deformation

The change of the stresses caused by external loads is:

$$[\sigma^{\alpha\beta} + \Delta\sigma^{\alpha\beta}] \cdot [b_{\alpha\beta} + \Delta b_{\alpha\beta}] = p^3 \tag{2}$$

Multiplication of the terms and neglecting terms of high order result in:

$$\Delta\sigma^{\alpha\beta} b_{\alpha\beta} + \Delta b_{\alpha\beta}\sigma^{\alpha\beta} = p^3 \tag{3}$$

Linear elastic behaviour of the material lead to

$$\Delta\sigma^{\alpha\beta} = n^{\alpha\beta\delta\gamma} \cdot \Delta\varepsilon_{\delta\gamma} \tag{4}$$

The deviation of the curvature can be approximately seen as the displacement in vertical direction

$$\Delta b_{\alpha\beta} = u^3_{|\alpha,\beta} \tag{5}$$

The change in the elastic strain is approximately multiplication of the curvature with the vertical displacement:

$$\Delta\varepsilon_{\delta\gamma} = b_{\delta\gamma} \cdot u^3 \tag{6}$$

(6) substituted (4) and with (5) is the change of the stresses

$$\Delta\sigma^{\alpha\beta} = \frac{n^{\alpha\beta\delta\gamma} \cdot b_{\delta\gamma}}{n^{\alpha\beta\delta\gamma} \cdot b_{\alpha\beta} \cdot b_{\delta\gamma}} \cdot p^3$$

with (4), (5) and (6) in (2) follows

$$n^{\alpha\beta\delta\gamma} b_{\delta\gamma} u^3 b_{\alpha\beta} + u^3_{|\alpha,\beta} \cdot \sigma^{\alpha\beta} = p^3$$

Assuming only vertical loads, allows setting the 2. term to zero and the vertical displacement is

$$u^3 = \frac{1}{n^{\alpha\beta\delta\gamma} \cdot b_{\alpha\beta} \cdot b_{\delta\gamma}} \cdot p^3$$

In both equations, the change of the stresses and the vertical displacement, the denominator is the same and a product of the elastic stiffness and the curvature of the surface. The lower this product is the higher the vertical deformations will be. Therefore this term describes the stiffness of the surface and is named as

$$D = n^{\alpha\beta\delta\gamma} \cdot b_{\alpha\beta} \cdot b_{\delta\gamma}$$

Expanded and the orientation of the coordinate system in direction of the principle stresses ($\sigma^{12} = \sigma^{21} = 0$) or principle curvature ($b_{12} = b_{21} = 0$) leads to:

$$D = \frac{n^{1111}}{R_1^2} + \frac{n^{2222}}{R_2^2} = \frac{R_2^2 \cdot n^{1111} + R_1^2 \cdot n^{2222}}{R_1^2 \cdot R_2^2}$$

Flexibility ellipsoids

Two aspects have to be mentioned considering the load bearing behaviour of cable nets or membrane structures, the stiffness of a three-dimensional shape and the possibility of pretensioning the structure in relation to the material behaviour and the stiffness. There exists an analogy between net calculation in geodesy and the analysis of membranes [22,23] and leads to new aspects describing the load carrying capacity of structures. Flexibility can be seen as the deformation of each node loaded by a rotating unit load and leads to flexibility ellipsoids showing the three dimensional deformation of the node.

The in plane stiffness of the cable net or membranes has a large influence to the possibility of pretension because this allows to change plane two dimensional flat strips into a three dimensional surface without wrinkles. The advantage of cinematic cable nets and membranes is the ability to distribute the forces during the process of pretensioning nearly homogeneous by tensioning only boundary cables or lifting high points. The ability of a double curved cable net distributing forces which are acting at the edges or boundaries homogeneous through the net can be described by redundancy.

The comparison between three different types of cable nets will give an example for the application of flexibility ellipsoids in evaluation of the structural behaviour. In geometry three homogeneous nets are exiting which can be transformed in double curved cable nets. Each net has the same tension forces and the stiffness per meter. The net with hexagonal meshes has only nodes with three links and leads to an equilibrium of each node under pretension only if the forces in all links are the same. The shape is then comparable with a minimal surface. The high degree of kinematics makes these nets very flexible and the stiffness can be mostly influenced by the height of the pretension forces. The net with square meshes has still no in plane shear stiffness but if the cables are arranged in the direction of the main curvature this net has even less deformation for a uniformly distributed load compared to the net with triangle meshes. The net with hexagonal meshes is the one with the highest flexibility; the flexibility of the net with triangle meshes is in direction normal to the net surface nearly the same compared to the net with square meshes. The in plane stiffness reduces theses deformation of the net with triangle meshes.

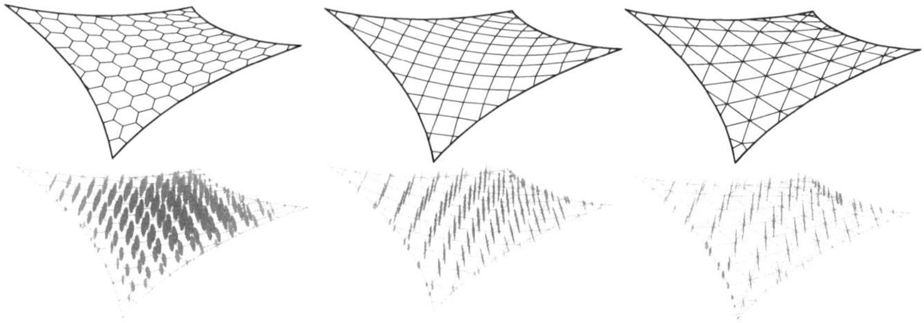

Fig. 16. Flexibility ellipsoids of pretensioned nets with different meshes

Redundancy

A common definition of redundancy is the availability of more functional system components than necessary for meeting the requirements. For example the statically indeterminacy is a measure for redundancy of structures by the meaning of safety against failure. In cinematic structures such as cable nets this definition has to be extended by taking into account the geometrical stiffness. Well known is the fact that each node has 3 independent possibilities of movement in space structure, the normal force of each element and the boundary forces are giving the necessary equations for equilibrium. If the number of independent movements of the nodes is equal to the number of equations resulting from the element forces and boundary forces the structure is statically determined.

Redundancy is more than the statically indeterminacy, which can be shown by the considerations of Dieter Stroebel [23] The number of indeterminacy is distributed among all elements in relation the geometry of the structure, the stiffness of the elements and the relatively elongation of each element within the structure. The elastic elongation of each element is influenced by stiffness of the connected elements. Focusing on one element means an extremely stiff surrounding structure compared to the element the change of length causes directly an increase of the force. The opposite is a very flexible structure which allows the elongation of the element without influencing the forces.

This behaviour can be described by the *elastic redundancy* in terms between 0 and 1. An elastic redundancy of 0 means no forces arise if the length an single element is changed. An elastic redundancy of 1 says changing the length leads to a force depending on the elastic stiffness.

Cinematic structures can be stabilised by tension forces. The tension forces built up a resistance if the structure is deformed under load known as geometrical stiffness. The three components of the stabilising tension forces lead to three additional

Elastic redundancy $r_E = 2.06$ / Geometric redundancy $r_G = 0,02$ | Elastic redundancy $r_E = 2,33$ / Geometric redundancy $r_G = 0,20$ | Elastic redundancy $r_E = 2,68$ / Geometric redundancy $r_G = 0,33$

Fig. 17. Change of forces by shortening one link of 0,5% of its length

equations for each element in a pretensioned structure and an additional geometric redundancy. The geometric redundancy of 3 says no influence of the geometric stiffness is required; the geometric redundancy lower than 3 describes the part of the geometric stiffness necessary for stabilisation and higher than 3 means the element is unstable and needed to be stabilized

For the three cable nets the elastic and geometric redundancy can be analysed for the links and gives information to the influence of manufacturing errors, the possibility of pretensioning and the height of the tension forces related to the deformation. The change of the 0,5% of the length of one link causes in the net with the hexagonal meshes no changes in the forces, shown by the elastic redundancy close to zero. The opposite can be seen in the net with the triangle meshes, the change in the length causes in that element an increasing force.

References

1. Gründig L (1976) Die Berechnung von vorgespannten Seilnetzen und Hängenetzen unter Berücksichtigung ihrer topologischen und physikalischen Eigenschaften und der Ausgleichsrechnung. 1. SFB 64 Mitteilungen 34/1975, 2. Dissertation DGK Reihe C, Nr. 216.
2. Singer P (1995) Die Berechnung von Minimalflächen, Seifenblasen, Membranen und Pneus aus geodätischer Sicht. Dissertation, Technische Universität Stuttgart, Deutsche Geodätische Kommission – Reihe C, Heft Nr. 448, München.
3. Bletzinger K-U (2002) Formfindung von leichen Flächentragwerken. in Baustatik-Baupraxis 8, Institut für Statik, TU Braunschweig.
4. Bellmann J (1998) Membrantragwerke und Seifenhaut Unterschiede in der Formfindung. Bauingenieur, 3/98.
5. Lewis WJ, Lewis TS (1996) Application of Forminan and Dynamic Relaxation to the Form Finding of Minimal Surfaces. Journal of International Association of space and shell structures 37(3):165–186.
6. Barnes M (1999) Form Finding and Analysis of Tension Structures by Dynamic Relaxation. Int. Journal of Space Structures 14:89–104.

7. Bletzinger K-U and Wüchner R (2001) Form finding of anisotropic pre-stressed membrane structures. Proc. of Trends in Computational Structural Mechanics, Bregenz, Austria, May 20–23, 2001, WA Wall, K-U Bletzinger, K Schweizerhof (Eds.), CIMNE, Barcelona, Spain.
8. Moncrieff E and Topping BHV (1990) Computer Methods for the generation of membrane cutting pattern. Computers & Structures 37(4):441–450.
9. Ishii K (1999) Form Finding Analysis in Consideration of Cutting Patterns of Membrane Structures. International Journal of Space Structures 14:105–119.
10. Maurin B and Motro R (1999) Cutting Pattern of Fabric membranes with Stress composition Method. International Journal of Space Structures 14:121–129.
11. Draft 2003. Research Project: Vorgespannte Membranen – eine interaktive Methodik zur Auslegung von Membranen in der Luftfahrt und im Bauwesen, ISD, University of Stuttgart, Labor Blum, Stuttgart and Femscope GmbH, Sigmaringen.
12. Linkwitz K, Gründig L, Hangleiter U and Bahndorf J (1985) Mathemaisch-numerische Netzberechnung. In: SFB 64, Weitgespannte Flächentragwerke, Universität Stuttgart, Mitteilungen 72, Abschlu*β*bericht Teilprojekt F2, Stuttgart.
13. Blum R (1974) Beitrag zur nichtlinearen Membrantheorie. SFB 64 Weitgespannte Flächentragwerke, Universität Stuttgart, Mitteilungen 73/1985, Werner-Verlag, Düsseldorf.
14. Seifenblasen/Forming Bubbles (1988) Unter d. Leitung von Klaus Bach, Frei Otto, Mitteilungen des Instituts für Leichte Flächentragwerke IL 18, Universität Stuttgart, Krämer Verlag Stuttgart.
15. Reimann K (2003) Numerische Berechnung von Minimalflächen. Internal Report, Femscope GmbH, Sigmaringen.
16. EASY – Software, Form finding, analysis and cutting pattern, Technet GmbH, Berlin.
17. Wagner R, Blum R (2004) Spannung im Tragwerk – Bauen mit Seilen und Membranen, (to be published).
18. Meffert B (1978) Mechanische Eigenschaften PVC – beschichteter Polyestergewebe, Disseration, Aachen.
19. Blum R and Bidmon W (1987) Spannungs-Dehnungs-Verhalten von Bautextilien. SFB 64, Weitgespannte Flächentragwerke, Universität Stuttgart, Mitteilungen 74/1987, Werner-Verlag, Düsseldorf.
20. Reimann K (2003) Zur numerischen Berechnung des Spannungs-Dehnungs-Verhalten von Geweben nach Blum/Bidmon. Internal report, Femscope GmbH, Sigmaringen.
21. Zimmermann M (2000) Untersuchungen zum Unterschied zwischen der Formfindung und dem aus Bahnen zusammengefügten System bei Membrantragwerken. Diplomarbeit, Institut für Konstruktion und Entwurf II, Universität Stuttgart.
22. Linkwitz K (1988) Einige Bemerkungen zur Fehlerellipse und zum Fehlerellipsoid. Vermessung, Photogrammetrie, Kulturtechnik, Schweizerischer Verein für Vermessungs- und Kulturtechnik (SVVK), S. 345-364. 86. Jahrgang, Heft 7.
23. Ströbel D (1997) Die Anwendung der Ausgleichungsrechnung auf elastomechanische Systeme. Dissertation, Technische Universität Stuttgart, Deutsche Geodätisceh Kommission – Reihe C, Heft Nr. 478, München.

Systems for Lightweight Structure Design: the *State-of-the-Art* and Current Developments

Erik Moncrieff[1]

Kurvenbau
Emdener Str. 39
D-10551 Berlin, Germany
erik.moncrieff@kurvenbau.com
http://www.kurvenbau.com

Summary. *This paper deals with the design of lightweight structures. In particular the role of computational modelling software in this process is discussed. The state-of-the-art is first described paying close attention to the requirements for industrially effective solutions. Some of the less well understood aspects of the modelling processes are discussed. In particular the load analysis, form-finding and cutting pattern generation processes are covered. The modelling of textile is addressed in detail. Approaches to the design of software design systems for lightweight structure design are discussed in the context of system flexibility and effectiveness. Finally, interesting applications in the field of lightweight structures arising from design system developments are highlighted.*

Key words: design, lightweight structure, modelling, simulation, textile, element type, form-finding, geometrically non-linear structural analysis, elastically non-linear structural analysis, cutting pattern generation, crimp, pneumatic structure, hybrid structure, adaptive design

1 Introduction

To most structural engineers and architects the design of lightweight structures is mysterious. The objective of this paper is to summarise the *state-of-the-art* in lightweight structure design systems in order to highlight several important concepts. Emphasis is directed to the requirements of industrial procedures.

2 Lightweight Structure Design

2.1 Design Process

As with conventional structural engineering projects, the design process for lightweight structures involves three key players. These are the *Client*, the *Architect* and the

E. Oñate and B. Kröplin (eds.), Textile Composites and Inflatable Structures, 17–28.

Structural Engineer. The client commissions the project and invites tenders from architects. The architects prepare conceptual designs working in collaboration with structural engineers. The client chooses a conceptual design and appoints the architect. The architect, again working in close collaboration with a structural engineer proceeds to refine the conceptual design into a production design. Finally the design is fabricated and installed. The critical path of this design process is shown in Fig. 1. In reality there are several design modification cycles operating.

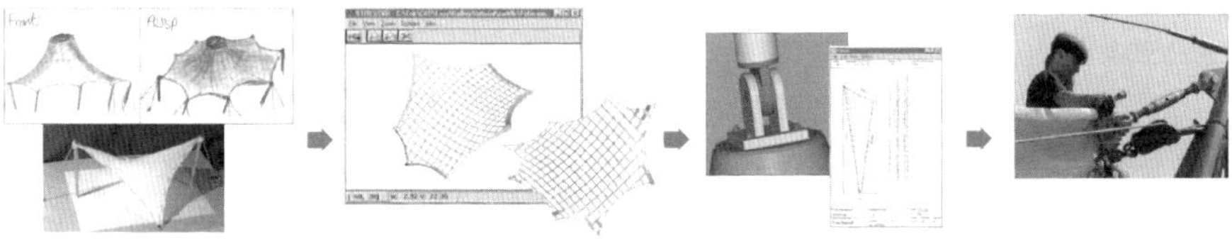

Conceptual design   Form-finding/Load analysis   Patterning/Detailing   Installation

Fig. 1. The phases of the design process critical path

2.2 Design deliverables

The deliverables can be conveniently divided between those for the conceptual and production designs.

Conceptual design	
• Pre-stress surface geometry	***Form – finding***
• Reaction, support and cable forces	***Load Analysis***
• Textile stresses	

Production design	
• Pre-stress surface geometry	***Form – finding***
• Reaction, support and cable forces	***Load Analysis***
• Textile stresses	
• Cloth pattern system line geometry	***Cutting Pattern Generation***
• Support structure design	***Detailing***
• Cable dimensions	
• Connection design	
• Cloth and reinforcement cutting patterns	

Detailing is a critical process and highly integrated with the other processes. It will, however be the *Form-finding*, *Load Analysis* and *Cutting Pattern Generation* processes which will be mainly considered here.

2.3 Load Analysis and Form-Finding

The tasks

Load analysis and form-finding require the determination of Force Equilibrant models. In a *Force Equilibrant* model the residual forces acting on any degree of freedom after summing the internal and external loads acting there is zero. In the case of computational load analysis the elemental forces may be calculated using several elastic models. Similarly several methods may be used to define the elemental forces in computational form-finding. The load intensity distribution must be estimated.

Historical development
Before the development of computational structural modelling, textile structures were form-found using physical models and load analysis was performed using hand calculations. The development of linear structural analysis software had little applicability for the design of textile roofs due to their strong geometrical non-linearity Non-linear systems have been developed since the 1970's and are now routinely used.

Current system configurations
Today industrial systems are broadly based on three main solver algorithms.

- *Conjugate Gradient (CG)/Force Density (FD)*
- *Dynamic Relaxation (DR)*
- *Modified Stiffness (MS)*

Developments in mainland Europe have mostly used *CG/FD* solvers, Britain has concentrated on *DR*, and Japan and the USA have mainly used the *MS* method.

Two element types are commonly used to model textile roofs. Cable net models using link elements have been popular in *CG/FD* systems, while triangular continuum elements have been typically used in *DR* and *MS* systems. It is important to highlight that the prevalence of using particular elements with particular solver algorithms does not have a theoretical or computational basis. *CG/FD* systems with triangular continuum elements are used when appropriate, and *MS* and *DR* systems can also use link elements to model textile. Appropriate element types for the modelling of lightweight structures will be discussed in Section 3.2 below.

2.4 Cutting Pattern Generation

The tasks
Cutting Pattern Generation is the process where two dimensional unstressed cloth polygons are created from three-dimensional doubly curved stressed surfaces. This involves the specification of seam line locations, transformation of the stressed 3D surfaces to stressed 2D surfaces, and compensating the stressed 2D surfaces to unstressed 2D surfaces.

Historical development
Before the advent of computer modelling, textile roofs were patterned using physical models. Simple triangle strip development between computer model seam lines were next implemented and have been used successfully for medium to large structures. Distortion minimisation techniques have been adapted from map making to cope with the demands of smaller and more sensitive configurations.

Seam generation
Regardless of whether physical or computational modelling is used, patterning based on geodesic seam lines is the preferred approach. This is because geodesic lines

are, by definition, straight when developed to a plane. Cloths patterned between geodesic seam lines will therefore be straighter than non-geodesic patterns. Non-geodesic patterns typically have banana or "S" shaped cloths which cause larger cloth wastage.

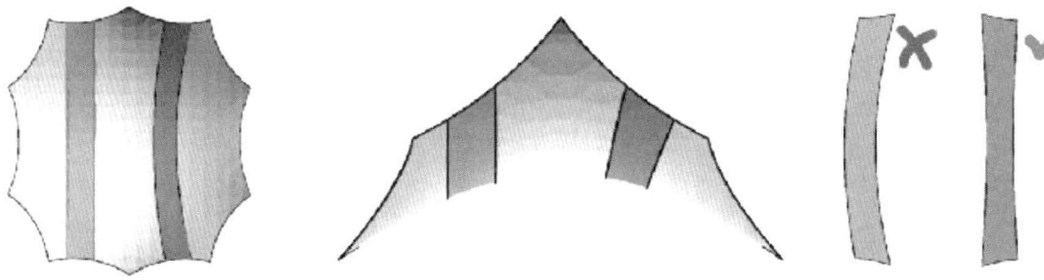

Fig. 2. Non-geodesic (lighter cloth), and geodesic (darker cloth) patterns

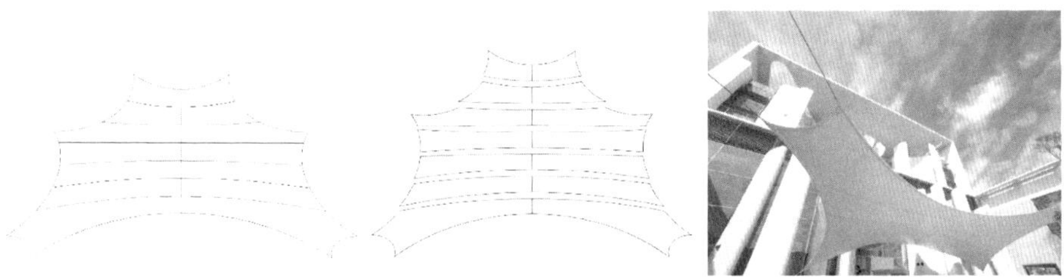

Fig. 3. Architecturally mandated semi-geodesic seam pattern for the German Chancellory

In some situations architectural requirements mandate non-geodesic seam lines. A prominent example of such a situation is the *German Chancellory* [1]. In such situations the use of line generation algorithms which are curvature based rather than length-minimising is helpful [2].

Cloth planarisation

The process of transforming a doubly curved surface into planar cloth sub-surfaces requires the introduction of distortion. The most basic approach taken to solve this problem is to define the cloths in terms of developable triangle strips. This works entirely adequately for large structures, but small structures are more difficult because cloth roll width does not limit pattern widths. This results in a wish to have fewer cloths for economic reasons. Meshes for small structures based on simple triangle development fail to reliably model the surface [3]. This can be seen in Figure 4. In such cases the use of more sophisticated distortion minimisation algorithms is very effective.

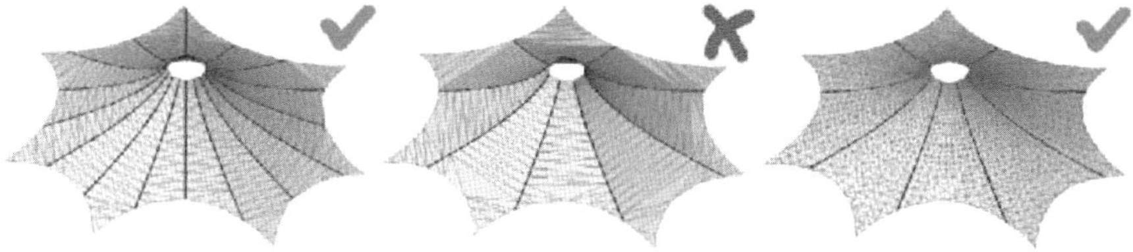

Fig. 4. Planarisation: (a) Large structure simple triangle development, (b) Small structure triangle development, (c) Small structure deformation minimising flattening

2.5 Design Methodologies

Textile structures have been designed in three general ways.

Non-computational: Physical models are used to form-find the pre-stress surface geometry and create the cutting patterns. Simplified "hand calculations" are used to predict structural response.

Non-specialised software: Non-equilibrium computational modelling software, such as *3ds max*, is used to generate the pre-stress surface geometry and cutting pattern generation. Standard FE structural analysis software is used to perform load analysis.

Specialised software: Lightweight structure task-specific equilibrium based computational modelling software is used to perform form-finding, load analysis and cutting pattern generation.

The non-computational method has the advantages that it is intuitive, the form can be realised, it can be implemented with low initial investment, and modification of conceptual forms is quick and simple. It suffers from its lack of computational non-linear structural analysis, low precision and lack of computational mesh for rendering. Its slowness, particularly with respect to making modifications to the production form and cutting patterns, makes it operationally expensive.

Using the non-specialised software method leverages existing CAD and analysis software skills and provides many sophisticated geometric tools. With few exceptions, the forms generated are, however, not force equilibrant. Consequently they can not necessarily be realised with a tensile surface. Lack of integration between the mesh generation and analysis leads to slow design modification cycles. Conventional FE software is often inappropriate for use with textile models. In particular, convergence problems are usually experienced by standard FE systems when dealing with textile slackening on-off non-linearity.

Specialist textile structure software systems quickly provides high confidence, high precision, integrated solutions. Initial investment is higher but when design volume is adequate, per-design costs are low. It is therefore the recommended method for production design. Having said that it is important to stress that the continued use of physical modelling during the conceptual modelling phase should always be encouraged.

3 Modelling Textile

The materials most commonly used for lightweight structures are PVC coated polyester and PTFE coated glass. Despite different production methods these textiles are similar in their structural configuration. A woven base cloth is coated on both sides as shown in Fig. 5. The warp threads are typically less crimped than the weft. The warp and weft crimp are more similar with *Precontraint* textile due to the weft stressing during coating. All textiles exhibit extremely complex structural behaviour. In addition to pronounced bi-axial non-linearity, they have thermally and load history dependent relaxation. Consequently they are very difficult to model.

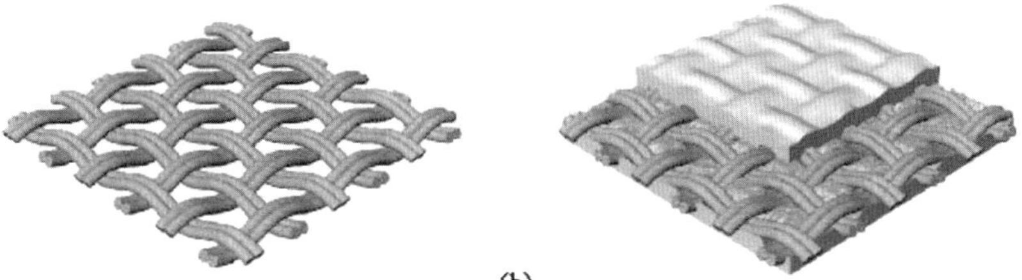

(a) (b)

Fig. 5. Schematics of coated textile composition. (a) Base cloth. (b) Base cloth with coating

Elastic modelling is needed for both "what if ..." prediction and production dimensioning. The relative difficulty of these processes is radically different with lightweight compared to conventional structures. Lightweight structures are inherently safer structures but are more susceptible to aesthetic failure due to patterning errors.

3.1 Modelling and Simulation

It is helpful to consider the terms *Modelling and Simulation*. In common usage these terms are relatively synonymous. In the field of structural engineering design I endorse the following distinction.

- *Structural Modelling* is the general use of a structural model to predict a structural response to a loading condition.
- *Structural Simulation* is Structural Modelling using calibrated models.

Models have different levels of complexity, as well as different levels of accuracy. It is important to recognise that models must be appropriate to the task they are being used for. There is no "best" model for all situations. Many people concentrate on absolute levels of model complexity. There is no doubt that the highest levels of model predictive accuracy will usually be achieved with a model of high complexity. Such accuracy will, however, require extensive quality calibration. Without such calibration the extra sophistication of the model becomes a liability. Some people express the view that a model is only as good as its accuracy. My opinion is that a model is only as good as its relevance to what it is being used to model.

It is a remarkable fact that so little measurement of lightweight structures has been conducted. In particular almost no textile surface measurements have been performed. We advocate the use of non-contact photogrammetry for strain measurement of in-situ textile structures [4].

3.2 Element Types for Textile Modeling

Before considering the problem of modelling elastically non-linear textile, consider the modelling of simple 1 DOF ties. Fig. 6 shows the load deflection behaviour for a steel bar and a low stiffness rope. These illustrate elastically linear and elastically non-linear behaviour respectively. In practice with both linear and non-linear ties stiffness values for load analysis are typically linearised according to narrow bracketing based on the expected and model observed strain. Cable compensation needs to consider in detail the hysteretic relaxation behaviour.

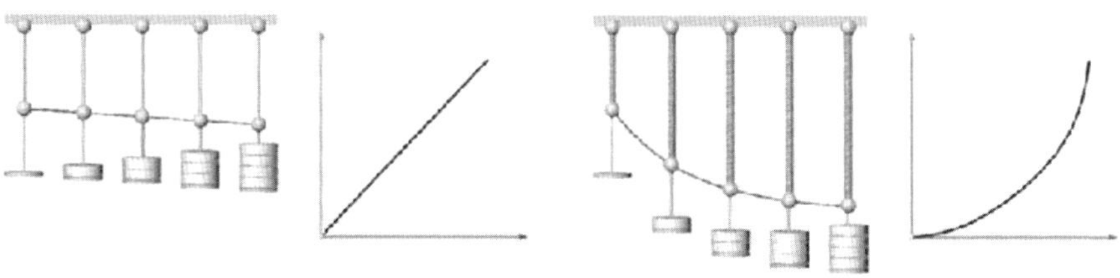

Fig. 6. DOF elastically linear and elastically non-linear ties

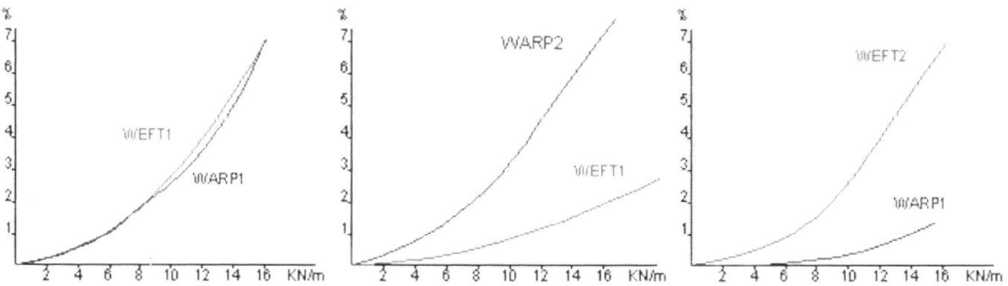

Fig. 7. Stiffness relationships of coated textile (Ferrari [5] *Precontraint* 502)

Modelling textile is much more complicated. Sanitised stiffness relationships from biaxial stress tests are shown in Fig. 7. Typically such graphs are generated for both the warp and weft directions under several fixed warp to weft stress ratios. These are often 1:1, 2:1, 1:2, 1:0 and 0:1. Clearly such tests typically provide extremely small sample sets.

Due to the nature of the weaving process the warp and weft threads are crimped and interact in a complex way. Various composite crimp models have been developed which seek to model the micro structure of the textile within the coating matrix. Crimp models, when properly calibrated, can achieve very high accuracy. As such these models are the most suitable for applications where design duration and cost is secondary to the performance of the structure. Satellite and other space structures are examples. In the architectural textile structure industry, complex crimp models suffer from slow performance and the infeasibility of adequate calibration.

As stated earlier, two element types are mostly used in industrial lightweight structure design systems. These are the cable net (link) and continuum (usually triangulated) models. There is a controversy concerning the relative accuracy and appropriateness of these models. Both elements are used in our systems, as well as several others. We advocate the cable net model as being the most appropriate for current industrial practice on the following basis.

The shear resistance of textile is extremely low and is customarily ignored by everyone. *Poisson*'s ratio effects caused by the textile structure are usually small, but can be noticeable in some parts of some structures. Contrary to extensive belief the cable net model can, like the continuum model, be used to model this effect. It must be stressed, however, that for both methods the sophistication is primitive and limited by the difficulty in choosing *Poisson*'s ratio values. Looking back to the biaxial test data in Fig. 7 above it should be clear that the extraction of a single representative *Poisson*'s ratio parameter is pure fantasy. Another element type which we have developed for textile structure modelling is cruciform based with link forces a function of the two cruciform link strains. We believe that in the medium future the use of automatically calibrated cruciform elements will become feasible. This will require the development of an integrated system linking the producers, test laboratories, software developers and design engineers. The fundamental test will be whether it is possible to use an automatically calibrated model to confidently predict any real biaxial test.

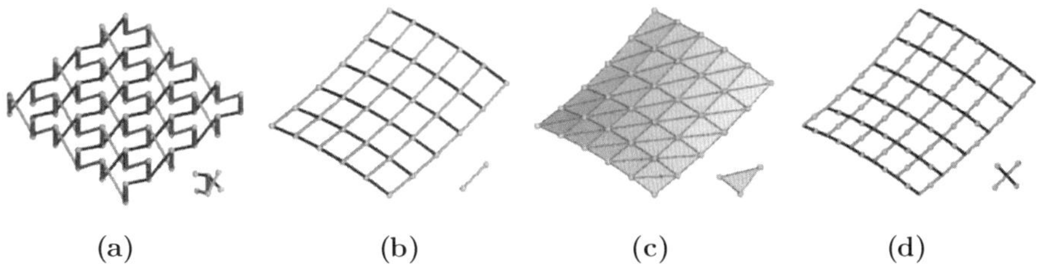

Fig. 8. Element types: (a) Composite crimp, (b) Cable net link, (c) Continuum triangle and (d) Cruciform

It is important to stress again that extra complexity does not necessarily imply extra accuracy. In particular, complex models require many difficult to measure parameters. Which are usually unavailable. When parameters are unavailable, less complex models are no less accurate and maintain contact between the designer and the physical behaviour.

4 Design System Architecture

4.1 Design Sensitivity

Structural configurations vary in their vulnerability to design, fabrication and installation errors. Some configurations are very tolerant and can be successfully installed with large pattern design flaws or installation errors. Other configurations are so sensitive that design and production tolerances have to be extremely narrow. To the novice designer it is not obvious whether a configuration is sensitive or tolerant. Contrast the simple saddles and the *Mina* [6] roof structures shown in Fig. 9.

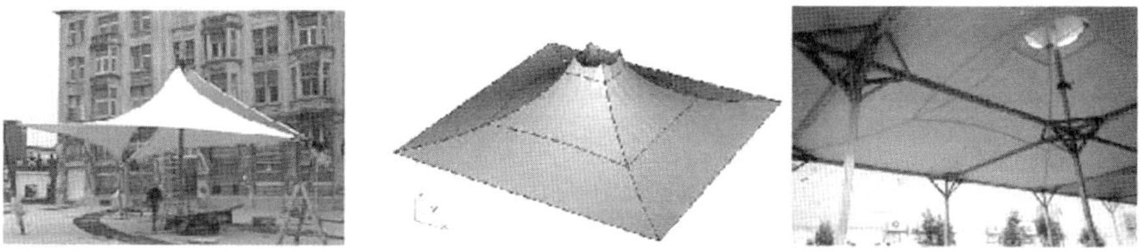

Fig. 9. (a) Tolerant tensile border saddle configuration. (b,c) Sensitive *Mina* valley tent

With the simple saddles we have maximum adjustability. Errors in the compensation of the patterns would mean that the installed system line geometry will differ significantly from the computer model. Such an outcome will usually still be completely acceptable. With the *Mina* roofs we have one of the most sensitive configurations possible. There is a single adjustable DOF, the mast top height. We also have the additional difficulties of a very high stiffness PTFE coated glass textile, and the production constraint requiring very few cloths. Despite the apparent large unit volume of the project, the need for roll optimised patterning meant pattern refinement using scale 1:1 prototypes could not be effective. The *Mina* requirements were clear and therefore had to be accommodated. In many cases, however, avoidably sensitive configurations are regularly designed. The use of parametric search strategies is extremely useful for determining the sensitivity of particular structural configurations.

4.2 System Objectives

Due to the variability of lightweight structure configurations, the use of configurable self-contained tools is a sound basic philosophy. Each of the design modification cycles appropriate to the specifics of the project can be automated. The provision of multiple levels of design abstraction also leads to enhancements of both design quality and productivity. Higher level design parameters, such as boundary sags or textile pre-stress ratios, should be used as much as possible during the design modification cycles. Consider the examples of mesh fineness and mesh boundary sag variation shown in Fig. 10.

Full access to lower level parameters, such as link lengths, should be provided for maximum power. Such low level power functionality should be provided before high level optimisations are introduced. It is also worth highlighting that we favour streamlining the critical design iteration paths before performing such productivity optimizations to "once only" tasks.

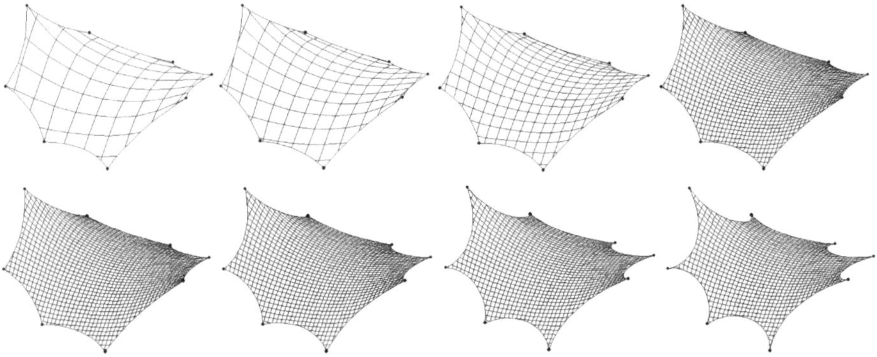

Fig. 10. Advantages of high-level mesh fineness and mesh border parameter variation

5 Current Developments

Dramatic developments are being made to computational design systems. In addition to enhancements in the power of the fundamental tools, emphasis is being focused on task based control interfaces, post-processing detailing and reporting. These system developments are enabling the feasible industrial exploitation of hitherto overly complex hybrid configurations. By integrating sophisticated technology for "as built" geometry feedback in *Adaptive Design* systems, installation accuracy and confidence can be greatly enhanced.

5.1 Complex Hybrid Structure Designs

One of the most interesting developments in lightweight structure design is the increasing use of constant volume pneumatics and flexible battens for primary and secondary load bearing or stiffening. *Tensegrity* structural configurations are also increasing in popularity. The use of active structural systems offers perhaps the most exciting prospects. Active control have been implemented with simple airhalls for many years at a very primitive level. The *Festo Airtecture* [7] developments extended the scope of pneumatically stressed structures significantly. Current developments include much more adventurous configurations, especially those for automatic deployability and even fully autonomous mobility. The actively controlled *Baba Yaga* walking shelter, which brings together many of these concepts, is shown in Fig. 11.

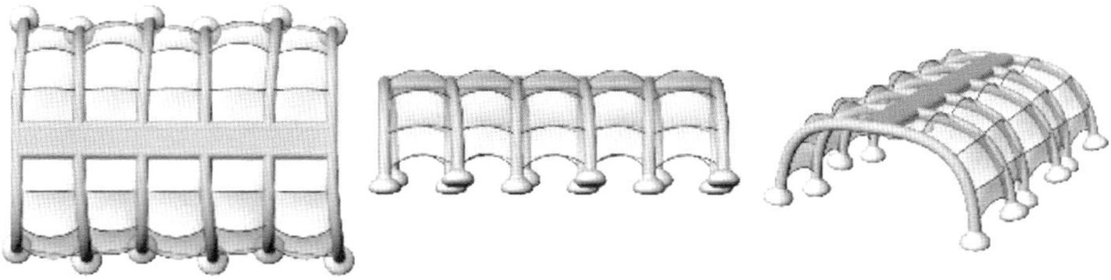

Fig. 11. Baba Yaga pneumatic mobile shelter during walk cycle

5.2 Adaptive Design

Lightweight structures are characterised by their complex doubly curved geometry. Designs are consequently usually composed of many precisely dimensioned components. Achieving an adequate level of installation success can therefore become difficult. With the increased use of CNC CAM more and more adventurous designs are being realised.

In conventional civil engineering CAD practice, design dimensions are mostly defined in absolute terms. Contractors produce structures which are within defined tolerances of these absolute values. Under such a system the "as built" geometry measurement serves only a policing role. If the measurement of "as built" geometry is progressively, and automatically, monitored during installation, the contractor is provided with valuable early warning of installation problems. Moreover, if the feedback geometry is integrated into the core design system, the geometry and location of the remaining components can be adjusted to match those already installed. This shift in paradigm brings the advantages of traditional highly skilled craft based construction methodologies to CAD/CAM.

An excellent traditional example of an *Adaptive Design* strategy is the construction of an igloo. Each of the ice blocks used is cut to fit the existing structure with high precision. Other pertinent examples from history are the leaning tower of Pisa and the building of wooden boats.

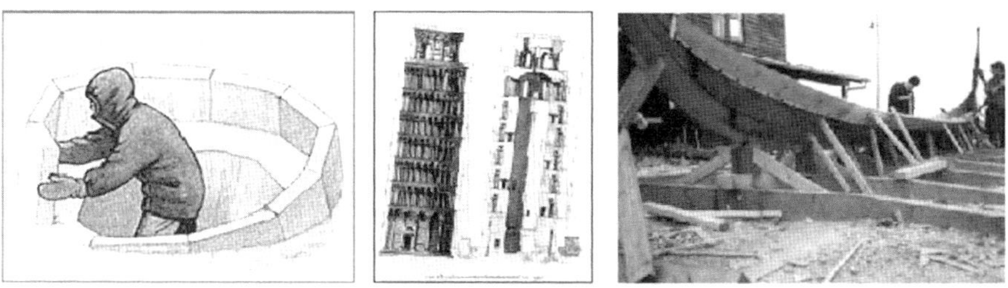

(a) Igloo construction. (b) Tower of Pisa. (c) Wooden boat.

Fig. 12. Traditional examples of adaptive design

Developments in close range photogrammetry and machine vision are now on the verge of providing the necessary automated real-time geometry aquisition systems to enable practical adaptive design systems. These systems look likely to prove very effective in all fields of complex geometric object design. The design of multi-cellular pneumatic structures is clearly of direct applicability.

References

1. TensiNet (2002) German Federal Chancellory. TensiNews 2:4, www.tensinet.com
2. Gründig L, Ekert L and Moncrieff E (1996) Geodesic and Semi-Geodesic Line Algorithms for Cutting Pattern Generation of Architectural Textile Structures. in T.T. Lan (Ed.), Proc. ASIA-PACIFIC Conference on Shell and Spatial Structures, Beijing
3. Moncrieff E and Topping BHV (1990) Computer Methods for the Generation of Membrane Cutting Patterns. Computers and Structures 37(4):441–450
4. Gründig L, Moncrieff E and Schewe H (1995) On the feasibility of using large scale photogrammetry to accurately determine in service strain distribution across three-dimensional textile roofs. 311-326 in W.D. De Wilde and P. Hamelin (Eds.), Euromech 334 Textile Composites and Textile Structures, Lyon
5. Ferrari Textiles (2003). www.ferrari-textiles.com
6. Moncrieff E (2003) Extreme Patterning: Lessons from the cutting pattern generation of the Mina valley tent city and the German federal chancellery projects. in M. Mollaert (Ed.), Designing Tensile Architecture: Proc. TensiNet Symposium, Vrije Universiteit Brussel, September 19th and 20th 2003, Brussels
7. Festo AG (2003). www.festo.com

Recent Developments in the Analytical Design of Textile Membranes

Lothar Gründig[1], Dieter Ströbel[2] and Peter Singer[2]

[1] Technical University Berlin
Sekr. H20, Strasse des 17. Juni 135
10623 Berlin, Germany
gruendig@inge3.bv.tu-berlin.de
[2] technet GmbH Stuttgart
Pestalozzistrasse 8, 70563 Stuttgart, Germany
Dieter.Stroebel@technet-gmbh.com
http:/www:technet-gmbh.de

Summary. *The task of determining appropriate forms for stressed membrane surface structures is considered. Following a brief introduction to the field, the primitive form-finding techniques which were traditionally used for practical surface design are described. The general concepts common to all equilibrium modelling systems are presented next, and then a more detailed exposition of the Force Density Method follows. The extension of the Force Density Method to geometrically non-linear elastic analysis is described. A brief overview of the* Easy *lightweight structure design system is given with particular emphasis paid to the formfinding and statical analysis suite. Finally, some examples are used to illustrate the flexibility and power of* Easy*'s formfinding tools.*
The task of generating planar cutting patterns for stressed membrane surface structures is considered next. Following a brief introduction to the general field of cutting pattern generation, the practical constraints which influence textile surface structures are presented. Several approaches which have been used in the design of practical structures are outlined. These include the physical paper strip modelling technique, together with geodesic string relaxation and flattening approaches. The combined flattening and planar sub-surface regeneration strategy used in the Easy *design system is described in detail. Finally, examples are given to illustrate the capabilities of* Easys *cutting pattern generation tools.*

1 Introduction

Contrary to the design of conventional structures a form finding procedure is needed with respect to textile membrane surfaces because of the direct relationship between the geometrical form and the force distribution. A membrane surface is always in the state of equilibrium of acting forces, and is not defined under unstressed conditions. In general there are two possibilities to perform the formfinding procedures: the

E. Oñate and B. Kröplin (eds.), Textile Composites and Inflatable Structures, 29–45.

physical formfinding procedure and the analytical one. The physical modelling of lightweight structures is characterized by stretchening a soft rubber type material between the chosen boundary positions in order to generate a physically feasible geometry. It has limitations with respect to an accurate description due the small scale of the model. The computational model allows for a proper description by discretizing the surface by a large number of points: a scale problem does not exist any more. Therefore the computational modelling of lightweight structures becomes more and more important; without this technology advanced lightweight structures cannot be built.

2 Analytical Formfinding

The analytical formfinding theories are based on *Finite Element Methods* in general: the surfaces are divided into a number of small finite elements like link elements or triangular elements for example. In such a way all possible geometries can be calculated. There are two theories established in practice: The linear Force Density Approach which uses links as finite elements and the nonlinear *Dynamic Relaxation Method* based on finite triangles.

The Force Density Method

The *Force Density Method* was first published in [1] and extended in [2-3, 9]. It is a mathematical approach for solving the equations of equilibrium for any type of cable network, without requiring any initial coordinates of the structure. This is achieved through the exploitation of a mathematical trick. The essential ideas are as follows. Pin-jointed network structures assume the state of equilibrium when internal forces s and external forces p are balanced.

In the case of node i in Fig. 1

$$\begin{aligned} s_a\cos(a,x)+s_b\cos(b,x)+s_c\cos(c,x)+s_d\cos(d,x)&=p_x\\ s_a\cos(a,y)+s_b\cos(b,y)+s_c\cos(c,y)+s_d\cos(d,y)&=p_y\\ s_a\cos(a,z)+s_b\cos(b,z)+s_c\cos(c,z)+s_d\cos(d,z)&=p_z \end{aligned}$$

where s_a, s_b, s_c and s_d are the bar forces and f.i. $\cos(a,x)$ is the normalised projection length of the cable a on the x-axis. These normalised projection lengths can also be expressed in the form $(x_m-x_i)/a$. Substituting the above cos values with these coordinate difference expressions results in

$$\begin{aligned} \frac{s_a}{a}(x_m-x_i)+\frac{s_b}{b}(x_j-x_i)+\frac{s_c}{c}(x_k-x_i)+\frac{s_d}{d}(x_l-x_i)&=p_x\\ \frac{s_a}{a}(y_m-y_i)+\frac{s_b}{b}(y_j-y_i)+\frac{s_c}{c}(y_k-y_i)+\frac{s_d}{d}(y_l-y_i)&=p_y\\ \frac{s_a}{a}(z_m-z_i)+\frac{s_b}{b}(z_j-z_i)+\frac{s_c}{c}(z_k-z_i)+\frac{s_d}{d}(z_l-z_i)&=p_z \end{aligned}$$

In these equations, the lengths a, b, c and d are nonlinear functions of the coordinates. In addition, the forces may be dependent on the mesh widths or on areas of partial surfaces if the network is a representation of a membrane. If we now apply

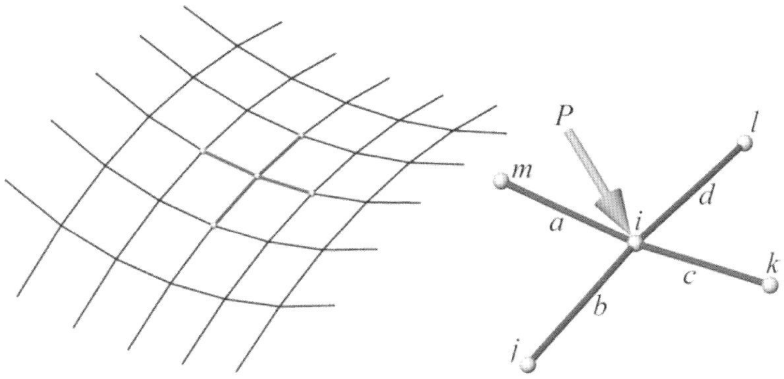

Fig. 1. Part of a cable network

the trick of fixing the force density ratio $s_a/a = q_a$ for every link, linear equations result.

These read

$$\begin{aligned} q_a(x_m - x_i) + q_b(x_j - x_i) + q_c(x_k - x_i) + q_d(x_l - x_i) &= p_x \\ q_a(y_m - y_i) + q_b(y_j - y_i) + q_c(y_k - y_i) + q_d(y_l - y_i) &= p_y \\ q_a(z_m - z_i) + q_b(z_j - z_i) + q_c(z_k - z_i) + q_d(z_l - z_i) &= p_z \end{aligned}$$

The force density values q have to be choosen in advance depending on the desired prestress. The procedure results in practical networks which are reflecting the architectural shapes and beeing harmonically stressed. The system of equations assembled is extremely sparse and can be efficiently solved using the Method of Conjugate Gradients as described in [3].

3 Analytical Formfinding with Technet's Easy Software

The 3 main steps of the Analytical Formfinding of Textile Membrane with the technet's EASY Software are described as follows:

1. Definition of all design parameters, of all boundary conditions as: the coordinates of the fixed points, the warp -and weft direction, the mesh-size and mesh-mode (rectangular or radial meshes), the prestress in warp- and weft direction, the boundary cable specifications (sag or force can be chosen).
2. The linear Analytical Formfinding with Force Densities is performed: the results are: the surface in equilibrium of forces, described by all coordinates of points on the surface, the stress in warp- and weft direction, the boundary cable-forces, the reaction forces of the fixed points. The stresses in warp and weft-direction and the boundary forces may differ in a small range with respect to the desired one from Step 1.

3. Evaluation and visualization tools in order to judge the result of the formfinding. The stresses and forces can be visualized, layer reactions can be shown, contour-lines can be calculated and visualized, cut-lines through the structure can be made.

4 Force Density Statical Analysis

The *Force Density Method* can be extended efficiently to perform the elastic analysis of geometrically non-linear structures. The theoretical background is described in detail in [3] where it was also compared to the *Method of Finite Elements*. It was shown that the *Finite Element Method*'s formulae can be derived directly from the *Force Density Method*'s approach. In addition, the Force Density Method may be seen in a more general way. According to [3] it has been proven to be numerically more stable for the calculation of structures subject to large deflections, where sub-areas often become slack. The nonlinear force density method shows powerful damping characteristics.

Prior to any statical analysis, the form-found structure has to be materialized. Applying Hooke's law the bar force sa is given by:

$$s_a = EA\frac{a - a_0}{a_0}$$

where A is the area of influence fore bar a, E is the modulus of elasticity, and a_0 is the unstressed length of bar a. Substituting s_a by q_a according to $q_a = s_a/a$ results in

$$a_0 = \frac{EAa}{q_a a + EA}$$

Because of a being a function of the coordinates of the bar end points, the materialized unstressed length is a function of the force density q_a, the stressed length a and the element stiffness EA.

In order to perform a statical structural analysis subject to external load, the unstressed lengths have to be kept fixed. This can be achieved mathematically by enforcing the equations of materialization together with the equations of equilibrium. This system of equations is no longer linear. The unknown variables of the enlarged system of equations are now the coordinates x, y, z and the force density values q. Eliminating q from the equations of equilibrium, by applying the formula above to each bar element, leads to a formulation of equations which are identically to those resulting from the *Finite Element Method*. Directly solving the enlarged system has been shown to be highly numerically stable, as initial coordinates for all nodes are available, and positive values or zero values for q can be enforced through the application of powerful damping techniques.

The usual relationship between stress and strain for the orthotropic membrane material is given by:

$$\begin{bmatrix} \sigma_{uu} \\ \sigma_{vv} \end{bmatrix} = \begin{bmatrix} e_{1111} & 0 \\ 0 & e_{2222} \end{bmatrix} \begin{bmatrix} \varepsilon_{uu} \\ \varepsilon_{vv} \end{bmatrix}$$

The warp-direction u and the weft-direction v are independent from each other; this means: the stress in warp-direction σ_{uu} f.i. is only caused by the modulus of elasticity e_{1111} and the strain ε_{uu} in this direction. Because of this independency cable net theories can be used also for Textile membranes.

In [4] the *Force Density Method* has been applied very favorably to triangular surface elements. This triangle elements allow the statical analysis taking into consideration a more precise material behavior in case of Textile membranes. Actually the both material directions u and v are depending from each other; a strain ε_{uu} leads not only to a stress in u-direction but also to a stress σ_{vv} in v-direction caused by the modulus of elasticity e_{1122}. The fact that shear-stress depends on a shear-stiffness e_{1212} seems not to be important for membranes because of its smallness

$$\begin{bmatrix} \sigma_{uu} \\ \sigma_{vv} \\ \sigma_{uv} \end{bmatrix} = \begin{bmatrix} e_{1111} & e_{1122} & 0 \\ e_{2211} & e_{2222} & 0 \\ 0 & 0 & e_{1212} \end{bmatrix} \begin{bmatrix} \varepsilon_{uu} \\ \varepsilon_{vv} \\ \varepsilon_{uv} \end{bmatrix}$$

Using these constitutive equations *Finite Element Methods* should be applied. We are using in this case the finite triangle elements.

5 Further Extensions of the Force Density Approach

The force density approach can be favorably exploited for further applications.

According to [3] the following system of equations of equilibrium is valid:

$$\mathbf{C}^t\mathbf{QCx} = \mathbf{p}$$

$\mathbf{C}$ is the matrix describing the topology of the system, $\mathbf{Q}$ is the diagonal matrix storing the force density values, $\mathbf{x}$ contains the coordinates of the nodes of the figure of equilibrium and $\mathbf{p}$ the external forces acting on the structure. For linear formfinding $\mathbf{C}$, $\mathbf{Q}$ and $\mathbf{p}$ are given, $\mathbf{x}$ is the result of the above equation.

In some applications it might be of interest to know, how close a given geometrical surface will represent a figure of equilibrium. In this case $\mathbf{C}$, $\mathbf{x}$ and $\mathbf{p}$ are given and $\mathbf{q}$ is searched for. As there might be no exact solution to the task described above the best approximating solution is achieved allowing for minimal corrections to the external forces. The system now reads:

$$\mathbf{C}^t\mathbf{Uq} = \mathbf{p} + \mathbf{v}$$

$\mathbf{U}$ now represents a diagonal matrix of coordinate differences $(\mathbf{C}\ \mathbf{x})$, $\mathbf{q}$ the force density values, and $\mathbf{v}$ the residuals of the systems to be minimal.

Solving the system of equations, applying the method of least squares, results in best approximating force density values for any given surface under external loads or subject to internal prestress, if some force density values are chosen as fixed in the structure.

As shown in [3] the system can be extended even further, by choosing $\mathbf{q}$ and $\mathbf{x}$ as observables and enforcing the equations of equilibrium, according to the method of least squares condition equations. In this case an architectural design can be best approximated computationally, enforcing the necessary conditions. This extension proves to be a powerful optimization strategy.

6 Statical Analysis with Technet's Easy Software

The statical Analysis of lightweight structures under external loads can be performed after three introducing steps:

1. To define stiffness values to all finite membrane and cable elements.
2. To calculate the unstressed link lengths by using the assigned stiffness valueses and the prestress of the membrane and the forces in the cables of the formfinding result.
3. To check if the result of the statical analysis with the loads of the formfinding procedure is identical with the formfinding result.

After these three steps, the statical analysis without beam elements under external loads can be achieved very easily.

1. To calculate the external load vectors as for example snow, wind or normal loads.
2. To perform the nonlinear statical analysis: the approximate values, which are needed in this nonlinear process, are given by the formfinding result.
3. Evaluation- and visualization tools in order to judge the result of the statical analysis. The stresses and forces can be visualized and compared with the maximum possible values. Stresses, forces and layer reactions can be shown, contour-lines can be calculated and visualized, cut-lines through the structure can be made, deflection of the nodes can be calculated.

If beam elements are included, the statical analysis under external loads has to be done as follows: all data for the beam-elements as cross-section areas, moments of inertia, local coordinate systems, joints, etc. have to be defined firstly. In order to set all these values in a convenient way the user is supported by a Beam-Editor in EasyBeam.

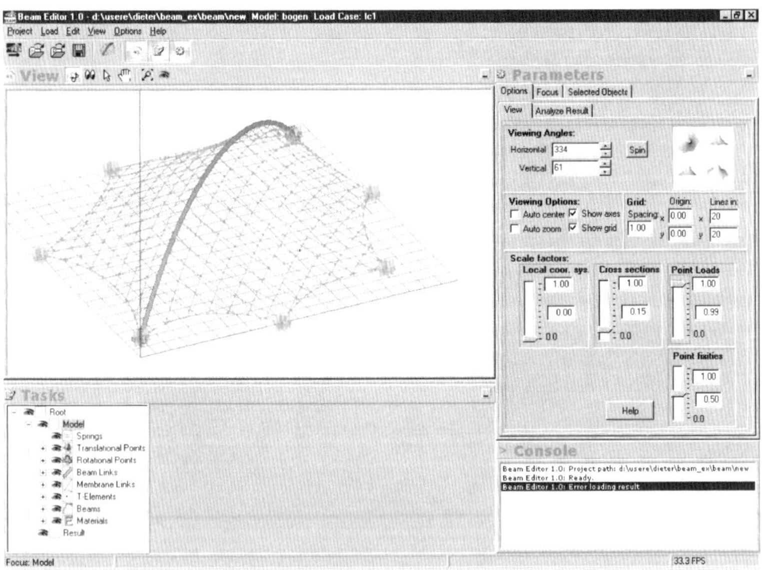

Fig. 2. The Easy beam editor

Then -see above- the steps 1-3 follow. The Beam Editor is also used for checking the results as internal forces and moments, layer-reactions, flexibility-ellipsoids, etc.

7 The Complete Easy Lightweight Structure Design System

The Easy system is composed of a number of program suites. These are represented schematically in Fig. 3.

EasyForm Formfinding of lightweight structures
EasySan Nonlinear Statical Load Analysis (without Beam elements)
EasyCut Cutting pattern generation
EasyBeam Nonlinear hybride Membrane structures including Beam elements
EasyVol Formfinding and Load Analysis of pneumatic constructions

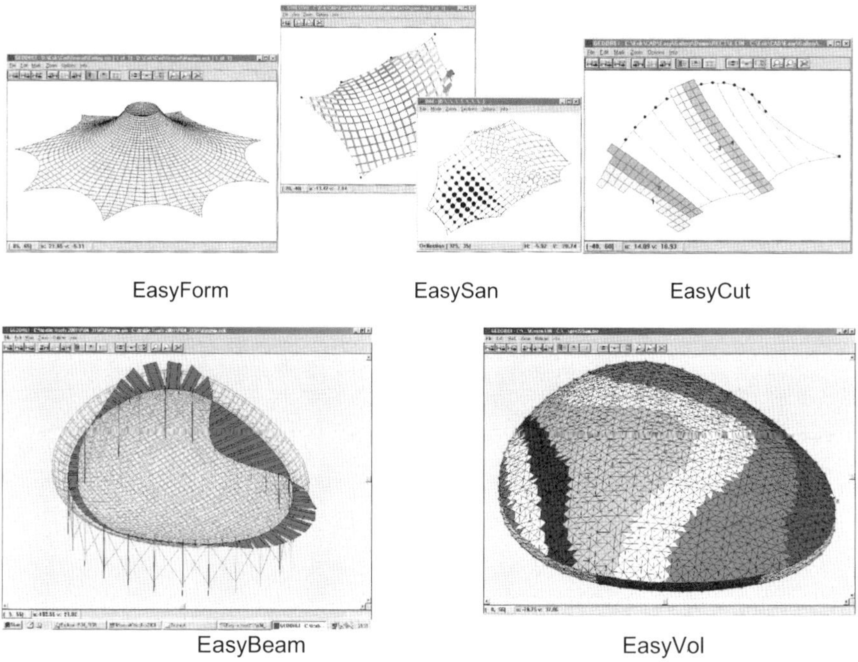

Fig. 3. The Easy program suites

EasyForm comprises the programs used for data generation together with force density form-finding. When the EasySan programs are additionally installed statical structural analysis of non-linear structures becomes possible. The EasyCut programs enable the generation of high quality planar cutting patterns from EasyForm output.

In most situations the incorporation of geometrically non-linear bending elements to lightweight structure models is not economically appropriate. Rather, it is more convenient to treat the beam supports as fully fixed points. The resulting reaction forces on these points are then exported to conventional rigid frame design packages as applied loads. Such a decoupled analysis is appropriate if the resulting deflections are low. However for a sensitive structure decoupling may not be adequate, due the strong interaction of forces causing geometry changes of the membrane surface and the beam elements. In this case the EasyBeam add-on module permits the incorporation of geometrically non-linear frame elements [5].

EasyForm and EasySan together can deal with all standard pneumatic structural configurations which have defined internal pressure prestress. In situations with closed volumes, such as high pressure air beams, this assumption is not valid. It becomes necessary to use more sophisticated algorithms which constrain the cell volumes to prescribed values, and vary the internal pressure accordingly.

Fig. 4. Formfinding and statical analysis under inner pressure and buoyancy

8 Cutting Pattern Generation of Textile Structures

The theory, being used to project a 2D surface in 3 dimensional space to a 2D surface in a plane is very old; it is part of the mathematical field named map projection theory. For example the Mercator Projection dates back to the 17th century.

The surfaces, which are used in practical membrane structure design are in general not developable without distortions. In addition there does not exist a geometrical shape without prestress. The map projection theories -used for the flattening of textile membranes- minimize the distortions with respect to lengths, angles and areas respectively.

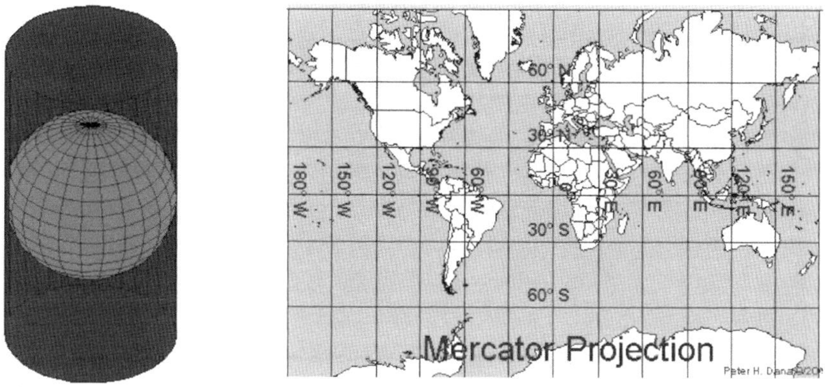

Fig. 5. Mercator projection

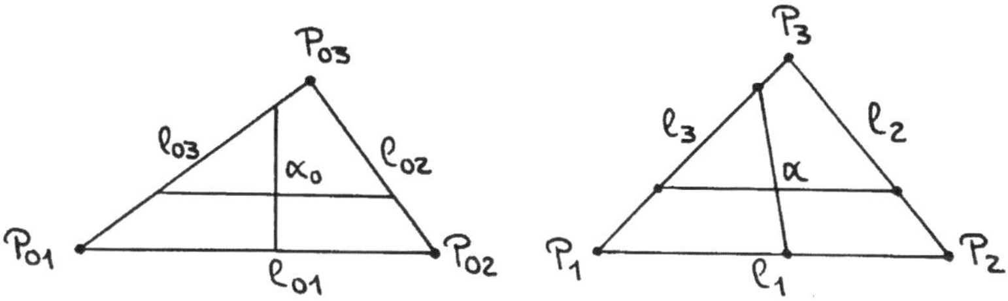

Fig. 6. Triangles non deformed (3D) and deformed (2D)

The theory to be applied optimizes the total distortion energy by means of the adjustment thoory.

The surface, which has to be flattened is described using finite triangles. The distortion between the non deformed and deformed situation can be calculated and has to be minimized for all triangles.

The paper strip method is exactly described in [10]. Practical examples are described in [6-8].

Fig. 7 illustrates the paper strip method. A paper is pressed on the surface of the physical model in such a way, that the seam line and the border of the paper are touching each other as good as (in general the paper strip will touch the surface only in one common line, with an increasing distance from this line the difference between paper strip and surface becomes larger.) In the next step a needle is used to perforate the paper strip in a certain number of equidistant points so that the neighboring seam or the boundary line is reached on the shortest way. In doing so the direction of the needle has to be perpendicular to the surface. The connection of the needle holes by straight lines on the flat paper strip leads to the patterns.

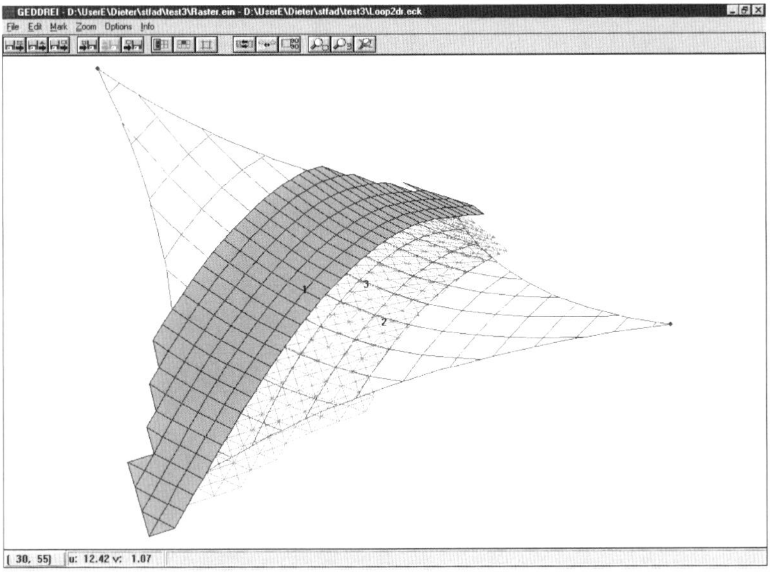

Fig. 7. Paper strip method

9 Cutting Pattern Generation with Technet's Easy Software

The Cutting pattern generation can be performed in the following steps:

1. Geodesic lines are created as seam lines.
2. Cutting procedures are used to cut the surface into different sub-surfaces according to these geodesic lines.
3. Ways of flattening are achieved: map projection, paper strip method.
4. Spline algorithms are applied to create equidistant points on the planar circumference.
5. Boundary adjustment is performed in order to produce identical seam lengths.
6. Compensation values are defined to compensate the strips.
7. Job-drawings are produced.

10 Flexibility Ellipsoids for the Evaluation of Mechanical Structures

The geodetic network adjustment determines the geometrical position of points connected by link observables. In general the adjustment theory was invented by C.F. GAUSS by mimizing the residuals of the observations. The process is well known as Least-Squares-Method.

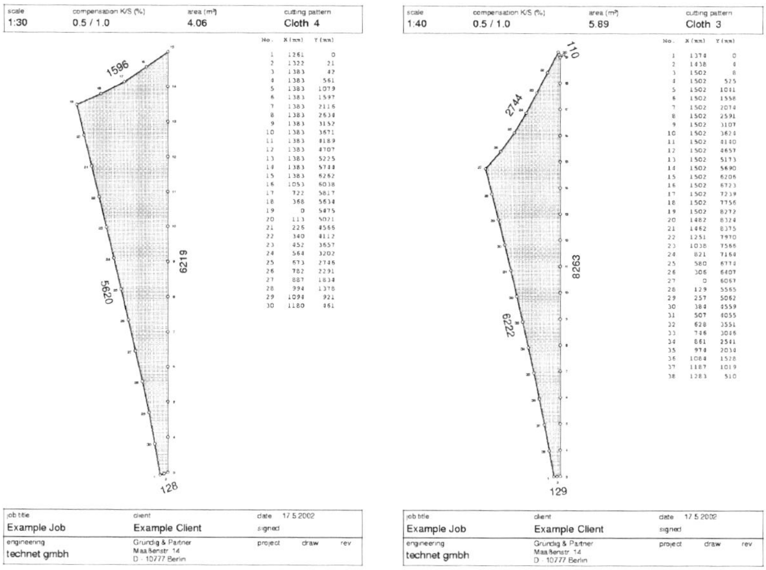

Fig. 8. Cutting patterns

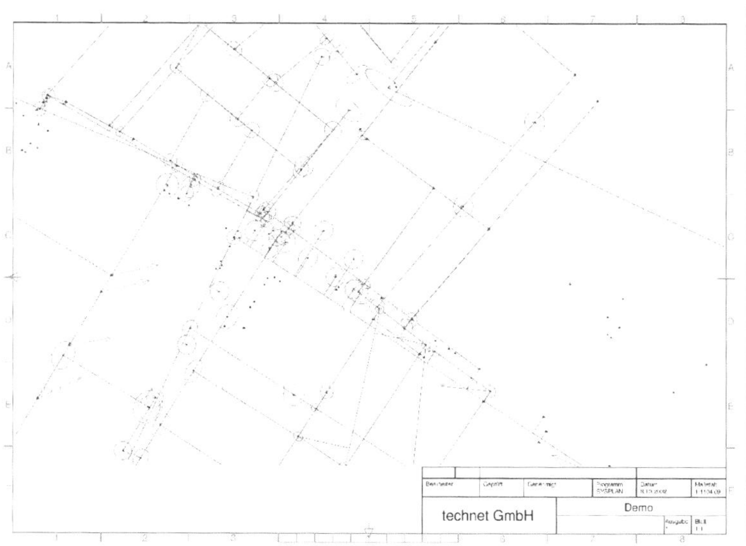

Fig. 9. Error-ellipses in a geodetic network

Analogical relationships between this adjustment theory and the calculation of mechanical structures exist due to the fact that in mechanical structures minimal-principles are valid too; the total energy is minimal.[5] The error-ellipsoids of the geodetic network adjustment have been an important tool for the evaluation of the point quality for more than 100 years, they show very descriptively the accuracy of the points [14].

The error-ellipsoids of the geodetic network adjustment are an essential tool used for the evaluation of mechanical structures too. The relation between the deflection v and the external loads p in the global coordinate system is given using the stiffness matrix $\mathbf{S}$

$$\mathbf{Sv} = \mathbf{p}$$

The inverse relation is

$$\mathbf{S}^{-1}\mathbf{p} = \mathbf{Fp} = \mathbf{v}$$

The inverse stiffness matrix $\mathbf{F}$ is usually called flexibility matrix; the matrix defines the relation between the external loads at a node and the according deflections. In detail, we receive for a node with the coordinates (x, y, z):

$$\begin{bmatrix} \ddots & & \vdots & & \vdots \\ & \begin{bmatrix} f_{xx} & f_{xy} & f_{xz} \\ & f_{yy} & f_{yz} \\ sym. & & f_{zz} \end{bmatrix} & & & \vdots \\ sym. & & & & \ddots \end{bmatrix} \begin{bmatrix} \vdots \\ \begin{bmatrix} p_x \\ p_y \\ p_z \end{bmatrix} \\ \vdots \end{bmatrix} = \begin{bmatrix} \vdots \\ \begin{bmatrix} v_x \\ v_y \\ v_z \end{bmatrix} \\ \vdots \end{bmatrix}$$

For the following deductions the submatrix $\mathbf{F}$ as 3×3 matrix has been introduced for any point P and the 3×1 vectors p and v referring also to this point. $\mathbf{F}$ is now not any more the total flexibility matrix, but only a sub matrix with 9 elements, p are the 3 components of the external load and v the deflections of the point. Let us assume, that all point loads are zero with the exception of the considered point, then the sub matrix $\mathbf{F}$ describes the relation between the point load and the deflections of the point. All other elements in $\mathbf{F}$ are not important in this context, because they are multiplied with zero. Hence, we receive for a free point:

$$\mathbf{Fp} = \mathbf{v}$$

In detail

$$\begin{bmatrix} f_{xx} & f_{xy} & f_{xz} \\ & f_{yy} & f_{yz} \\ sym. & & f_{zz} \end{bmatrix} \begin{bmatrix} p_x \\ p_y \\ p_z \end{bmatrix} = \begin{bmatrix} v_x \\ v_y \\ v_z \end{bmatrix}$$

How the deflections v and the external loads p are transformed by an orthogonal transformation $\mathbf{R}$ (orthogonal means: the transposed matrix equals the inverse matrix $\mathbf{R}^t = \mathbf{R}^{-1}$), the deflections are in the transformed system u and the loads q

$$\mathbf{u} = \mathbf{R}^t\mathbf{v} \rightarrow \mathbf{v} = \mathbf{Ru}$$
$$\mathbf{q} = \mathbf{R}^t\mathbf{p} \rightarrow \mathbf{p} = \mathbf{Rq}$$

The relation between the deflections u and the loads q is after some conversions:

$$\mathbf{u} = \mathbf{R}^t\mathbf{FRq} = \mathbf{Dq}$$

Now the question arises, if there is a matrix $\mathbf{R}$ generating a matrix $\mathbf{D}$ having only diagonal elements. The answer is yes, if $\mathbf{F}$ is symmetric with real values. $\mathbf{F}$ is always real symmetric in the described applications and also positive definite if there is a stable equilibrium; therefore all eigenvalues are real and all eigenvectors are orthogonal to each other. The diagonal elements of the matrix $\mathbf{D}$ are the eigenvalues of $\mathbf{F}$, the matrix of rotations $\mathbf{R}$ is generated by the eigenvectors.

This can easily be proofed. We suppose the eigenvalues of 3×3. Matrix $\mathbf{F}$ to be d_1, d_2 and d_3 and the corresponding eigenvectors to be $\mathbf{r}_1$, $\mathbf{r}_2$ and $\mathbf{r}_3$. The unit matrix is $\mathbf{E}$. We have at the beginning for the calculations of the eigenvalues:

$$\begin{aligned} \mathbf{F}r_1 = d_1 r_1 &\rightarrow (\mathbf{F} - d_1\mathbf{E})r_1 = 0 \\ \mathbf{F}r_2 = d_2 r_2 &\rightarrow (\mathbf{F} - d_2\mathbf{E})r_2 = 0 \\ \mathbf{F}r_3 = d_3 r_3 &\rightarrow (\mathbf{F} - d_3\mathbf{E})r_3 = 0 \end{aligned}$$

We define the diagonal matrix $\mathbf{D}$ with the diagonal elements d_1, d_2 and d_3 and the rotation matrix $\mathbf{R} = (\mathbf{r}_1, \mathbf{r}_2, \mathbf{r}_3)$. Doing so, we receive from the upper equations:

$$\mathbf{FR} = \mathbf{RD}$$

The eigenvectors $\mathbf{R}$ are pair wise orthogonal to each other, therefore the orthogonality of $\mathbf{R}$ is obvious, hence $\mathbf{RR}^t = \mathbf{E}$. Multiplication from the left hand side with $\mathbf{R}^t$ leads to

$$\mathbf{R}^t\mathbf{FR} = \mathbf{R}^t\mathbf{RD} = \mathbf{ED} = \mathbf{D}$$

We have seen, that the eigenvalues of the matrix $\mathbf{F}$ are the diagonal elements of the diagonal matrix $\mathbf{D}$. This matrix give the relation in a rotated coordinate system between the deflection u and the external loads q

$$\mathbf{u} = \mathbf{Dq}$$

whereby we receive the new coordinate system by creating the rotation matrix $\mathbf{R}$ using the eigenvectors. The upper equation reads in detail.

Now we are investigating which surface is created by the deflections $u = (u, v, w)$, if the external load

$$\begin{bmatrix} u \\ v \\ w \end{bmatrix} = \begin{bmatrix} d_1 & 0 & 0 \\ 0 & d_2 & 0 \\ 0 & 0 & d_3 \end{bmatrix} \begin{bmatrix} q_u \\ q_v \\ q_w \end{bmatrix}$$

vector $q = (q_u, q_v, q_w)$ is showing in any direction having the size 1. We receive:

$$q_u q_u + q_v q_v + q_w q_w = 1$$

In general the matrix $\mathbf{F}$ is positive definite; all eigenvalues are therefore positive. We get immediately with $q_u = u/d_1$, $q_v = v/d_2$ and $q_w = w/d_3$ the equation of an ellipsoid

$$q_u q_u + q_v q_v + q_w q_w = \frac{u^2}{d_1^2} + \frac{v^2}{d_2^2} + \frac{w^2}{d_3^2} = 1$$

as a surface being created by the deflections, if the unit load vector is rotating around a point

$$u_F = fq$$

We see, that the direction of the external load q and the direction of the deflections are only identical in case of a sphere, if $d_1 = d_2 = d_3$. In general we do not have a sphere, therefore we investigate the question, which geometrical figure is created, if the size of the deflection is accorded to the direction of the external load q; the point on the load vector having the size of the total deflection is called root point F; its coordinates are:

$$\begin{bmatrix} u_F \\ v_F \\ w_F \end{bmatrix} = f \begin{bmatrix} q_u \\ q_v \\ q_w \end{bmatrix}$$

The load q leads to the deflections $\mathbf{u} = [\, u \quad v \quad w\,]$. Because of the fact that the load has the size 1, is f the total length of the deflection, hence:

$$f^2 = u_F^2 + v_F^2 + w_F^2 = u^2 + v^2 + w^2$$

Also

$$\begin{bmatrix} u \\ v \\ w \end{bmatrix} = \begin{bmatrix} d_1 q_u \\ d_2 q_v \\ d_3 q_w \end{bmatrix} = \begin{bmatrix} d_1 u_F f^{-1} \\ d_2 v_F f^{-1} \\ d_3 w_F f^{-1} \end{bmatrix}$$

We receive by substituting

$$f^2 = u_F^2 d_1^2 f^{-2} + v_F^2 d_2^2 f^{-2} + w_F^2 d_3^2 f^{-2} = u^2 + v^2 + w^2$$

By a simple conversion we end up with the equation of a so-called Booth-Lemniscate

$$f^4 - u_F^2 d_1^2 - v_F^2 d_2^2 - w_F^2 d_3^2 = 0$$

In general written as follows

$$(u_F^2 + v_F^2 + w_F^2) - u_F^2 d_1^2 - v_F^2 d_2^2 - w_F^2 d_3^2 = 0$$

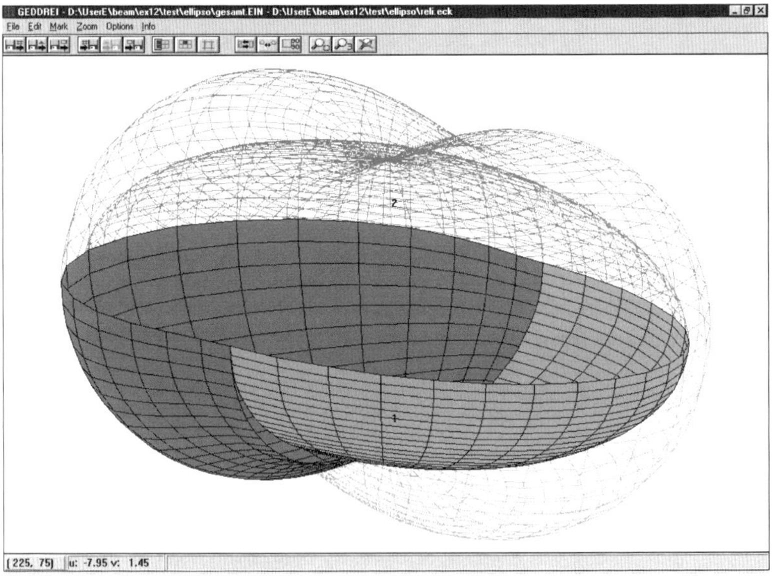

Fig. 10. Flexibility ellipsoide and lemniscate

The lemniscate has 6 common points with the ellipsoid of the deflections, this are the points on the local coordinate axis of the ellipsoid. In this 6 point the direction of the external loads and the deflections of the point are identical.

The example in Fig. 12 clarifies that the points being far from the fixed points are generally more flexible than the closed ones.

In Fig. 13 we see immediately that the points in the upper layer are very flexible in tangential directions. In order to change this behavior diagonals should be introduced [11].

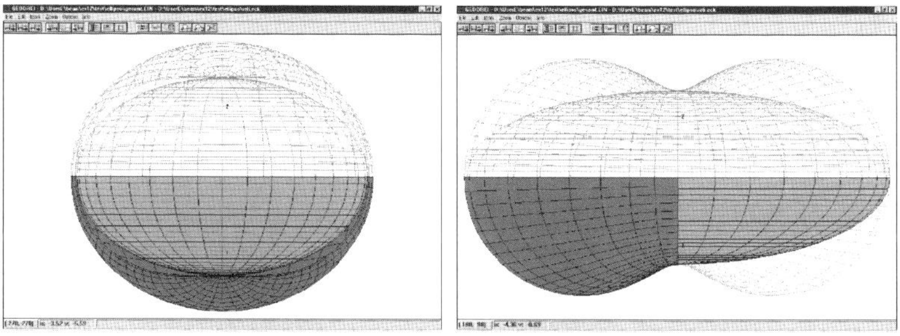

Fig. 11. Side views

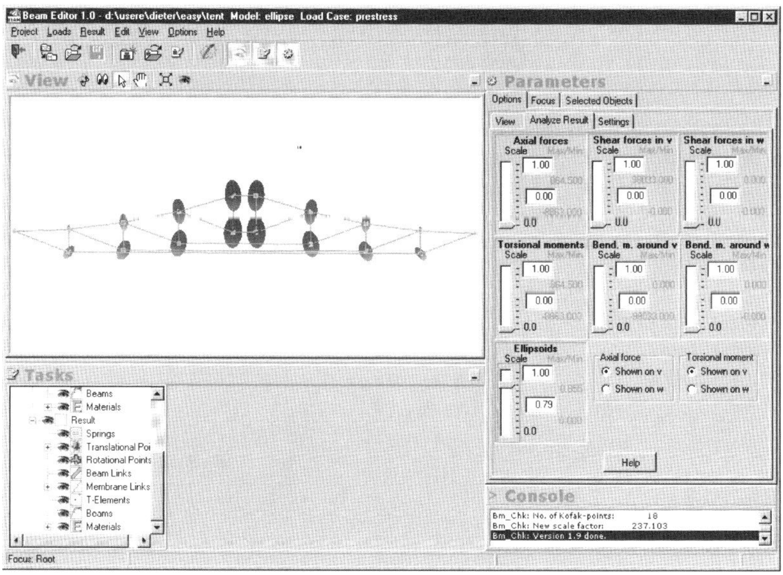

Fig. 12. 2-dimensional structure

Some remarks to the practical realization of the flexibility ellipsoids in computer software.

Due to the fact, that the inverse stiffness matrix is generally not calculated; conventional methods to solve this matrix are very time consuming and also intensive with respect to storage. In those cases the ellipsoids cannot be calculated. However, we are using in so-called hyper sparse algorithms, which are storing only the non-zero elements [12],[13] and generates the inverse for only those elements needed. Without such a strategy the calculation of those ellipsoids is not possible for large structures.

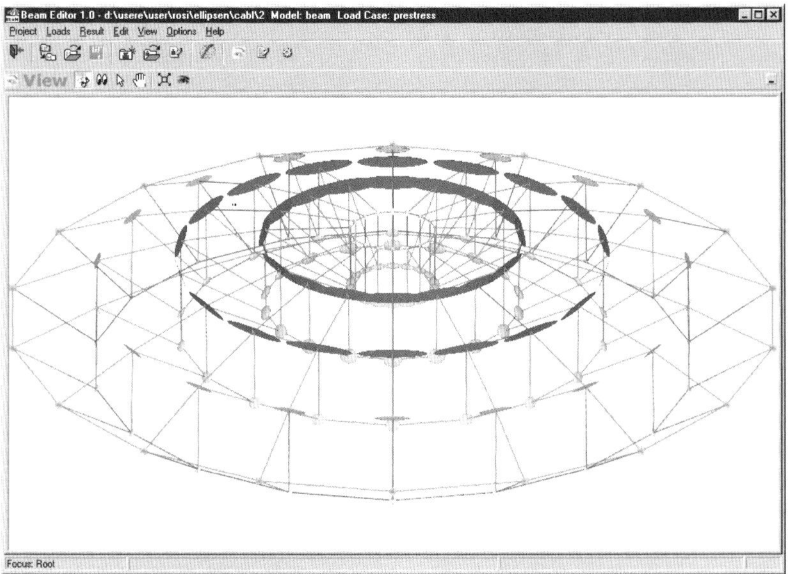

Fig. 13. 3-dimensional space-structure

11 Conclusions

It has been shown that, by using a modular approach for the design of membrane structure surfaces, the resulting system is extremely powerful and flexible. The very large number of structures which have been built using the Easy tools (many thousands) prove the validity of this strategy.

References

1. Linkwitz K and Schek H-J (1971) Einige Bemerkungen zur Berechnung von vorgespannten Seilnetzkonstruktionen. Ingenieur-Archiv 40:145–158.
2. Schek HJ (1974) The force density method for form finding and computation of general net- works. Computer Methods in Applied Mechanics and Engineering 3:115–134.

3. Gründig L (1975) Die Berechnung von vorgespannten Seilnetzen und Hängenetzen unter Berücksichtigung ihrer topologischen und physikalischen Eigenschaften und der Ausgleichungs- rechnung. Dissertationsschrift, DGK Reihe C, Nr. 216, 1976 and SFB 64-Mitteilungen 34/1976.
4. Singer P (1995) Die Berechnung von Minimalflächen, Seifenblasen, Membrane und Pneus aus geodätischer Sicht. Dissertationsschrift, DGK Reihe C, Nr. 448.
5. Ströbel D (1997) Die Anwendung der Ausgleichungsrechnung auf elastomechanische Systeme. DGK, Reihe C, Nr. 478.
6. Gründig L and Moncrieff E (1998) Formfinding, Analysis and Patterning of Regular and Irregular-Mesh Cablenet Structures. in Hough R and Melchers R (Eds.), LSA98: Lightweight Structures in Architecture Engineering and Construction Proceedings IASS 39th Congress, October, Sydney, Australia, IASS/LSAA.
7. Moncrieff E, Gründig L and Ströbel D (1999) The Cutting Pattern Generation of the Pilgrim's Tents for Phase II of the Mina Valley Project. in Astudillo, R. and Madrid, A. J. (Eds.), Proc. IASS 40th Anniversary Congress, September 20-24, 1999, Madrid, Spain, IASS/CEDEX.
8. Gründig L, Moncrieff E, Singer P and Ströbel D (2000) High-performance cutting pattern generation of architectural textile structures. in Papadrakakis E (Ed.), Proc. IASS-IACM 2000 Fourth International Colloquium on Computation of Shell & Spatial Structures, June 4-7, Chania-Crete, Greece.
9. Gründig L, Moncrieff E, Singer P and Ströbel D (2000) A history of the principal developements and applications of the force density method in Germany 1970-1999. in Papadrakakis, E. (Ed.), Proc. IASS-IACM 2000 Fourth International Colloquium on Computation of Shell & Spatial Structures, June 4-7, Chania-Crete, Greece.
10. Moncrieff E and Gründig L (2000) Computation modelling of lightweight structures. 5th international workshop on the design and practical realization of architecture membrane structures, Technical University Berlin, June 22-24, Germany.
11. Wagner R (2202) Tensegrity and Cable Domes. In: The Seventh International Workshop on the Design and Realisation of Architectural Membrane Structures, Technische Hochschule Berlin.
12. Ströbel D (1985) Entwicklung von Hypersparsetechniken zur Gleichungslösung für Probleme in der Ausgleichungsrechnung und Statik. Diplomarbeit, Institut für Anwendungen der Geodäsie im Bauwesen, Technische Universität Stuttgart.
13. Gründig L, Bahndorf J und Ströbel D (1985) Hyper Sparse Techniques Applied to Geodetic Networks. In: 7th International Symposium on Geodetic Computations, Krakau.
14. Linkwitz K (1988) Einige Bemerkungen zur Fehlerellipse und zum Fehlerellipsoid. Vermessung, Photogrammetrie, Kulturtechnik, Schweizerischer Verein fr Vermessungs- und Kulturtechnik (SVVK), S. 345–364, 86. Jahrgang, Heft 7.

http://www.technet-gmbh.com

Finite Element Analysis of Membrane Structures

Robert L. Taylor*[1], Eugenio Oñate[2] and Pere-Andreu Ubach[2]

[1] Department of Civil and Environmental Engineering
University of California at Berkeley, USA
rlt@ce.vulture.berkeley.edu
[2] International Center for Numerical Methods in Engineering
Edificio C1, Gran Capitán s/n
08034 Barcelona - Spain
onate@cimne.upc.es

Summary. *This paper summarizes the development for a large displacement formulation of a membrance composed of three-node triangular elements. A formulation in terms of the deformation gradient is first constructed in terms of nodal variables. In particular, the use of the right Cauchy-Green deformation tensor is shown to lead to a particulary simple representation in terms of nodal quantities. This may then be used to construct general models for use in static and transient analyses.*

Key words: Membrane structures, finite element

1 Introduction

The behavior of curved, thin bodies can be modeled by a membrane theory of shells. In such a theory only the in-plane stress resultants are included. The deformation state for a membrane may be represented by the position of points on the two-dimensional surface. General theories for shells may be specialized to those for a membrane by ignoring the resultant couples and associated changes in curvature deformations as well as any transverse shearing effects. A numerical approximation to the shell may then be constructed using a finite element approach. Examples for general shell theory and finite element solution may be found for small deformations in standard books.[1] Theory for large deformation can proceed following the presentations of Simo *et al.*[2, 3, 4, 5] or Ramm et al.[6, 7, 8, 9, 10, 11, 12, 13]

For the simplest shape finite element composed of a 3-node triangular form with displacement parameters at each vertex (a 9-degree of freedom element) it is far simpler to formulate the membrane behavior directly. This is especially true for large displacement response. Here the initially flat form of a simple triangular response is maintained throughout all deformation states. Consequently, one may proceed

* Visiting Professor, CIMNE, UPC, Barcelona, Spain.

E. Oñate and B. Kröplin (eds.), Textile Composites and Inflatable Structures, 47–68.

directly with the construction of the kinematic behavior, even in the presence of large strains. This approach is followed in the present work.

The loading of a membrane is often by pressures which remain normal to the surface throughout all deformations. Such *follower* loading generally leads to a form which yields an unsymmetric tangent matrix. Such formulation has been presented in works by Schweizerhof and Ramm[14] and by Simo *et al.*[15] The general approach presented in the last work is used for the special case of the flat triangular element used in this work.

The formulation included in the present study includes inertial and damping terms based on second and first time derivatives of the motion. These are discretized in time using standard techniques (e.g., the Newmark method[16, 17]). Both explicit and implicit schemes are presented together with all linearization steps needed to implement a full Newton type solution. The inclusion of the damping term permits solution of the first order form in order to obtain a final static solution. Generally, the first order form is used until the final state is reached at which point the rate terms are deleted and the full static solution achieved using a standard Newton iterative method.

The work presented is implemented in the general purpose finite element solution system *FEAP*[18] and used to solve example problems. The solution to some basic example problems are included to show the behavior of the element and solution strategies developed.

2 Governing Equations

Reference configuration coordinates in the global Cartesian frame are indicated in upper case by $\boldsymbol{X}$ and current configuration in lower case by $\boldsymbol{x}$. The difference between the coordinates defines a displacement $\boldsymbol{u}$.

Using standard interpolation for a linear triangle positions in the element may be specified as

$$\boldsymbol{X} = \xi_\alpha \tilde{\boldsymbol{X}}^\alpha \tag{1}$$

in the reference configuration and

$$\boldsymbol{x} = \xi_\alpha \tilde{\boldsymbol{x}}^\alpha \tag{2}$$

for the current configurations. If necessary, the displacement vector may be deduced as

$$\boldsymbol{u} = \xi_\alpha \tilde{\boldsymbol{u}}^\alpha \tag{3}$$

In the above $\tilde{\boldsymbol{X}}^\alpha$, $\tilde{\boldsymbol{x}}^\alpha$, $\tilde{\boldsymbol{u}}^\alpha$ denote *nodal* values of the reference coordinates, current coordinates and displacement vector, respectively. Furthermore, the natural (*area*) coordinates satisfy the constraint

$$\xi_1 + \xi_1 + \xi_3 = 1 \tag{4}$$

It is convenient to introduce a surface coordinate system denoted by Y_1, Y_2 with normal direction N in the reference state and y_1, y_2 with normal direction n in the current state (see Fig. 1).

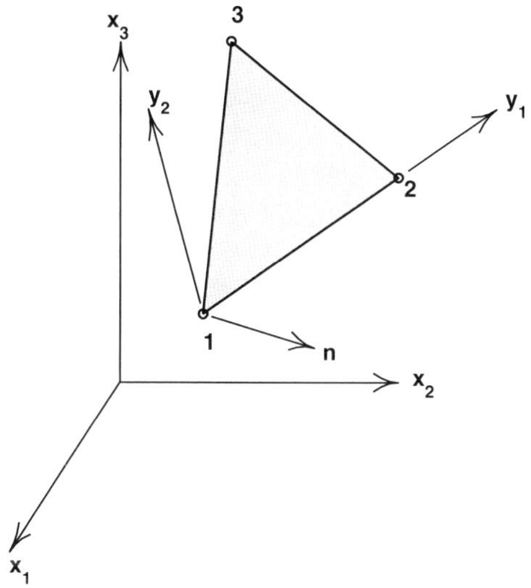

Fig. 1. Description of coordinates for triangular element

Placing the origin of the Y_I and y_i coordinates at nodal location $\tilde{\boldsymbol{X}}^1$ and $\tilde{\boldsymbol{x}}^1$, respectively, the unit base vectors may be constructed from the linear displacement triangle by aligning the first vector along the $1-2$ side. Accordingly, we define the first unit vector as

$$\boldsymbol{v}_1 = \frac{\tilde{\boldsymbol{x}}^2 - \tilde{\boldsymbol{x}}^1}{\|\tilde{\boldsymbol{x}}^2 - \tilde{\boldsymbol{x}}^1\|} = \frac{\Delta\tilde{\boldsymbol{x}}^{21}}{\|\Delta\tilde{\boldsymbol{x}}^{21}\|} \tag{5}$$

where

$$\Delta\tilde{\boldsymbol{x}}^{ij} = \tilde{\boldsymbol{x}}^i - \tilde{\boldsymbol{x}}^j \quad \text{and} \quad \|\boldsymbol{a}\| = \left(\boldsymbol{a}^T\boldsymbol{a}\right)^{1/2} \quad .$$

Next a vector *normal* to the triangle is constructed as

$$\boldsymbol{v}_3 = \Delta\tilde{\boldsymbol{x}}^{21} \times \Delta\tilde{\boldsymbol{x}}^{31} \tag{6}$$

and normalized to a unit vector as

$$\boldsymbol{n} = \frac{\boldsymbol{v}_3}{\|\boldsymbol{v}_3\|} \tag{7}$$

The vector $\boldsymbol{v}_3$ plays a special role in later development of nodal forces for *follower pressure loading* as it is twice the area of the triangle times the unit normal vector $\boldsymbol{n}$.

Finally, a second orthogonal unit vector in the plane of the triangle is be computed as

$$\boldsymbol{v}_2 = \boldsymbol{n} \times \boldsymbol{v}_1 \quad . \tag{8}$$

The above developments have been performed based on the current configuration. However, reference quantities may be deduced by replacing lower case letters by upper case ones (e.g., $\boldsymbol{v}_1 \rightarrow \boldsymbol{V}_1$, etc.).

With the above base vectors defined, positions in the plane of the triangle may be given directly as

$$y_i = \left(\boldsymbol{x} - \tilde{\boldsymbol{x}}^1\right) \cdot \boldsymbol{v}_i \tag{9}$$

In general, an interpolation may be given as

$$\boldsymbol{y} = \xi_\alpha \tilde{\boldsymbol{y}}^\alpha \tag{10}$$

We note from Eq. (9) that $\tilde{\boldsymbol{y}}^1$ is identically zero hence Eq. (10) reduces to

$$\boldsymbol{y} = \xi_2\, \tilde{\boldsymbol{y}}^2 + \xi_3\, \tilde{\boldsymbol{y}}^3 \quad . \tag{11}$$

However, the constraint (4) still restricts the admissible values for ξ_2 and ξ_3.

2.1 Deformation Gradient

From the above description of the motion of the triangle it is now possible to deduce the deformation gradient in the plane of the triangle as

$$\boldsymbol{F} = \frac{\partial \boldsymbol{y}}{\partial \boldsymbol{Y}} = \mathbf{1} + \frac{\partial \boldsymbol{u}}{\partial \boldsymbol{Y}} \tag{12}$$

Using the parametric representations (11) we can compute the deformation gradient from

$$\frac{\partial \boldsymbol{y}}{\partial \boldsymbol{Y}} \frac{\partial \boldsymbol{Y}}{\partial \boldsymbol{\xi}} = \boldsymbol{F} \frac{\partial \boldsymbol{Y}}{\partial \boldsymbol{\xi}} = \frac{\partial \boldsymbol{y}}{\partial \boldsymbol{\xi}} \tag{13}$$

If we define the arrays $\boldsymbol{J}$ and $\boldsymbol{j}$ as

$$\boldsymbol{J} = \frac{\partial \boldsymbol{Y}}{\partial \boldsymbol{\xi}} \quad \text{and} \quad \boldsymbol{j} = \frac{\partial \boldsymbol{y}}{\partial \boldsymbol{\xi}} \tag{14}$$

then the deformation gradient is given by

$$\boldsymbol{F} = \boldsymbol{j}\, \boldsymbol{J}^{-1} \tag{15}$$

In the above $\boldsymbol{J}$ is the Jacobian transformation for the reference frame and $\boldsymbol{j}$ that for the current frame. Expanding the relations for each Jacobian we obtain

$$\boldsymbol{J} = \begin{bmatrix} \left(\Delta \tilde{\boldsymbol{X}}^{21}\right)^T \boldsymbol{V}_1 & \left(\Delta \tilde{\boldsymbol{X}}^{31}\right)^T \boldsymbol{V}_1 \\ \left(\Delta \tilde{\boldsymbol{X}}^{21}\right)^T \boldsymbol{V}_2 & \left(\Delta \tilde{\boldsymbol{X}}^{31}\right)^T \boldsymbol{V}_2 \end{bmatrix} \tag{16}$$

and

$$\boldsymbol{j} = \begin{bmatrix} \left(\Delta \tilde{\boldsymbol{x}}^{21}\right)^T \boldsymbol{v}_1 & \left(\Delta \tilde{\boldsymbol{x}}^{31}\right)^T \boldsymbol{v}_1 \\ \left(\Delta \tilde{\boldsymbol{x}}^{21}\right)^T \boldsymbol{v}_2 & \left(\Delta \tilde{\boldsymbol{x}}^{31}\right)^T \boldsymbol{v}_2 \end{bmatrix} \tag{17}$$

By noting that $\tilde{\boldsymbol{X}}^2 - \tilde{\boldsymbol{X}}^1$ is orthogonal to $\boldsymbol{V}_2$ and similarly for the current configuration that $\tilde{\boldsymbol{x}}^2 - \tilde{\boldsymbol{x}}^1$ is orthogonal to $\boldsymbol{v}_2$ and in addition using the definition for $\boldsymbol{V}_i$

and $\boldsymbol{v}_i$ the above simplify to

$$\boldsymbol{J} = \begin{bmatrix} \|\Delta\tilde{\boldsymbol{X}}^{21}\|, & \left[\left(\Delta\tilde{\boldsymbol{X}}^{21}\right)^T \left(\Delta\tilde{\boldsymbol{X}}^{31}\right)\right] / \|\Delta\tilde{\boldsymbol{X}}^{21}\| \\ 0 \quad , & \|\boldsymbol{V}_3\| / \|\Delta\tilde{\boldsymbol{X}}^{21}\| \end{bmatrix} \tag{18}$$

and

$$\boldsymbol{j} = \begin{bmatrix} \|\Delta\tilde{\boldsymbol{x}}^{21}\|, & \left[\left(\Delta\tilde{\boldsymbol{x}}^{21}\right)^T \left(\Delta\tilde{\boldsymbol{x}}^{31}\right)\right] / \|\Delta\tilde{\boldsymbol{x}}^{21}\| \\ 0 \quad , & \|\boldsymbol{v}_3\| / \|\Delta\tilde{\boldsymbol{x}}^{21}\| \end{bmatrix} \tag{19}$$

Using these definitions, the right Cauchy-Green deformation tensor may be expanded as

$$\boldsymbol{C} = \boldsymbol{F}^T \boldsymbol{F} = \boldsymbol{J}^{-T} \boldsymbol{j}^T \boldsymbol{j} \boldsymbol{J}^{-1} = \boldsymbol{G}^T \boldsymbol{g} \boldsymbol{G} \tag{20}$$

where $\boldsymbol{G}$ is used to denote the inverse of $\boldsymbol{J}$. In component form we have

$$\boldsymbol{C} = \frac{1}{J_{11}^2 J_{22}^2} \begin{bmatrix} J_{22} & 0 \\ -J_{12} & J_{11} \end{bmatrix} \begin{bmatrix} g_{11} & g_{12} \\ g_{12} & g_{22} \end{bmatrix} \begin{bmatrix} J_{22} & -J_{12} \\ 0 & J_{11} \end{bmatrix} \tag{21}$$

in which the terms in the kernel array involving $\boldsymbol{j}$ may be expressed in the particularly simple form

$$\begin{aligned} g_{11} &= \quad j_{11}^2 = \left(\Delta\tilde{\boldsymbol{x}}^{21}\right)^T \Delta\tilde{\boldsymbol{x}}^{21} \\ g_{12} &= \quad j_{12} j_{11} = \left(\Delta\tilde{\boldsymbol{x}}^{21}\right)^T \Delta\tilde{\boldsymbol{x}}^{31} \\ g_{22} &= j_{12}^2 + j_{22}^2 = \left(\Delta\tilde{\boldsymbol{x}}^{31}\right)^T \Delta\tilde{\boldsymbol{x}}^{31} \end{aligned} \tag{22}$$

2.2 Material Constitution - Elastic Behavior

In the present work we assume that a simple St.Venant-Kirchhoff material model may be used to express the stresses from the deformations. Stresses are thus given by

$$\boldsymbol{S} = \boldsymbol{\mathcal{D}}\, \boldsymbol{E} \tag{23}$$

where $\boldsymbol{\mathcal{D}}$ are constant elastic moduli. and the Green-Lagrange strains $\boldsymbol{E}$ are given in terms of the deformation tensor as

$$\boldsymbol{E} = \frac{1}{2} \left(\boldsymbol{C} - \boldsymbol{I}\right) \quad . \tag{24}$$

In each triangular element the deformation may be computed from (19) to (22), thus giving directly the stress.

3 Weak Form for Equations of Motion

A weak form for the membrane may be written using a virtual work expression given by

$$\begin{aligned} \delta\Pi = & \int_\Omega \delta x_i\, \rho_0\, h\, \ddot{x}_i\, h\, \mathrm{d}\Omega + \int_\Omega \delta x_i\, c_0\, \dot{x}_i\, \mathrm{d}\Omega + \int_\Omega \delta E_{IJ} S_{IJ}\, h\, \mathrm{d}\Omega \\ & - \int_\Omega \delta x_i b_i\, \mathrm{d}\omega - \int_{\gamma_t} \delta x_i\, \bar{t}_i\, \mathrm{d}\gamma = 0 \end{aligned} \tag{25}$$

in which ρ_0 is mass density in the reference configuration, c_0 is a linear damping coefficient in the reference configuration, h is membrane thickness, S_{IJ} are components of the second Piola-Kirchhoff stress, b_i are components of loads in global coordinate directions (e.g., gravity), and $\bar{t}_i$ are components of specified membrane force per unit length. Upper case letters refer to components expressed on the reference configuration, whereas, lower case letters refer to current configuration quantities. Likewise, Ω and ω are surface area for the reference and current configurations, respectively. Finally, γ_t is a part of the current surface contour on which traction values are specified.

The linear damping term is included only for purposes in getting initially stable solutions. That is, by ignoring the inertial loading based on $\ddot{\boldsymbol{x}}$ only first derivatives of time will occur. This results in a transient form which is critically damped - thus permitting the reaching of a static loading state in a more monotonic manner.

We note that components for a normal pressure loading may be expressed as

$$b_i = p\, n_i \tag{26}$$

where p is a specified pressure and n_i are components of the normal to the surface.

Writing Eq. (20) in component form we have

$$C_{IJ} = G_{iI}\, g_{ij}\, G_{jJ} \quad \text{for} \quad i,j = 1,2 \;\; I,J = 1,2 \tag{27}$$

where

$$G_{11} = \frac{1}{J_{11}} \;\; ; \;\; G_{22} = \frac{1}{J_{22}} \;\; ; \;\; G_{12} = \frac{-J_{12}}{J_{11}\, J_{22}} \;\; ; \;\; G_{21} = 0 \tag{28}$$

The integrand of the first term in (25) may be written as

$$\delta C_{IJ}\, S_{IJ} = G_{iI}\, \delta g_{ij}\, G_{jJ}\, S_{IJ} = \delta g_{ij}\, s_{ij} \tag{29}$$

where the stress like variable s_{ij} is defined by

$$s_{ij} = G_{iI}\, G_{jJ}\, S_{IJ} \tag{30}$$

The transformation of stress given by (30) may be written in matrix form as

$$\boldsymbol{s} = \boldsymbol{Q}^T \boldsymbol{S} \tag{31}$$

in which

$$Q_{ab} \leftarrow G_{iI}\, G_{jJ}$$

where the index map is performed according to Table 1, yielding the result

$$\boldsymbol{Q} = \begin{bmatrix} G_{11}^2 & 0 & 0 \\ G_{12}^2 & G_{22}^2 & G_{12}\, G_{22} \\ 2\, G_{11}\, G_{12} & 0 & G_{11}\, G_{22} \end{bmatrix} \tag{32}$$

Since the deformation tensor is constant over each element, the results for the stresses are constant when h is taken constant over each element and, thus, the surface integral for the first term leads to the simple expression

$$\int_\Omega h\, \delta E_{IJ}\, S_{IJ}\, \mathrm{d}\Omega = \int_\Omega \frac{h}{2} \delta C_{IJ}\, S_{IJ}\, \mathrm{d}\Omega = \int_\Omega \frac{h}{2} \delta g_{ij}\, s_{ij}\, \mathrm{d}\Omega = \frac{h}{2} \delta g_{ij}\, s_{ij}\, A \tag{33}$$

Table 1. Index map for $\boldsymbol{Q}$ array

Indices	Values		
a	1	2	3
I,J	1,1	2,2	1,2 & 2,1
b	1	2	3
i,j	1,1	2,2	1,2 & 2,1

where A is the reference area for the triangular element.

The variation of g_{ij} results in the values

$$\begin{aligned} \delta g_{11} &= 2\left(\delta\tilde{\boldsymbol{x}}^2 - \delta\tilde{\boldsymbol{x}}^1\right)^T \Delta\tilde{\boldsymbol{x}}^{21} \\ \delta g_{12} &= \left(\delta\tilde{\boldsymbol{x}}^2 - \delta\tilde{\boldsymbol{x}}^1\right)^T \Delta\tilde{\boldsymbol{x}}^{31} + \left(\delta\tilde{\boldsymbol{x}}^3 - \delta\tilde{\boldsymbol{x}}^1\right)^T \Delta\tilde{\boldsymbol{x}}^{21} \\ \delta g_{22} &= 2\left(\delta\tilde{\boldsymbol{x}}^3 - \delta\tilde{\boldsymbol{x}}^1\right)^T \Delta\tilde{\boldsymbol{x}}^{31} \end{aligned} \tag{34}$$

At this stage it is convenient to transform the second order tensors to matrix form and write

$$\frac{1}{2}\,\delta C_{IJ} S_{IJ} = \delta E_{IJ} S_{IJ} = \left[\,\delta E_{11}\ \delta E_{22}\ 2\,\delta E_{12}\,\right] \begin{bmatrix} S_{11} \\ S_{22} \\ S_{12} \end{bmatrix} = \delta \boldsymbol{E}^T \boldsymbol{S} \tag{35}$$

or for the alternative form

$$\frac{1}{2}\,\delta g_{ij} s_{ij} = \frac{1}{2}\left[\,\delta g_{11}\ \delta g_{22}\ 2\,\delta g_{12}\,\right] \begin{bmatrix} s_{11} \\ s_{22} \\ s_{12} \end{bmatrix} = \frac{1}{2}\,\delta \boldsymbol{g}^T \boldsymbol{s} \tag{36}$$

Using (34) we obtain the result directly in terms of global cartesian components as

$$\begin{aligned} \frac{1}{2}\,\delta \boldsymbol{g}^T \boldsymbol{s} &= \left[\,\delta(\tilde{\boldsymbol{x}}^1)^T\ \delta(\tilde{\boldsymbol{x}}^2)^T\ \delta(\tilde{\boldsymbol{x}}^3)^T\,\right] [\boldsymbol{b}]^T\ \boldsymbol{s} \\ &= \left[\,\delta(\tilde{\boldsymbol{x}}^1)^T\ \delta(\tilde{\boldsymbol{x}}^2)^T\ \delta(\tilde{\boldsymbol{x}}^3)^T\,\right] [\boldsymbol{b}]^T\ \boldsymbol{Q}^T \boldsymbol{S} = \delta \boldsymbol{E}^T \boldsymbol{S} \end{aligned} \tag{37}$$

where the *strain-displacement matrix* $\boldsymbol{b}$ is given by

$$\boldsymbol{b} = \underbrace{\begin{bmatrix} -(\Delta\tilde{\boldsymbol{x}}^{21})^T & (\Delta\tilde{\boldsymbol{x}}^{21})^T & \boldsymbol{0} \\ -(\Delta\tilde{\boldsymbol{x}}^{31})^T & \boldsymbol{0} & (\Delta\tilde{\boldsymbol{x}}^{31})^T \\ -(\Delta\tilde{\boldsymbol{x}}^{21} + \Delta\tilde{\boldsymbol{x}}^{31})^T & (\Delta\tilde{\boldsymbol{x}}^{31})^T & (\Delta\tilde{\boldsymbol{x}}^{21})^T \end{bmatrix}}_{3\times 9} \tag{38}$$

Thus, directly we have in each element

$$\delta \boldsymbol{E} = \boldsymbol{Q}\,\boldsymbol{b}\,\delta\tilde{\boldsymbol{x}} = \frac{1}{2}\,\delta \boldsymbol{C} \tag{39}$$

where $\tilde{\boldsymbol{x}}$ denotes the three nodal values on the element. It is immediately obvious that we can describe a strain-displacement matrix for the variation of $\boldsymbol{E}$ as

$$\boldsymbol{B} = \boldsymbol{Q}\,\boldsymbol{b} \tag{40}$$

A residual form for each element may be written as

$$\begin{Bmatrix} \boldsymbol{R}^1 \\ \boldsymbol{R}^2 \\ \boldsymbol{R}^3 \end{Bmatrix} = \begin{Bmatrix} \boldsymbol{f}^1 \\ \boldsymbol{f}^2 \\ \boldsymbol{f}^3 \end{Bmatrix} - [\boldsymbol{M}_e] \begin{Bmatrix} \ddot{\tilde{\boldsymbol{x}}}^1 \\ \ddot{\tilde{\boldsymbol{x}}}^2 \\ \ddot{\tilde{\boldsymbol{x}}}^3 \end{Bmatrix} - [\boldsymbol{C}_e] \begin{Bmatrix} \dot{\tilde{\boldsymbol{x}}}^1 \\ \dot{\tilde{\boldsymbol{x}}}^2 \\ \dot{\tilde{\boldsymbol{x}}}^3 \end{Bmatrix} - h\, A\, [\boldsymbol{B}]^T \begin{Bmatrix} S_{11} \\ S_{22} \\ S_{12} \end{Bmatrix} \tag{41}$$

where $[\boldsymbol{M}_e]$ and where $[\boldsymbol{C}_e]$ are the element mass and damping matrices given by

$$[\boldsymbol{M}_e] = \begin{bmatrix} \boldsymbol{M}^{11} & \boldsymbol{M}^{12} & \boldsymbol{M}^{13} \\ \boldsymbol{M}^{21} & \boldsymbol{M}^{22} & \boldsymbol{M}^{23} \\ \boldsymbol{M}^{31} & \boldsymbol{M}^{32} & \boldsymbol{M}^{33} \end{bmatrix} \quad \text{and} \quad [\boldsymbol{C}_e] = \begin{bmatrix} \boldsymbol{C}^{11} & \boldsymbol{C}^{12} & \boldsymbol{C}^{13} \\ \boldsymbol{C}^{21} & \boldsymbol{C}^{22} & \boldsymbol{C}^{23} \\ \boldsymbol{C}^{31} & \boldsymbol{C}^{32} & \boldsymbol{C}^{33} \end{bmatrix} \tag{42}$$

with

$$\boldsymbol{M}^{\alpha\beta} = \int_{\Omega} \rho_0\, h\, \xi_\alpha\, \xi_\beta\, \mathrm{d}\Omega\, \boldsymbol{I} \quad \text{and} \quad \boldsymbol{C}^{\alpha\beta} = \int_{\Omega} c_0\, h\, \xi_\alpha\, \xi_\beta\, \mathrm{d}\Omega\, \boldsymbol{I} \tag{43}$$

3.1 Pressure Follower Loading

For membranes subjected to internal pressure loading, the finite element nodal forces must be computed based on the *deformed* current configuration. Thus, for each triangle we need to compute the nodal forces from the relation

$$\delta\tilde{\boldsymbol{x}}^{\alpha,T}\, \boldsymbol{f}^\alpha = \delta\tilde{\boldsymbol{x}}^{\alpha,T} \int_\omega \xi_\alpha\, (p\, \boldsymbol{n})\, \mathrm{d}\omega \tag{44}$$

For the constant triangular element and constant pressure over the element, denoted by p_e, the normal vector $\boldsymbol{n}$ is also constant and thus the integral yields the nodal forces

$$\boldsymbol{f}^\alpha = \frac{1}{3}\, p_e\, \boldsymbol{n}\, A_e \tag{45}$$

We noted previously from Eq. (6) that the cross product of the incremental vectors $\Delta\tilde{\boldsymbol{x}}^{21}$ with $\Delta\tilde{\boldsymbol{x}}^{31}$ resulted in a vector normal to the triangle with magnitude of twice the area. Thus, the nodal forces for the pressure are given by the simple relation

$$\boldsymbol{f}^\alpha = \frac{1}{6}\, p_e\, \Delta\tilde{\boldsymbol{x}}^{21} \times \Delta\tilde{\boldsymbol{x}}^{31} \tag{46}$$

Instead of the cross products it is convenient to introduce a matrix form denoted by

$$\Delta\tilde{\boldsymbol{x}}^{21} \times \Delta\tilde{\boldsymbol{x}}^{31} = \left[\widehat{\Delta\tilde{\boldsymbol{x}}}^{21}\right] \Delta\tilde{\boldsymbol{x}}^{31} \tag{47}$$

where

$$\left[\widehat{\Delta\tilde{\boldsymbol{x}}}^{ij}\right] = \begin{bmatrix} 0 & -\Delta\tilde{x}_3^{ij} & \Delta\tilde{x}_2^{ij} \\ \Delta\tilde{x}_3^{ij} & 0 & -\Delta\tilde{x}_1^{ij} \\ -\Delta\tilde{x}_2^{ij} & \Delta\tilde{x}_1^{ij} & 0 \end{bmatrix} . \tag{48}$$

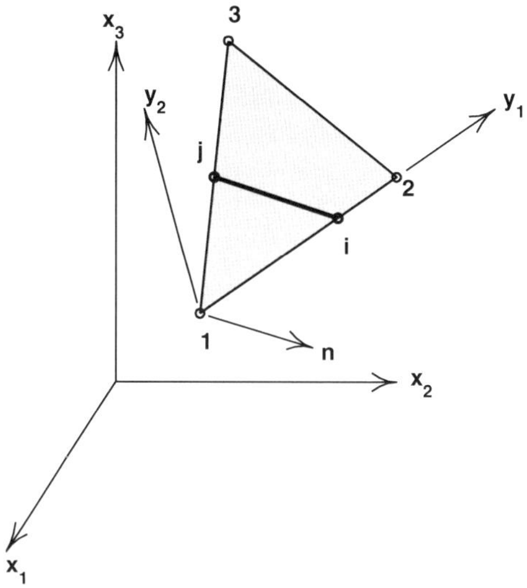

Fig. 2. Cable reinforced membrane element at(i-j)

4 Reinforcement Cables

It is common to place reinforcing cables in membranes to provide added strength or shape control. The cables are generally very strong in axial load capacity (generally tension) and weak in bending. Accordingly, they may be modeled by a *truss* type member. In the form admitted here it is not necessary for the reinforcement to be placed at the edges of membrane elements - they may pass through an element as shown in Fig 2.

The ends of a typical reinforcement are denoted as i and j in the figure and have reference coordinates $\boldsymbol{X}^i$ and $\boldsymbol{X}^j$,respectively. These points may be referred to the nodal values of the membrane by computing the values of the natural coordinates so that

$$\boldsymbol{X}^k = \xi_\alpha^k \, \boldsymbol{X}^\alpha \quad \text{for} \;\; k = i, j \tag{49}$$

The solution for the two points may be trivially constructed from linear interpolation on the edges. The results are (for the points intersecting the edges shown in the figure)

$$\begin{aligned} \xi_2^i &= \frac{\|\boldsymbol{X}^i - \boldsymbol{X}^1\|}{\|\boldsymbol{X}^2 - \boldsymbol{X}^1\|} \quad ; \quad \xi_1^i = 1 - \xi_2^i \quad ; \quad \xi_3^i = 0 \\ \xi_3^j &= \frac{\|\boldsymbol{X}^j - \boldsymbol{X}^1\|}{\|\boldsymbol{X}^3 - \boldsymbol{X}^1\|} \quad ; \quad \xi_1^i = 1 - \xi_3^i \quad ; \quad \xi_2^i = 0 \end{aligned} \tag{50}$$

Using these values the deformed position of the reinforcement cable may be written as

$$\begin{Bmatrix} \boldsymbol{x}^i \\ \boldsymbol{x}^j \end{Bmatrix} = \begin{Bmatrix} \xi_\alpha^i \, \boldsymbol{I} \\ \xi_\alpha^j \, \boldsymbol{I} \end{Bmatrix} \tilde{\boldsymbol{x}}^\alpha \tag{51}$$

4.1 Deformation of Cable

The deformation of the reinforcement cable may be expressed in terms of the Green-Lagrange strain given by

$$E_{ij} = \frac{1}{2}\left[\frac{(\boldsymbol{x}^j - \boldsymbol{x}^i)^T(\boldsymbol{x}^j - \boldsymbol{x}^i)}{(\boldsymbol{X}^j - \boldsymbol{X}^i)^T(\boldsymbol{X}^j - \boldsymbol{X}^i)} - 1\right] = \frac{1}{2}\left[\frac{\|\Delta\boldsymbol{x}^{ji}\|^2}{\|\Delta\boldsymbol{X}^{ji}\|^2} - 1\right] \tag{52}$$

where $\Delta\boldsymbol{X}^{ji} = \boldsymbol{X}^j - \boldsymbol{X}^i$. The variation of the strain is then expressed as

$$\delta E_{ij} = \frac{(\delta\boldsymbol{x}^j - \delta\boldsymbol{x}^i)^T(\Delta\boldsymbol{x}^{ji})}{\|\Delta\boldsymbol{X}^{ji}\|^2} \tag{53}$$

4.2 Material Constitution

For simplicity we again assume that the material is elastic and may be represented by a one-dimensional form of a St.Venant-Kirchhoff model expressed as

$$S_{ij} = \mathsf{E}\, E_{ij} \tag{54}$$

where S_{ij} is the constant second Piola-Kirchhoff stress in the cable and E is an elastic modulus.

4.3 Weak Form for Reinforcement

A weak form for an individual reinforcement cable in an element may be written as

$$\delta\Pi_r = \delta\boldsymbol{x}^k\left(\boldsymbol{M}_{kl}\ddot{\boldsymbol{x}}^l + \boldsymbol{C}_{kl}\dot{\boldsymbol{x}}^l\right) - \delta E_{ij}\, S_{ij}\, A_{ij}\, L_{ij} \quad ; \quad k,l = i,j \tag{55}$$

where A_{ij} is the cross sectional area of the reinforcement; L_{ij} the length of the cable (i.e., $\|\Delta\boldsymbol{X}^{ji}\|$); $\boldsymbol{M}_{kl}$ is the mass matrix; and $\boldsymbol{C}_{kl}$ is the damping matrix.

The variation of the strain is rewritten from Eq. (53) as

$$\delta E_{ij} = \left[\delta\boldsymbol{x}^{i,T} \;\; \delta\boldsymbol{x}^{j,T}\right]\begin{Bmatrix} -\dfrac{\Delta\boldsymbol{x}^{ji}}{L_{ij}^2} \\ \dfrac{\Delta\boldsymbol{x}^{ji}}{L_{ij}^2} \end{Bmatrix} \tag{56}$$

Equation (55) is appended to the other terms from the membrane by replacing variations of end displacements and the rate terms by their representation in terms of the membrane nodal parameters as given by Eq. (51). The result is:

$$\delta\Pi_r = \delta\tilde{\boldsymbol{x}}^{\alpha,T}\left\{\boldsymbol{M}_r^{\alpha\beta}\,\ddot{\tilde{\boldsymbol{x}}}^\beta + \boldsymbol{C}_r^{\alpha\beta}\,\dot{\tilde{\boldsymbol{x}}}^\beta - \boldsymbol{P}_r^\alpha\right\} \tag{57}$$

where

$$\boldsymbol{P}_r^\alpha = \Delta\xi_\alpha^{ji}\,\Delta\xi_\beta^{ji}\,\tilde{\boldsymbol{x}}^\beta\,\frac{S_{ij}\,A_{ij}}{L_{ij}} \quad ; \tag{58}$$

$$\boldsymbol{M}_r^{\alpha\beta} = \xi_\alpha^i \boldsymbol{M}_{ii} \xi_\beta^i + \xi_\alpha^i \boldsymbol{M}_{ij} \xi_\beta^j + \xi_\alpha^j \boldsymbol{M}_{ji} \xi_\beta^i + \xi_\alpha^j \boldsymbol{M}_{jj} \xi_\beta^j \tag{59}$$

and

$$\boldsymbol{C}_r^{\alpha\beta} = \xi_\alpha^i \boldsymbol{C}_{ii} \xi_\beta^i + \xi_\alpha^i \boldsymbol{C}_{ij} \xi_\beta^j + \xi_\alpha^j \boldsymbol{C}_{ji} \xi_\beta^i + \xi_\alpha^j \boldsymbol{C}_{jj} \xi_\beta^j \tag{60}$$

In the definition of $\boldsymbol{P}_r^\alpha$ the incremental ξ denote

$$\Delta \xi_\alpha^{ji} = \xi_\alpha^j - \xi_\alpha^i \tag{61}$$

Multiple reinforcement strands within any element may be simply considered by summing over all the ij-pairs of intersection points.

5 Solution Methods

5.1 Explicit Solution

In an explicit time integration the lumped mass is commonly used. Furthermore, we shall assume that the damping is negligible and thus may be ignored. A diagonal (lumped) mass is usually also constructed where

$$\boldsymbol{M}^{\alpha\beta} = \begin{cases} \frac{1}{3} \int_\Omega \rho_0 \, h \, \mathrm{d}\Omega \, \boldsymbol{I} \; ; & \text{for } \alpha = \beta \\ \boldsymbol{0} \; ; & \text{for } \alpha \neq \beta \end{cases} \tag{62}$$

Diagonalization of the mass matrix in the presence of reinforcement is more difficult and, if performed, must be based on purely physical lumping arguments as no clear mathematical justification is available.[17]

A solution is then computed for each discrete time t_n from

$$\ddot{\boldsymbol{x}}_n = \boldsymbol{M}^{-1} \left[\boldsymbol{f}_n - \sum_e \left(\boldsymbol{B}_{n,e}^T \boldsymbol{S}_{n,e} A_e h_e + \boldsymbol{P}_{r,e} \right) \right] \tag{63}$$

where subscripts e refer to individual element quantities and subscript n to the time step. The inverse of the mass is trivial due to its diagonal form, hence the method is directly proportional to the number of nodes in the mesh.

The solution state may now be advanced in time using any time integration procedure. For example using a Newmark method

$$\begin{aligned} \dot{\boldsymbol{x}}_n &= \dot{\boldsymbol{x}}_{n-1} + \frac{1}{2} \Delta t_{n-1} \left(\ddot{\boldsymbol{x}}_{n-1} + \ddot{\boldsymbol{x}}_n \right) \\ \boldsymbol{x}_{n+1} &= \boldsymbol{x}_n + \Delta t_n \, \dot{\boldsymbol{x}}_n + \frac{1}{2} \Delta t_n^2 \, \ddot{\boldsymbol{x}}_n \end{aligned} \tag{64}$$

in which $\Delta t_n = t_n - t_{n-1}$.

An explicit solution is conditionally stable and requires

$$\Delta t_n \le \Delta t_{cr} \tag{65}$$

for all time steps. The critical time step depends on element size and the maximum wave speed for the element material. The resulting time increment is often much too small for practical considerations in computer effort and for the response necessary to model slowly varying loads. In these situations it is more expedient to use in implicit time integration procedure in which inertial, stress, and loading matrices also depend on position and velocity at the current time.

5.2 Implicit Solution

In an implicit solution case it is necessary to use an iterative solution scheme at each time step which solves a sequence of linear, coupled algebraic problems. Here we only present results for the St.Venant-Kirchhoff material model and the normal *follower* pressure loading. We assume that the transient problem will be integrated using the Newmark method, however, other schemes may also be utilized with minor modifications.

The Newmark method may be written for implicit solutions as

$$\begin{aligned} \boldsymbol{x}_n &= \boldsymbol{x}_{n-1} + \Delta t_n \, \dot{\boldsymbol{x}}_{n-1} + \left(\frac{1}{2} - \beta\right) \Delta t_n^2 \, \ddot{\boldsymbol{x}}_{n-1} + \beta \, \Delta t_n^2 \, \ddot{\boldsymbol{x}}_n \\ \dot{\boldsymbol{x}}_n &= \dot{\boldsymbol{x}}_{n-1} + (1-\gamma) \, \Delta t_n \, \ddot{\boldsymbol{x}}_{n-1} + \gamma \, \Delta t_n \, \ddot{\boldsymbol{x}}_n \end{aligned} \tag{66}$$

The equations to be solved at each time step may be expressed as

$$\begin{aligned} \boldsymbol{R}_n^\alpha = \boldsymbol{f}_n^\alpha &- \sum_e \left(\boldsymbol{M}_e^{\alpha\beta} + \boldsymbol{M}_r^{\alpha\beta}\right) \ddot{\boldsymbol{x}}_n^\beta - \sum_e \left(\boldsymbol{C}_e^{\alpha\beta} + \boldsymbol{C}_r^{\alpha\beta}\right) \dot{\boldsymbol{x}}_n^\beta \\ &- \sum_e \left(h_e \, A_e \, \boldsymbol{B}_e^{\alpha,T} \, \boldsymbol{S}_e + \boldsymbol{P}_{r,e}^\alpha\right)_n \end{aligned} \tag{67}$$

In an implicit method Eq. (67) may be solved iteratively using a Newton method. In this process the nonlinear residual equations are linearized about a given set of nodal positions $\tilde{\boldsymbol{x}}_n^k$ corresponding to known values at some iteration stage k. The result is written as

$$\boldsymbol{R}_n^{k+1} \approx \boldsymbol{R}_n^k + \left.\frac{\partial \boldsymbol{R}_n}{\partial \tilde{\boldsymbol{x}}}\right|^k d\tilde{\boldsymbol{x}}_n^k = \boldsymbol{0} \tag{68}$$

If we define the tangent (jacobian) matrix $\boldsymbol{A}$ as

$$\boldsymbol{A} = -\frac{\partial \boldsymbol{R}}{\partial \tilde{\boldsymbol{x}}} \tag{69}$$

the set of linear algebraic equations to be solved at each iteration may be expressed as

$$\boldsymbol{A}_n^k \, d\tilde{\boldsymbol{x}}_n^k = \boldsymbol{R}_n^k \tag{70}$$

The solution may then be updated using

$$\boldsymbol{x}_n^{k+1} = \boldsymbol{x}_n^k + d\boldsymbol{x}_n^k \tag{71}$$

and iteration continued until convergence is achieved.

The Newton scheme will have a quadratic asymptotic rate of convergence provided a careful derivation of the tangent matrix $\boldsymbol{A}$ is constructed. Typically such jacobians are referred to as the *consistent tangent matrix*. For transient applications the use of the specified time stepping algorithm is required to compute the tangent matrix. The computation for the transient term is performed as

$$\boldsymbol{A} = -\frac{\partial \boldsymbol{R}}{\partial \tilde{\boldsymbol{x}}} - \frac{\partial \boldsymbol{R}}{\partial \dot{\tilde{\boldsymbol{x}}}} \frac{\partial \dot{\tilde{\boldsymbol{x}}}}{\partial \tilde{\boldsymbol{x}}} - \frac{\partial \boldsymbol{R}}{\partial \ddot{\tilde{\boldsymbol{x}}}} \frac{\partial \ddot{\tilde{\boldsymbol{x}}}}{\partial \tilde{\boldsymbol{x}}} \tag{72}$$

The result may be written as

$$\boldsymbol{A} = c_1 \, \boldsymbol{K} + c_2 \, \boldsymbol{C} + c_3 \, \boldsymbol{M} \tag{73}$$

where the c_i result from any differentiation of the nodal vectors with respect to the solution vector. For the Newmark method the result from (67) gives $c_1 = 1$ and from (66) we obtain

$$\frac{\partial \tilde{\boldsymbol{x}}}{\partial \ddot{\tilde{\boldsymbol{x}}}_n} = \beta\,\Delta t_n^2 \boldsymbol{I} \quad \text{and} \quad \frac{\partial \dot{\tilde{\boldsymbol{x}}}}{\partial \ddot{\tilde{\boldsymbol{x}}}_n} = \gamma\,\Delta t_n \boldsymbol{I}$$

thus giving

$$c_2 = \frac{\gamma}{\beta\,\Delta t_n} \quad \text{and} \quad c_3 = \frac{1}{\beta\,\Delta t_n^2}$$

From (67) we find that

$$\begin{aligned} \boldsymbol{M}_n &= \sum_e (\boldsymbol{M}_e + \boldsymbol{M}_r) \\ \boldsymbol{C}_n &= \sum_e (\boldsymbol{C}_e + \boldsymbol{C}_r) \\ \boldsymbol{K}_n &= \sum_e (\boldsymbol{K}_e + \boldsymbol{K}_r + \boldsymbol{K}_f) \end{aligned} \tag{74}$$

where subscripts e, r and f denote quantities computed from the membrane element, reinforcement cable and pressure follower loading terms, respectively.

Membrane tangent matrix

To compute the element stiffness matrix it is necessary to determine the change in stress due to an incremental change in the motion. Accordingly, for the St.Venant-Kirchhoff model we obtain

$$d\boldsymbol{S}_e = \boldsymbol{\mathcal{D}}\, d\boldsymbol{E}_e \tag{75}$$

where

$$d\boldsymbol{E}_e = \boldsymbol{Q}_e\, \boldsymbol{b}_e\, d\tilde{\boldsymbol{x}}_e = \boldsymbol{B}_e\, d\tilde{\boldsymbol{x}}_e \tag{76}$$

The element stiffness matrix is given by

$$\boldsymbol{K}_e = \left(h\, A\, \boldsymbol{B}_n^T\, \boldsymbol{\mathcal{D}}_n\, \boldsymbol{B}_n + \boldsymbol{K}_g \right)_e \tag{77}$$

where $\boldsymbol{K}_g$ is a geometric stiffness computed from the term

$$\frac{h_e}{2} A_e\, d\,(\delta C_{IJ})\, S_{IJ} = \frac{h_e}{2} A_e\, d\,(\delta g_{ij})\, s_{ij} \tag{78}$$

From (34) we obtain

$$\begin{aligned} d(\delta g_{11}) &= 2\,\left(\delta\tilde{\boldsymbol{x}}^2 - \delta\tilde{\boldsymbol{x}}^1\right)^T \left(d\tilde{\boldsymbol{x}}^2 - d\tilde{\boldsymbol{x}}^1\right) \\ d(\delta g_{12}) &= \left(\delta\tilde{\boldsymbol{x}}^2 - \delta\tilde{\boldsymbol{x}}^1\right)^T \left(d\tilde{\boldsymbol{x}}^3 - d\tilde{\boldsymbol{x}}^1\right) + \left(\delta\tilde{\boldsymbol{x}}^3 - \delta\tilde{\boldsymbol{x}}^1\right)^T \left(d\tilde{\boldsymbol{x}}^2 - d\tilde{\boldsymbol{x}}^1\right) \\ d(\delta g_{22}) &= 2\,\left(\delta\tilde{\boldsymbol{x}}^3 - \delta\tilde{\boldsymbol{x}}^1\right)^T \left(d\tilde{\boldsymbol{x}}^3 - d\tilde{\boldsymbol{x}}^1\right) \end{aligned} \tag{79}$$

Using these expressions the geometric matrix may be written as

$$\boldsymbol{K}_g = h_e\, A_e \begin{bmatrix} (s_{11} + 2s_{12} + s_{22})\boldsymbol{I} & -(s_{11} + s_{12})\boldsymbol{I} & -(s_{22} + s_{12})\boldsymbol{I} \\ -(s_{11} + s_{12})\boldsymbol{I} & s_{11}\boldsymbol{I} & s_{12}\boldsymbol{I} \\ -(s_{22} + s_{12})\boldsymbol{I} & s_{12}\boldsymbol{I} & s_{22}\boldsymbol{I} \end{bmatrix} \tag{80}$$

Reinforcement tangent matrix

In a similar manner to that for the membrane, the tangent matrix for the reinforcement may be computed from

$$\begin{aligned}\boldsymbol{K}_r^{\alpha\beta} &= \frac{\partial \boldsymbol{P}_r^{\alpha}}{\partial \tilde{\boldsymbol{x}}^{\beta}} \\ &= \Delta\xi_{\alpha}^{ji}\left[\frac{\mathsf{E}A_{ij}}{L_{ij}}\left(\Delta\xi_{\gamma}^{ji}\tilde{\boldsymbol{x}}^{\gamma}\tilde{\boldsymbol{x}}^{\delta,T}\Delta\xi_{\delta}^{ji}\right)+\frac{S_{ij}A_{ij}}{L_{ij}}\boldsymbol{I}\right]\Delta\xi_{\beta}^{ji}\end{aligned} \tag{81}$$

Pressure follower loading tangent matrix

The pressure follower loading tangent matrix is obtained directly from (46) as

$$d\boldsymbol{f}^{\alpha} = \frac{1}{6}\,p_e\,\left[d(\Delta\tilde{\boldsymbol{x}}^{21})\times\Delta\tilde{\boldsymbol{x}}^{31}+\Delta\tilde{\boldsymbol{x}}^{21}\times d(\Delta\tilde{\boldsymbol{x}}^{31})\right] \tag{82}$$

Using (47) this may be written as

$$d\boldsymbol{f}^{\alpha} = \frac{1}{6}\,p_e\,\left([\widehat{\Delta\tilde{\boldsymbol{x}}}^{21}]\,d(\Delta\tilde{\boldsymbol{x}}^{31})-[\widehat{\Delta\tilde{\boldsymbol{x}}}^{31}]\,d(\Delta\tilde{\boldsymbol{x}}^{21})\right) \tag{83}$$

and gives the tangent matrix form

$$\delta\tilde{\boldsymbol{x}}^{\alpha,T}\boldsymbol{K}_f^{\alpha\beta}d\tilde{\boldsymbol{x}}^{\beta} = \delta\tilde{\boldsymbol{x}}^{\alpha,T}\left[\widehat{\Delta\boldsymbol{x}}^{32}\;\; -\widehat{\Delta\boldsymbol{x}}^{31}\;\; \widehat{\Delta\boldsymbol{x}}^{21}\right]\begin{bmatrix} d\tilde{\boldsymbol{x}}^1 \\ d\tilde{\boldsymbol{x}}^2 \\ d\tilde{\boldsymbol{x}}^3\end{bmatrix} \tag{84}$$

In general the tangent is unsymmetric and requires a solution scheme to solve equations which can include such effects.

5.3 Quasi-Static Solutions

Membrane structures typically have no stiffness during the initial phase of loading. Thus, it is necessary to perform some form of a transient analysis until an equilibrium position is neared, at which time it is possible to switch to a *static* loading in which no rate terms are included. To avoid oscillations during the solution process the equations of motion are treated here in a first order form as

$$\boldsymbol{R} = \boldsymbol{f} - \boldsymbol{C}\,\dot{\tilde{\mathbf{x}}} - \sum_e\left(h_e A_e \boldsymbol{B}^T\boldsymbol{S}+\boldsymbol{P}_{e,r}\right) = \mathbf{0} \tag{85}$$

and solved using an implicit *backward Euler* solution process in which discrete rates are approximated by

$$\dot{\tilde{\boldsymbol{x}}}_n = \frac{1}{\Delta t_n}\,(\tilde{\boldsymbol{x}}_n - \tilde{\boldsymbol{x}}_{n-1}) \tag{86}$$

A solution is computed until the rate terms are small at which time they are dropped and the solution is computed from the static form

$$\boldsymbol{R} = \boldsymbol{f} - \sum_e\left(h_e A_e \boldsymbol{B}^T\boldsymbol{S}+\boldsymbol{P}_{e,r}\right) = \mathbf{0} \tag{87}$$

In results reported in numerical examples, a diagonal (lumped) form of $\boldsymbol{C}$ is used. The diagonalization is performed identically to that for the lumped mass form.

6 Numerical Examples

6.1 Sphere Subjected to Internal Follower Pressure

The first problem presented is a sphere of initial (unstressed) radius of 10 units which is subjected to an internal follower pressure loading of 5. The material properties are

$$E = 1000 \quad ; \quad \nu = 0.25 \quad ; \quad \rho_0 = 10 \quad ; \quad h_e = 0.1$$

A mesh for one quadrant of the sphere (1/8 of the total) is constructed as shown in Fig. 3

Both the undeformed and deformed configurations are included to indicate the amount of deformation occurring.

The problem is solved using a backward Euler time integrator for the case where $\boldsymbol{M}$ is zero and a diagonal damping matrix $\boldsymbol{C}$ with $c_0 = \rho_0$ is used. The full internal pressure is applied during the first time step and held constant during subsequent steps. The problem is solved using 100 steps with a constant Δt of 0.001. Subsequently, the rate terms are ignored and a final static state determined during one additional step. Convergence to full machine precision is achieved in three iterations for all steps - indicating a correct implementation for the Newton strategy described in this study. A plot of the contours for the u_3 displacement is shown in Fig. 4 .

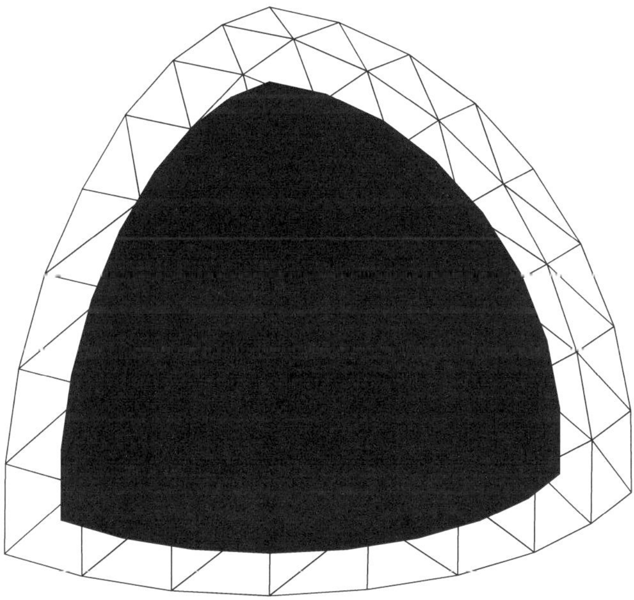

Fig. 3. Mesh for sphere problem symmetrically supported at edges

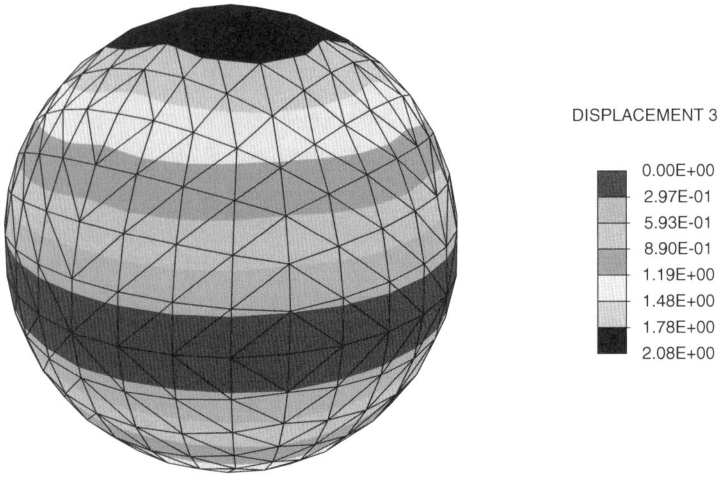

Fig. 4. Contours for u_3. Full sphere shown by reflections.

6.2 Corner Supported Sphere Quadrant Subjected to Internal Follower Pressure

The boundary conditions for the first problem are changed to ones in which the vertices of the quadrant are fully restrained (fixed in all directions). Loading is again by a follower loading with magnitude 5. The contours for the absolute displacement ($\|\boldsymbol{u}\|$) are shown on the deformed mesh in Fig. 5 . Solution is obtained in the same way as for the previous example. Convergence was achieved in all steps to full machine accuracy in 3-4 iterations each step.

Figure 6 presents a comparison for the different solution strategies used for this example. The same problem has been analyzed three (3) times including different terms in the backward Euler time integration. The first solution includes the damping term only, the second solution includes all the dynamic terms (mass and damping), and the third solution includes only the mass term of the equation. For each time step the figure shows the evolution of the displacement of the mesh's central point. Finally, each solution scheme is compared to the actual final static result.

6.3 Square Supported at 4-Corners

The third problem is presented to test the performance of the triangular membrane formulation described above under gravity loading. Here a square region of 20 units on a side is supported only at its four corners. Due to symmetry only one quadrant is analyzed with the origin of coordinates at the lower left corner and a uniform mesh of 20×20 spaces as shown in Fig. 7. The flat configuration lies in the $x_1 - x_2$ coordinate plane at $x_3 = 0$. A uniform body loading is applied in the x_3 direction. To permit a sag to occur the upper right corner node (which is fully restrained in all three directions) is displaced equally in the x_1 and x_2 by a negative 2 units (that is the deformed position of the node at this point has final coordinates $(8, 8, 0)$.

For the membrane problem considered here, the properties are taken as

$$E = 10000 \quad ; \quad \nu = 0.25 \quad ; \quad \rho_0 = 10 \quad ; \quad h_e = 0.1 \quad ; \quad b_3 = 1$$

The solution is computed using a backward Euler solution process with $\boldsymbol{M} = \mathbf{0}$ and a diagonal $\boldsymbol{C}$ matrix with $c_0 = \rho_0$. The time behavior is constructed using 1000 time steps of $\Delta t = 0.0001$ followed by an additional 1000 steps at $\Delta t = 0.001$ and a final 1000 steps at $\Delta t = 0.002$. The corner displacements are moved 2 units at a uniform rate until $t = 1$ after which they are held constant. At the final time of 3.1 the solution strategy is switched to a static algorithm and a converged solution achieved in 4 iterations. In general each time step converges at the quadratic rate in 3 to 4 iterations per step. Use of larger step sizes, however, resulted in divergence of the solution after a few steps. The final shape of the membrane is shown in Fig. 8 where the full problem is shown by reflecting results about the symmetry axes.

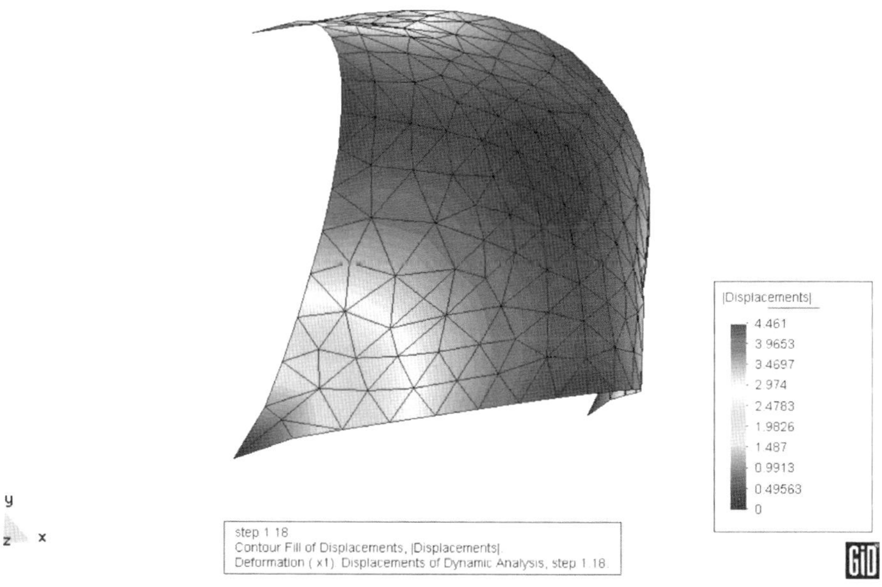

Fig. 5. Contours for $\|\boldsymbol{u}\|$. Corner supported sphere shown in deformed configuration

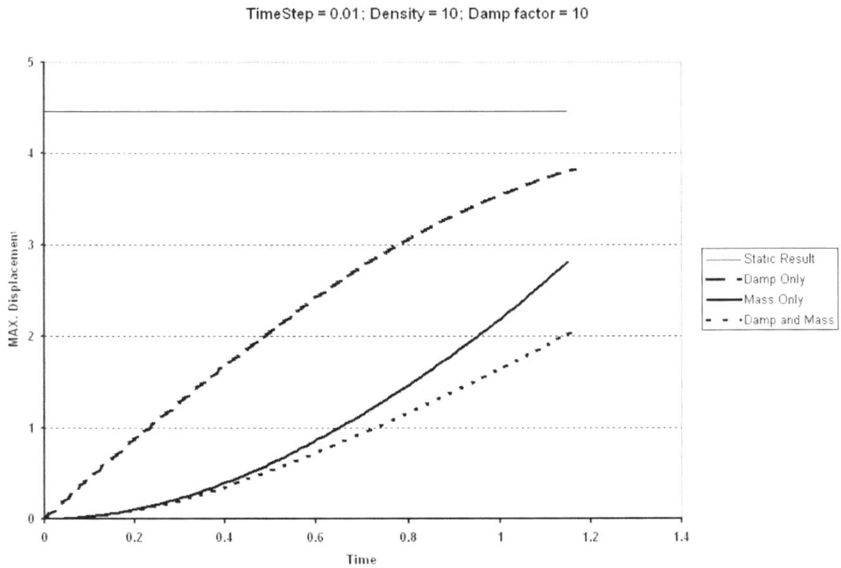

Fig. 6. Comparison of the evolution of the result for 3 different sets of dynamic terms. Representation of the absolute displacement $\|\boldsymbol{u}\|$ of the mesh's central point in the corner supported sphere

Fig. 7. Mesh for membrane problem supported at corners

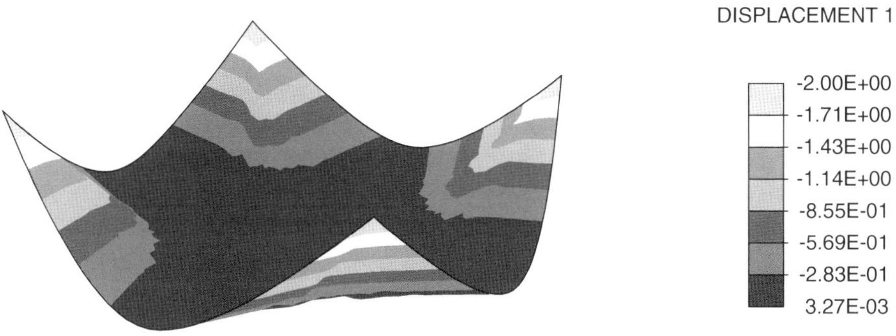

Fig. 8. Deformed configuration of membrane

6.4 Inflation of a Balloon

The fourth problem represents the inflation of a balloon and is presented to test the performance of the triangular membrane element and solution strategy on a very non-linear problem. The mesh geometry consists on two flat, horizontal surfaces connected to a vertical conical open end. Figure 9 shows the geometry of a balloon in its uninflated state. An internal pressure (follower) loading of 5 units is applied to bring the balloon to a final vertical configuration. The circular end of the cone has all its degrees of freedom fixed. The parameters for the material are

$$E = 1000 \quad ; \quad \nu = 0.25 \quad ; \quad \rho_0 = 10 \quad ; \quad h_e = 0.1$$

Fig. 9. Reference configuration of the balloon

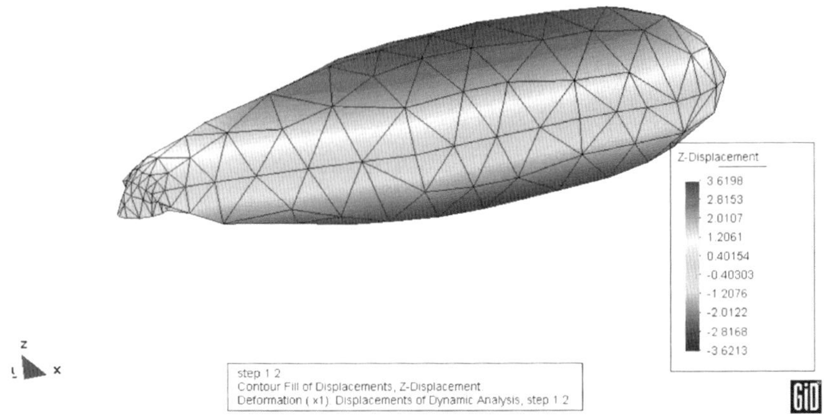

Fig. 10. Current configuration of the balloon after 120 time steps, displaying vertical displacement (u_3) contour

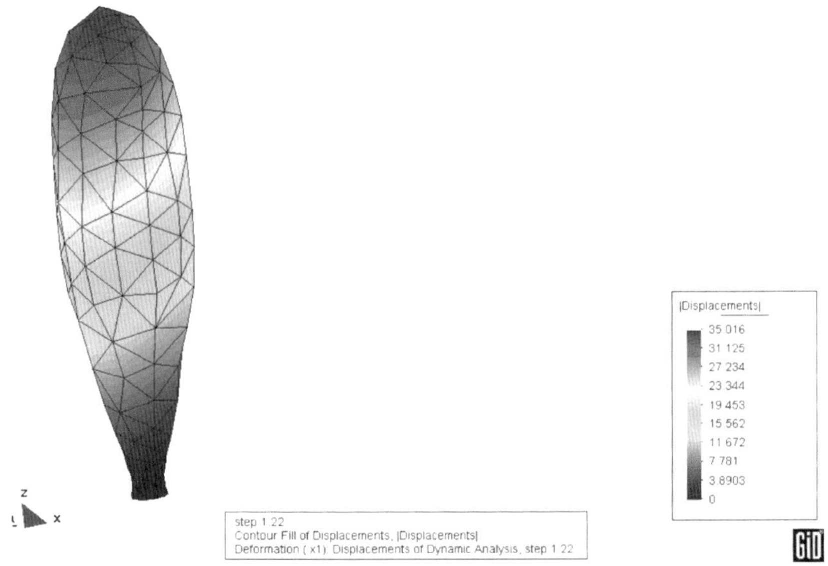

Fig. 11. Final configuration of the balloon, displaying absolute displacement $\|\boldsymbol{u}\|$

The solution is computed using a backward Euler strategy neglecting the inertia terms. The $\boldsymbol{C}$ matrix used is diagonal with $c_0 = \rho_0$. The transient solution is continued for 120 time steps using a $\Delta t = 0.01$. A final static step is then computed in which all the dynamic terms are dropped. Convergence for all the transient steps is achieved in 4-6 iterations, while the final static step takes 33 iterations. The solution state at the end of the transient steps is shown in Fig. 10 and the final state achieved after the static step in Fig. 11.

7 Closure

This paper has summarized the steps required to develop and implement a 3-node triangular membrane finite element which can undergo arbitrarily large finite motions. The material behavior included here is restricted to a St.Venant-Kirchhoff material, however, extension to other types of constitutive models for isotropic behavior is straight forward.

The formulation presented includes both inertia and damping effects leading to a general second-order semi-discrete form. Solution to these ordinary differential equations may be achieved using explicit or implicit transient forms or using a quasi-static form in which all rate effects are omitted. Use of a solution algorithm employing only first-order form is shown to lead to a simple means of solving the final static membrane state starting from an unstressed reference configuration.

References

1. Zienkiewicz OC, Taylor RL (2000) The Finite Element Method: Solid Mechanics. Volume 2. Butterworth-Heinemann, Oxford, 5th edition
2. Simo JC, Fox DD (1989) On a stress resultant geometrically exact shell model. Part I: Formulation and optimal parametrization. Computer Methods in Applied Mechanics and Engineering 72:267–304
3. Simo JC, Fox DD, Rifai MS (1989) On a stress resultant geometrically exact shell model. Part II: The linear theory; computational aspects. Computer Methods in Applied Mechanics and Engineering 73:53–92
4. Simo JC, Rifai MS, Fox DD (1990) On a stress resultant geometrically exact shell model. Part IV Variable thickness shells with through-the-thickness stretching. Computer Methods in Applied Mechanics and Engineering 81:91–126
5. Simo JC, Tarnow N (1994) A new energy and momentum conserving algorithm for the non-linear dynamics of shells. International Journal for Numerical Methods in Engineering 37:2527–2549
6. Büchter N, Ramm E (1992) Shell theory versus degeneration – a comparison in large rotation finite element analysis. International Journal for Numerical Methods in Engineering 34:39–59
7. Büchter N, Ramm E, Roehl D (1994) Three-dimensional extension of non-linear shell formulations based on the enhanced assumed strain concept. *International Journal for Numerical Methods in Engineering* 37:2551–2568
8. Braun M, Bischoff M, Ramm E (1994) Nonlinear shell formulations for complete three-dimensional constitutive laws include composites and laminates. Computational Mechanics 15:1–18
9. Bischoff M, Ramm E (1997) Shear deformable shell elements for large strains and rotations. International Journal for Numerical Methods in Engineering 40:4427–49
10. Bischoff M, Ramm E (1999) Solid-like shell or shell-like solid formulation? A personal view. In W. Wunderlich, editor, *Proc. Eur Conf on Comp. Mech (ECCM'99 on CD-ROM)*, Munich, September
11. Ramm E (1999) From Reissner plate theory to three dimensions in large deformation shell analysis. Z. Angew. Math. Mech. 79:1–8

12. Bischoff M, Ramm E (2000) On the physical significance of higher order kinematic and static variables in a three-dimensional shell formulation. International Journal of Solids and Structures 37:6933–60
13. Bletzinger K-U, Bischoff M, Ramm E (2000) A unified approach for shear-locking-free triangular and rectangular shell finite elements. Computers and Structures 75:321–34
14. Schweizerhof K, Ramm E (1984) Displacement dependent pressure loads in non-linear finite element analysis. Computers and Structures 18(6):1099–1114
15. Simo JC, Taylor RL, Wriggers P (1991) A note on finite element implementation of pressure boundary loading. Communications in Applied Numerical Methods 7:513–525
16. Newmark N (1959) A method of computation for structural dynamics. Journal of the Engineering Mechanics Division 85:67–94.
17. Zienkiewicz OC, Taylor RL (2000) The Finite Element Method: The Basis. Volume 1, Butterworth-Heinemann, Oxford, 5th edition
18. Taylor RL FEAP - A Finite Element Analysis Program, Programmer Manual. University of California, Berkeley, http://www.ce.berkeley.edu/~rlt

Applications of a Rotation-Free Triangular Element for Finite Strain Analysis of Thin Shells and Membranes

Fernando G. Flores[1] and Eugenio Oñate[2]

[1] Structures Department
National University of Córdoba
Casilla de Correo 916
5000 Córdoba - Argentina
fflores@efn.unc.edu.ar
Web page: http://www.efn.unc.edu.ar

[2] International Centre for Numerical
Methods in Engineering (CIMNE)
Universidad Politécnica de Cataluña
Campus Norte UPC, 08034 Barcelona, Spain
onate@cimne.upc.es
Web page: http://www.cimne.upc.es

Summary. *This paper shows applications of a recently developed shell element to the analysis of thin shell and membrane structures. The element is a three node triangle with only translational DOFs (rotation free) that uses the configuration of the three adjacent elements to evaluate the strains. This allows to compute (constant) bending strains and (linear) membrane strains. A total Lagrangian formulation is used. Strains are defined in terms of the principal stretches. This allows to consider rubber materials and other type of materials using the Hencky stress-strain pair. An explicit central difference scheme is used to integrate the momentum equations. Several examples, including inflation and deflation of membranes show the excellent convergence properties and robustness of the element for large strain analysis of thin shells and membranes.*

Key words: airbag inflation, deflation, shell triangular elements, rotation free, membranes

1 Introduction

The simulation of the inflation of membrane structures is normally performed with membrane finite elements, i.e. no bending stiffness included. The formulation of such elements is simple as they only require C^0 continuity [1], in contrast with elements based on thin shell theory where C^1 continuity implies important obstacles [2] in

E. Oñate and B. Kröplin (eds.), Textile Composites and Inflatable Structures, 69–88.

the development of conforming elements. Triangular elements are naturally preferred as they can easily adapt to arbitrary geometries and due to the robustness of the associated mesh generators.

When only the final configuration of the membrane is of interest implicit programs are normally used, including special algorithms due to the lack of stiffness of the membrane when no tensile stresses are yet present. When the inflation/deflation process is of interest, the use of programs with explicit integration of the momentum equations are largely preferred. In the latter case linear strain triangles are normally not effective, specially when contact between surfaces is present. This implies a fine discretization of constant strain triangles to capture the details, what makes simulation quite expensive due to the time increment limitations. In this paper a triangular finite element with similar convergence properties to the linear strain triangle, but without its drawbacks, is used.

Membrane structures components have some, although small, bending stiffness that in most of the cases is sensibly disregarded. However in many cases it may be convenient to include bending energy in the models due to the important regularization effect it supposes. Shell elements are of course more expensive due the increase in degrees of freedom (rotations) and integration points (through the thickness). In the last few years shell elements without rotation degrees of freedom have been developed (see [3]–[10] among others), which make shell elements more efficient for both implicit and explicit integrators.

The outline of this papers is as follows. Next section summarizes the rotation-free shell triangle used [10]. Sec. 3 shows convergence properties of the element in 2-d plane stress problems and 3-d linear bending/membrane problems. Sec. 4 presents examples of inflation/deflation of membranes with and without bending stiffness. Finally Sec. 5 summarizes some conclusions.

2 Formulation of the Rotation Free Shell Triangle

The rotation-free EBST (for Enhanced Basic Shell Triangle) element has three nodes with three displacement degrees of freedom at each node. An element patch is defined by the central triangle and the three adjacent elements (Fig. 1). This patch helps to define the membrane strains and curvature field within the central triangle (the EBST element) in terms of the displacement of the six patch nodes.

The node-ordering in the patch is the following (see Fig. 1.a)

- The nodes in the main element (M) are numbered locally as 1, 2 and 3. They are defined counter-clockwise around the positive normal
- The sides in the main element are numbered locally as 1, 2, and 3. They are defined by the local node opposite to the side
- The adjacent elements (which are part of the cell) are numbered with the number associated to the common side
- The extra nodes of the cell are numbered locally as 4, 5 and 6, corresponding to nodes on adjacent elements opposite to sides 1, 2 and 3 respectively
- The connectivities in the adjacent elements are defined beginning with the extra node.

Any convenient in plane $(\mathbf{t}_1, \mathbf{t}_2)$ local cartesian coordinate system can be defined for the patch, with $\mathbf{t}_3$ the unit normal to the plane.The main features of the element formulation are the following:

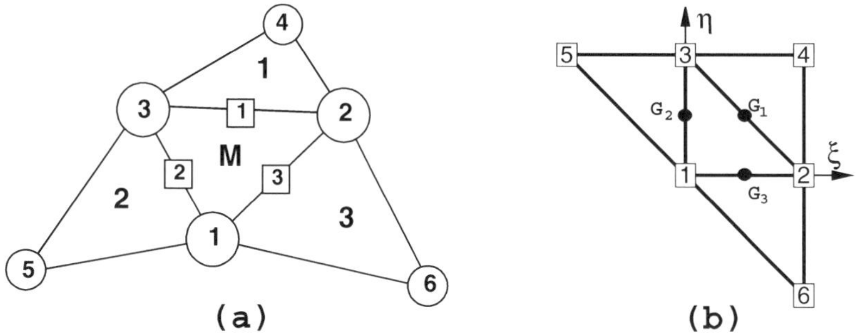

Fig. 1. Patch of elements for strain computation. (a)in spatial coordinates (b)in natural coordinates

1. The geometry of the patch formed by the central element and the three adjacent elements is *quadratically interpolated* from the position of the six nodes in the patch
2. The membrane strains are assumed to vary *linearly* within the central triangle and are expressed in terms of the (continuous) values of the deformation gradient at the mid side points of the triangle
3. An assumed *constant curvature* field within the central triangle is obtained using the values of the (continuous) deformation gradient at the mid side points.

Details of the derivation of the EBST element are given below.

2.1 Definition of the Element Geometry and Computation of Membrane Strains

As mentioned above a quadratic approximation of the geometry of the patch of four elements is chosen using the position of the six nodes. It is useful to define the patch in the isoparametric space using the nodal positions given in the Table 1 (see also Fig. 1.b).

Table 1. Isoparametric coordinates of the six nodes in the patch of Fig. 1.b

	1	2	3	4	5	6
ξ	0	1	0	1	-1	1
η	0	0	1	1	1	-1

The quadratic interpolation for the geometry is defined by

$$\varphi = \sum_{i=1}^{6} N_i \varphi_i \tag{1}$$

with φ_i being the position vector of node i, $\zeta = 1 - \xi - \eta$ and

$$\begin{array}{ll} N_1 = \zeta + \xi\eta & N_4 = \frac{\zeta}{2}(\zeta - 1) \\ N_2 = \xi + \eta\zeta & N_5 = \frac{\xi}{2}(\xi - 1) \\ N_3 = \eta + \zeta\xi & N_6 = \frac{\eta}{2}(\eta - 1) \end{array} \tag{2}$$

this interpolation allows to compute the displacement gradients at selected points in order to use an assumed strain approach. The computation of the gradients is performed at the mid side points of the central element (M) denoted by G_1, G_2 and G_3 in Fig. 1.b. This choice has the advantage that gradients at the three mid side points depend only on the nodes belonging to the two elements adjacent to each side. When gradients are computed at the common mid-side point of two adjacent elements, the same values are obtained, as the coordinates of the same four points are used. This in practice means that the gradients at the mid-side points are independent of the element where they are computed.

The deformation gradient at the mid-side points of the element are obtained from the quadratic interpolations (1) as

$$(\varphi'_\alpha)_{G_i} = \varphi'^i_\alpha = \left[\sum_{j=1}^{3} N^i_{j,\alpha}\varphi_j\right] + N^i_{i+3,\alpha}\varphi_{i+3} \quad , \quad \alpha = 1,2 \quad , \quad i = 1,2,3 \tag{3}$$

In Eq.(3) $(\cdot)^i$ denotes values computed at the ith mid-side point.

The cartesian derivatives of the shape functions are computed at the original configuration by the standard expression

$$\begin{bmatrix} N_{i,1} \\ N_{i,2} \end{bmatrix} = \mathbf{J}^{-1} \begin{bmatrix} N_{i,\xi} \\ N_{i,\eta} \end{bmatrix} \tag{4}$$

where the Jacobian matrix at the original configuration is

$$\mathbf{J} = \begin{bmatrix} \varphi'^0_\xi \cdot \mathbf{t}_1 & \varphi'^0_\eta \cdot \mathbf{t}_1 \\ \varphi'^0_\xi \cdot \mathbf{t}_2 & \varphi'^0_\eta \cdot \mathbf{t}_2 \end{bmatrix} \tag{5}$$

Once the deformation gradient is obtained, any convenient strain measure can be coupled. The membrane strains within the central triangle are now obtained using a linear assumed membrane strain field $\hat{\varepsilon}_m$, i.e.

$$\varepsilon_m = \hat{\varepsilon}_m \tag{6}$$

with

$$\hat{\varepsilon}_m = (1 - 2\zeta)\varepsilon^1_m + (1 - 2\xi)\varepsilon^2_m + (1 - 2\eta)\varepsilon^3_m = \sum_{i=1}^{3} \bar{N}_i \varepsilon^i_m \tag{7}$$

where ε^i_m are the membrane strains computed at the three mid side points G_i ($i = 1, 2, 3$ see Fig. 2). In (7)

$$\bar{N}_1 = (1 - 2\zeta) \quad , \quad \bar{N}_2 = (1 - 2\xi) \quad \text{and} \quad \bar{N}_3 = (1 - 2\eta) \tag{8}$$

If, for example, Green-Lagrange strains are used,

$$\varepsilon_{m_{ij}} = \frac{1}{2}(\varphi'_i \cdot \varphi'_j - \delta_{ij}) \tag{9}$$

substituting (3) into (9) and using the usual membrane strain vector [11]

$$\varepsilon_m = [\varepsilon_{m_{11}}, \varepsilon_{m_{12}}, \varepsilon_{m_{12}}]^T \tag{10}$$

equation (7) gives

$$\varepsilon_m = \sum_{i=1}^{3} \frac{1}{2} \bar{N}_i \begin{Bmatrix} \boldsymbol{\varphi}^i_{\prime 1} \cdot \boldsymbol{\varphi}^i_{\prime 1} - 1 \\ \boldsymbol{\varphi}^i_{\prime 2} \cdot \boldsymbol{\varphi}^i_{\prime 2} - 1 \\ 2\boldsymbol{\varphi}^i_{\prime 1} \cdot \boldsymbol{\varphi}^i_{\prime 2} \end{Bmatrix} \tag{11}$$

The virtual membrane strains are expressed by

$$\delta\varepsilon_m = \sum_{i=1}^{3} \bar{N}_i \begin{Bmatrix} \boldsymbol{\varphi}^i_{\prime 1} \cdot \delta\boldsymbol{\varphi}^i_{\prime 1} \\ \boldsymbol{\varphi}^i_{2} \cdot \delta\boldsymbol{\varphi}^i_{\prime 2} \\ \delta\boldsymbol{\varphi}^i_{\prime 1} \cdot \boldsymbol{\varphi}^i_{\prime 2} + \boldsymbol{\varphi}^i_{\prime 1} \cdot \delta\boldsymbol{\varphi}^i_{2} \end{Bmatrix} . \tag{12}$$

2.2 Computation of Curvatures

The curvatures (second fundamental form) of the middle surface are defined by []

$$\kappa_{\alpha\beta} = \frac{1}{2} (\boldsymbol{\varphi}_{\prime\alpha} \cdot \mathbf{t}_{3^\prime\beta} + \boldsymbol{\varphi}_{\prime\beta} \cdot \mathbf{t}_{3^\prime\alpha}) = -\mathbf{t}_3 \cdot \boldsymbol{\varphi}_{\prime\alpha\beta} \quad , \quad \alpha, \beta = 1, 2 \tag{13}$$

We will assume the following constant curvature field within each element

$$\kappa_{\alpha\beta} = \hat{\kappa}_{\alpha\beta} \tag{14}$$

where $\hat{\kappa}_{\alpha\beta}$ is the assumed constant curvature field obtained as

$$\hat{\kappa}_{\alpha\beta} = -\frac{1}{A^0_M} \int_{A^0_M} \mathbf{t}_3 \cdot \boldsymbol{\varphi}_{\prime\beta\alpha} \, dA^0 \tag{15}$$

and A^0_M is the area (in the original configuration) of the central element in the patch.

Substituting (15) into (14) and integrating by parts the area integral gives the curvature vector [11] within the element in terms of the following closed line integral

$$\kappa = \begin{Bmatrix} \kappa_{11} \\ \kappa_{22} \\ 2\kappa_{12} \end{Bmatrix} = \frac{1}{A^0_M} \int_{\Gamma^0_M} \begin{bmatrix} -n_1 & 0 \\ 0 & -n_2 \\ -n_2 & -n_1 \end{bmatrix} \begin{bmatrix} \mathbf{t}_3 \cdot \boldsymbol{\varphi}_{\prime 1} \\ \mathbf{t}_3 \cdot \boldsymbol{\varphi}_{\prime 2} \end{bmatrix} d\Gamma^0 \tag{16}$$

where n_i are the components (in the local system) of the normals to the element sides in the initial configuration Γ^0_M.

For the definition of the normal vector $\mathbf{t}_3$, the linear interpolation of the position vector over the central element is used.

$$\boldsymbol{\varphi}^M = \sum_{i=1}^{3} L^M_i \boldsymbol{\varphi}_i \tag{17}$$

where L^M_i are the standard linear shape functions of the central triangle (area coordinates) [11]. In this case the tangent plane components are

$$\boldsymbol{\varphi}^M_{\prime\alpha} = \sum_{i=1}^{3} L^M_{i,\alpha} \boldsymbol{\varphi}_i \quad , \quad \alpha = 1, 2 \tag{18}$$

$$\mathbf{t}_3 = \frac{\boldsymbol{\varphi}_{\prime 1}^M \times \boldsymbol{\varphi}_{\prime 2}^M}{|\boldsymbol{\varphi}_{\prime 1}^M \times \boldsymbol{\varphi}_{\prime 2}^M|} = \lambda\, \boldsymbol{\varphi}_1^M \times \boldsymbol{\varphi}_2^M \ . \tag{19}$$

From these expressions it is also possible to compute in the original configuration the element area A_M^0, the outer normals $(n_1, n_2)^i$ at each side and the side lengths l_i^M. Equation (19) also allows to evaluate the thickness ratio λ in the deformed configuration and the actual normal $\mathbf{t}_3$.

Direction $\mathbf{t}_3$ can be seen as a reference direction. If a different direction than that given by (19) is chosen, at an angle θ with the former, this has an influence of order θ^2 in the computation of the curvatures (see (23) below). This justifies (19) for the definition of $\mathbf{t}_3$ as a function exclusively of the three nodes of the central triangle, instead of using the 6-node isoparametric interpolation.

The numerical evaluation of the line integral in (16) results in a sum over the integration points at the element boundary which are, in fact, the same points used for evaluating the gradients when computing the membrane strains. As one integration point is used over each side, it is not necessary to distinguish between sides (i) and integration points (G_i).

The explicit form of the gradient evaluated at each side G_i (3) from the quadratic interpolation is

$$\begin{bmatrix} \boldsymbol{\varphi}_{\prime 1}^i \\ \boldsymbol{\varphi}_{\prime 2}^i \end{bmatrix} = \begin{bmatrix} N_{1,1}^i & N_{2,1}^i & N_{3,1}^i & N_{i+3,1}^i \\ N_{1,2}^i & N_{2,2}^i & N_{3,2}^i & N_{i+3,2}^i \end{bmatrix} \begin{bmatrix} \boldsymbol{\varphi}_1 \\ \boldsymbol{\varphi}_2 \\ \boldsymbol{\varphi}_3 \\ \boldsymbol{\varphi}_{i+3} \end{bmatrix} . \tag{20}$$

We note again that the gradient at each mid side point G_i depends only on the coordinates of the three nodes of the central triangle and on those of an additional node in the patch, associated to the side i where the gradient is computed.

In this way the curvatures can be computed by

$$\boldsymbol{\kappa} = 2 \sum_{i=1}^{3} \begin{bmatrix} L_{i,1}^M & 0 \\ 0 & L_{i,2}^M \\ L_{i,2}^M & L_{i,1}^M \end{bmatrix} \begin{bmatrix} \mathbf{t}_3 \cdot \boldsymbol{\varphi}_{\prime 1}^i \\ \mathbf{t}_3 \cdot \boldsymbol{\varphi}_{\prime 2}^i \end{bmatrix} \tag{21}$$

An alternative form to express the curvatures, which is useful when their variations are needed, is to define the vectors

$$\mathbf{h}_{ij} = \sum_{k=1}^{3} \left(L_{k,i}^M \boldsymbol{\varphi}_{\prime j}^k + L_{k,j}^M \boldsymbol{\varphi}_{\prime i}^k \right) \tag{22}$$

This gives

$$\kappa_{ij} = \mathbf{h}_{ij} \cdot \mathbf{t}_3 \tag{23}$$

The variation of the curvatures can be obtained as

$$\begin{aligned} \delta\boldsymbol{\kappa} = {} & 2 \sum_{i=1}^{3} \begin{bmatrix} L_{i,1}^M & 0 \\ 0 & L_{i,2}^M \\ L_{i,2}^M & L_{i,1}^M \end{bmatrix} \left\{ \sum_{i=1}^{3} \begin{bmatrix} N_{j,1}^i (\mathbf{t}_3 \cdot \delta\mathbf{u}_j) \\ N_{j,2}^i (\mathbf{t}_3 \cdot \delta\mathbf{u}_j) \end{bmatrix} + \begin{bmatrix} N_{i+3,1}^i (\mathbf{t}_3 \cdot \delta\mathbf{u}^{i+3}) \\ N_{i+3,2}^i (\mathbf{t}_3 \cdot \delta\mathbf{u}^{i+3}) \end{bmatrix} \right\} \\ & - \sum_{i=1}^{3} \begin{bmatrix} (L_{i,1}^M \varrho_{11}^1 + L_{i,2}^M \varrho_{11}^2) \\ (L_{i,1}^M \varrho_{22}^1 + L_{i,2}^M \varrho_{22}^2) \\ (L_{i,1}^M \varrho_{12}^1 + L_{i,2}^M \varrho_{12}^2) \end{bmatrix} (\mathbf{t}_3 \cdot \delta\mathbf{u}_i) \end{aligned} \tag{24}$$

where the projections of the vectors $\mathbf{h}_{ij}$ over the contravariant base vectors $\tilde{\boldsymbol{\varphi}}'^{M}_{\alpha}$ have been included

$$\varrho^{\alpha}_{ij} = \mathbf{h}_{ij} \cdot \tilde{\boldsymbol{\varphi}}'^{M}_{\alpha} \quad , \quad \alpha, i, j = 1,2 \tag{25}$$

with

$$\tilde{\boldsymbol{\varphi}}'^{M}_{1} = \lambda\, \boldsymbol{\varphi}'^{M}_{2} \times \mathbf{t}_3 \tag{26}$$

$$\tilde{\boldsymbol{\varphi}}'^{M}_{2} = -\lambda\, \boldsymbol{\varphi}'^{M}_{1} \times \mathbf{t}_3 \tag{27}$$

In above expressions superindexes in L^M_j and $\delta\mathbf{u}^k_j$ refer to element numbers whereas subscripts denote node numbers. As before the superindex M denotes values in the central triangle (Fig. 1.a). Note that as expected the curvatures (and their variations) in the central element are a function of the nodal displacements of the six nodes in the four elements patch.

Details of the derivation of (12) and (24) can be found in [10]. The explicit expressions of the membrane and curvature matrices can be found in [12]. The derivation of the element stiffnes matrix is described in [10, 12]. Also in [10, 12] details of the quasi-static formulation and the fully explicit dynamic formulation are given.

It must be noted that while the membrane strains are linear the curvature strains are constant. A full numerical integration of the stiffness matrix terms requires three points for the membrane part and one point for the bending part. Numerical experiments show that:

- when using one or three integration points the element is free of spurious energy modes and passes the patch test
- for initial curved surfaces the element with full (three point) integration leads to some membrane locking. This defect dissapears if one integration point is used for the membrane stiffness term.

It can also be observed that:

- for large strain elastic or elastic-plastic problems membrane and bending parts can not be integrated separately, and a numerical integration trought the thickness must be performed
- for explicit integrators (hydro codes) is much more effective to use only one integration point for both the membrane and bending parts.

Above arguments lead to reccomended the use of one integration point for both membrane and bending parts. This element is termed EBST1 to distinguish from the fully integrated one.

2.3 Boundary Conditions

Elements at the domain boundary, where an adjacent element does not exist, deserve a special attention. The treatment of essential boundary conditions associated to translational constraints is straightforward, as they are the degrees of freedom of the element. The conditions associated to the normal vector are crucial in this formulation for bending. For clamped sides or symmetry planes, the normal vector $\mathbf{t}_3$ must be kept fixed (clamped case), or constrained to move in the plane of symmetry (symmetry case). The former case can be seen as a special case of the latter, so we

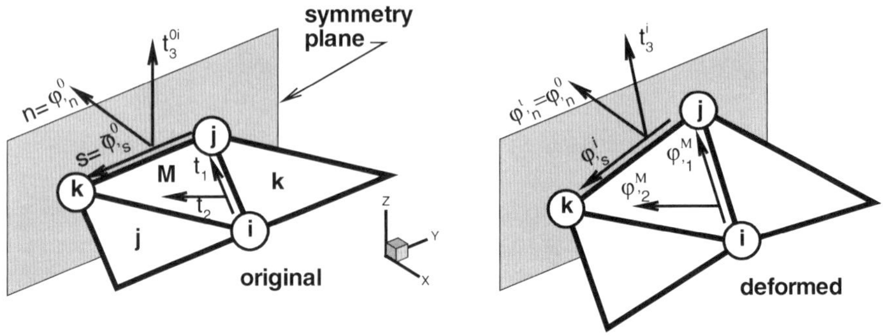

Fig. 2. Local cartesian system for the treatment of symmetry boundary conditions

will consider symmetry planes only. This restriction can be imposed through the definition of the tangent plane at the boundary, including the normal to the plane of symmetry $\varphi^0_{\prime n}$ that does not change during the process.

The tangent plane at the boundary (mid-side point) is expressed in terms of two orthogonal unit vectors referred to a local-to-the-boundary Cartesian system (see Fig. 2) defined as

$$\left[\varphi^0_{\prime n},\ \bar{\varphi}_{\prime s}\right] \tag{28}$$

where vector $\varphi^0_{\prime n}$ is fixed during the process while direction $\bar{\varphi}_{\prime s}$ emerges from the intersection of the symmetry plane with the plane defined by the central element (M). The plane (gradient) defined by the central element in the selected original convective Cartesian system $(\mathbf{t}_1, \mathbf{t}_2)$ is

$$\left[\varphi^M_{\prime 1},\ \varphi^M_{\prime 2}\right] \tag{29}$$

the intersection line (side i) of this plane with the plane of symmetry can be written in terms of the position of the nodes that define the side (j and k) and the original length of the side l^M_i, i.e.

$$\varphi^i_{\prime s} = \frac{1}{l^M_i}\left(\varphi_k - \varphi_j\right) \tag{30}$$

That together with the outer normal to the side $\mathbf{n}^i = [n_1, n_2]^T = [\mathbf{n}\cdot\mathbf{t}_1, \mathbf{n}\cdot\mathbf{t}_2]^T$ (resolved in the selected original convective Cartesian system) leads to

$$\begin{bmatrix} \varphi^{iT}_{\prime 1} \\ \varphi^{iT}_{\prime 2} \end{bmatrix} = \begin{bmatrix} n_1 & -n_2 \\ n_2 & n_1 \end{bmatrix} \begin{bmatrix} \varphi^{iT}_{\prime n} \\ \varphi^{iT}_{\prime s} \end{bmatrix} \tag{31}$$

where, noting that λ is the determinant of the gradient, the normal component of the gradient $\varphi^i_{\prime n}$ can be approximated by

$$\varphi^i_{\prime n} = \frac{\varphi^0_{\prime n}}{\lambda |\varphi^i_{\prime s}|} \tag{32}$$

In this way the contribution of the gradient at side i to vectors $\mathbf{h}_{\alpha\beta}$ (22) results in

$$\begin{bmatrix} \mathbf{h}_{11}^T \\ \mathbf{h}_{22}^T \\ 2\mathbf{h}_{12}^T \end{bmatrix}^i = 2 \begin{bmatrix} L_{i,1}^M & 0 \\ 0 & L_{i,2}^M \\ L_{i,2}^M & L_{i,1}^M \end{bmatrix} \begin{bmatrix} \boldsymbol{\varphi}'^{iT}_1 \\ \boldsymbol{\varphi}'^{iT}_2 \end{bmatrix} = 2 \begin{bmatrix} L_{i,1}^M & 0 \\ 0 & L_{i,2}^M \\ L_{i,2}^M & L_{i,1}^M \end{bmatrix} \begin{bmatrix} n_1 & -n_2 \\ n_2 & n_1 \end{bmatrix} \begin{bmatrix} \boldsymbol{\varphi}'^{iT}_n \\ \boldsymbol{\varphi}'^{iT}_s \end{bmatrix} \tag{33}$$

For a simple supported (hinged) side, the problem is not completely defined. The simplest choice is to neglect the contribution to the side rotations from the adjacent element missing in the patch in the evaluation of the curvatures via (16) [6, 8]. This is equivalent to assume that the gradient at the side is equal to the gradient in the central element. More precise changes can be however introduced to account for the different natural boundary conditions. One may assume that the curvature normal to the side is zero, and consider a contribution of the missing side to introduce this constraint. As the change of curvature parallel to the side is zero along the hinged side, both things lead to zero curvatures in both directions. For the case of a triangle with two sides associated to hinged sides, the normal curvatures to both sides must be set to zero.

For a free edge the same approximation can be used but due to Poisson's effect this will lead to some error.

For the membrane formulation of the EBST element, the gradient at the mid-side point of the boundary is assumed to be equal to the gradient of the main triangle.

2.4 Constitutive Models

In the numerical experiments presented below two constitutive models have been used. A standard linear elastic orthotropic material and a hyper-elastic material for rubbers.

For the case of rubbers, the Ogden [13] model extended to the compressible range is considered. The material behaviour is characterized by the strain energy density per unit undeformed volume defined as

$$\psi = \frac{K}{2}(\ln J)^2 + \sum_{p=1}^{N} \frac{\mu_p}{\alpha_p}\left[J^{-\frac{\alpha_p}{3}} \left(\sum_{i=1}^{3} \lambda_i^{\alpha_p - 1} \right) - 3 \right] \tag{34}$$

where K is the bulk modulus of the material, λ_i are the principal strain ratios, J is the determinant of the deformation gradient $\mathbf{F}$ ($J = \lambda_1\lambda_2\lambda_3$), N, μ_i and α_i are material parameters, μ_i, α_i are real numbers such that $\mu_i\alpha_i > 0 \quad (\forall i = 1, N)$ and N is a positive integer.

3 Convergence Studies

In this section three examples are presented to show the convergence properties and the performance of present element. Examples are solved with a implicit program capable of dealing static/dynamic problems with moderate non-linearities.

3.1 Cook's Membrane Problem

One of the main targets of the rotation-free triangular element is to obtain a membrane element with a behaviour similar to the linear strain triangle (LST). Such capacity is studied in this example [14] corresponding to a problem with an important amount of shear energy involved. This problem is also intended also to assess the ability of the element to distort. Figure 3.a shows the geometry of a tapered panel clamped on one side and with a uniformly distributed shear load in the opposite side. Figure 3.b presents the vertical displacement of point **C** (mid point of the loaded side) for the uniformly refined meshes considered as a function of the total number of nodes in the mesh.

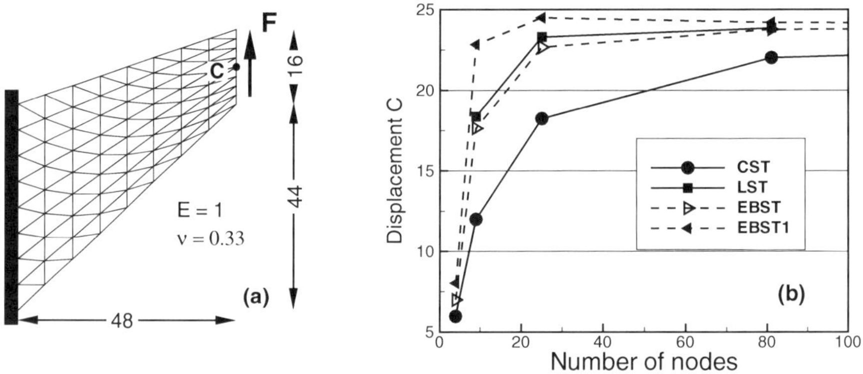

Fig. 3. Cook's membrane problem. (a) Geometry and load (b) Vertical displacement of point C for different meshes

For the EBST element with three integration points, it can be seen that for the coarsest mesh (two linear elements), the measured displacement is slightly superior than the constant strain triangle (CST); but when the mesh is refined, values rapidly catch up with those obtained with the linear strain triangle. The element with only one integration point (EBST1) shows excellent predictions for coarse meshes and fast convergence properties for the reported displacement.

3.2 Cylindrical Roof

In this example an effective membrane interpolation is of primary importance. Hence this is good test to assess the rotation-free element. The geometry is a cylindrical roof supported by a rigid diaphragm at both ends and it is loaded by a uniform dead weight (see Fig. 4.a). Only one quarter of the structure is meshed due to symmetry conditions. Unstructured and structured meshes are considered. In the latter case two orientations are possible (Fig. 4.a shows orientation B).

Figure 4.b shows the normalized vertical displacement of the midpoint of the free side (point-C) over both (A and B) structured and (U) unstructured meshes as

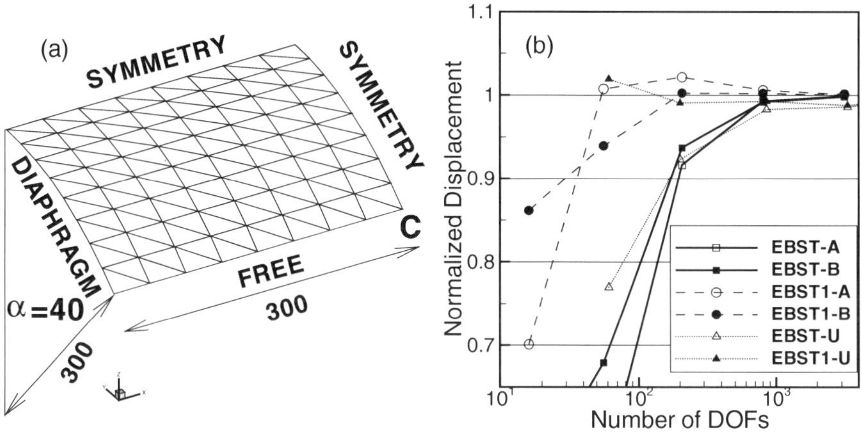

Fig. 4. Cylindrical roof under dead weight. $E = 3 \times 10^6$, $\nu = 0.0$, Thickness =3.0, shell weight =0.625 per unit area. (a)Geometry. (b)Normalized displacement of point "C" for different meshes

a function of the number of free degrees of freedom. Value used for normalization is $u_C = -3.610$ as quoted in [15].

An excellent convergence of the EBST element can be seen. The version with only one integration point (EBST1) presents a behaviour a little more flexible and converges from above. For non-structured meshes the result converges to the reference value but a bit more slowly.

3.3 Inflation of a Sphere

As the EBST element uses a quadratic interpolation of geometry, the existance of membrane locking must be assessed. For this example an originally curved surface is considered, where a standard linear strain triangle would lead to membrane locking. The example is the inflation of a spherical shell under internal pressure. An incompressible Mooney-Rivlin constitutive material have been considered. The Ogden parameters are $N = 2$, $\alpha_1 = 2$, $\mu_1 = 40$, $\alpha_2 = -2$, $\mu_2 = -20$. Due to the simple geometry an analytical solution exists [16] (with $\gamma = R/R^{(0)}$):

$$p = \frac{h^{(0)}}{R^{(0)}\gamma^2}\frac{dW}{d\gamma} = \frac{8h^{(0)}}{R^{(0)}\gamma^2}\left(\gamma^6 - 1\right)\left(\mu_1 - \mu_2\gamma^2\right)$$

In this numerical simulation the same geometric and material parameters used in [9] have been adopted: $R^{(0)} = 1$ and $h^{(0)} = 0.02$. The three meshes considered to evaluate convergence are shown in Fig. 5.a-c. The EBST1 element has been used. The value of the actual radius as a function of the internal pressure is plotted in Fig. 5.d for the different meshes and is also compared with the analytical solution. It can be seen that with a few degrees of freedom it is possible to obtain an excellent agreement for the range of strains considered. The final value corresponds to a thickness radius ratio of $h/R = 0.00024$. No membrane locking has therefore been detected in this problem.

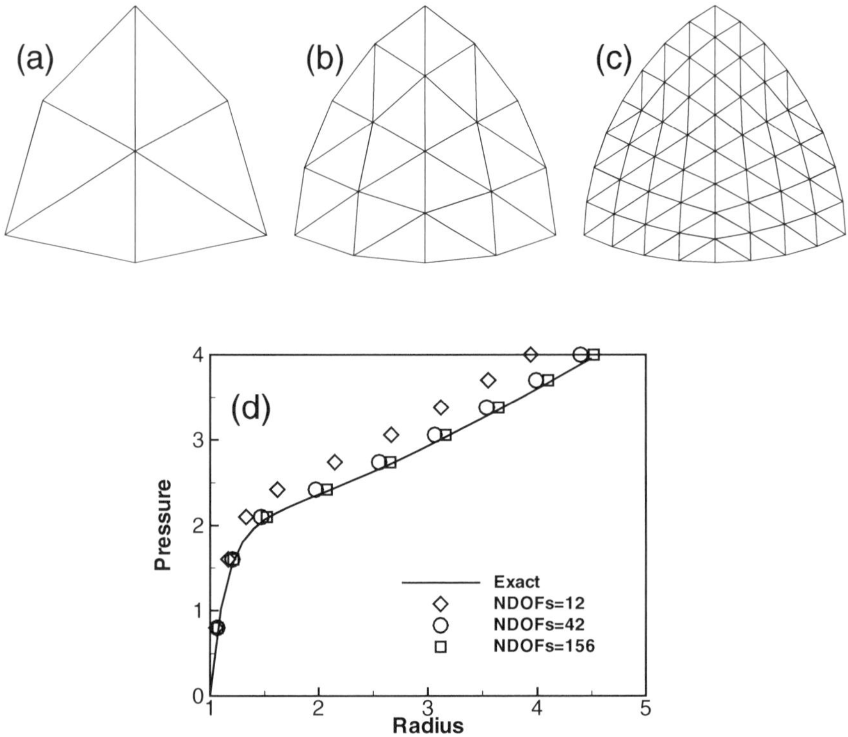

Fig. 5. Inflation of sphere of Mooney-Rivlin material. (a)-(c) EBST1 meshes used in the analysis (d)Radius as a function of the internal pressure

4 Thin Shells and Membranes

Results for examples with geometric and material non-linearities are presented next using the EBST1 element [10]. Due to the features of the modelled problems, with strong non linearities associated to instabilities and contact, a code with explicit integration of the dynamic equilibrium equations has been used [17]. This code allows to obtain pseudo-static solutions through dynamic relaxation. In most of the examples contact situations appear, including contact with walls, objects or self contact due to folds and wrinkles. A standard penalty formulation is used for contact assumed frictionless in the cases.

4.1 Inflation/Deflation of a Circular Airbag

This example has been taken from [9] where it is shown that the final configuration is mesh dependent due to the strong instabilities leading to a non-uniqueness of the solution. In [9] it is also discussed the important regularizing properties of the bending energy, that when disregarded leads to massive wrinkling in the compressed zones.

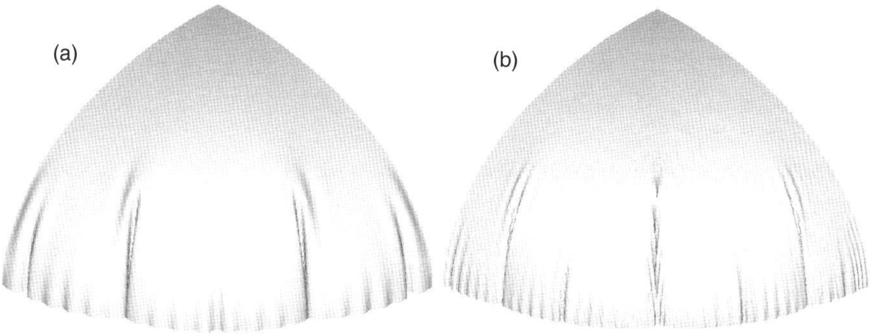

Fig. 6. Inflation of a circular airbag. Deformed configurations for final pressure. (a) bending effects included (b) membrane solution only

The airbag geometry is initially circular with an undeformed radius of 0.35. The constitutive material is a linear isotropic elastic one with modulus of elasticity $E = 6 \times 10^7$ and Poisson's ratio $\nu = 0.3$. Arbitrarily only one quarter of the geometry has been modelled. Only the normal displacement to the original plane is constrained along the boundaries. The thickness considered is $h = 0.0004$ and the inflation pressure is 5000. Using a density $\delta = 1000$, pressure is linearly increased from 0 to the final value in $t = 0.1$.

With comparative purposes and also to backup the comments in Ref. [9] two analyses have been performed, a purely membrane one and one including bending effects. Figure 6 shows the final deformed configurations for a mesh with 10201 nodes and 20000 elements. The figure on the left (a) corresponds to a full analysis including bending and the right figure (b) is a pure membrane analysis.

We note that when the bending energy is included a more regular final pattern is obtained. Also the final pattern is rather independent of the discretization (note that the solution is non unique due to the strong instabilities). On the other hand, the pure membrane solution shows a a noteworthy increment of "numerical" wrinkles.

Figure 7 shows the results obtained for the de-inflation process for three different times. Column on the left corresponds to the analysis with bending energy included. Note that the spherical membrane falls down due to the self weight. The final configuration is of course non-unique.

4.2 Inflation of a Square Air-bag

This example has also been taken from [9]. Again the final configuration is mesh dependent due to the strong instabilities leading to a non-uniqueness of the solution.

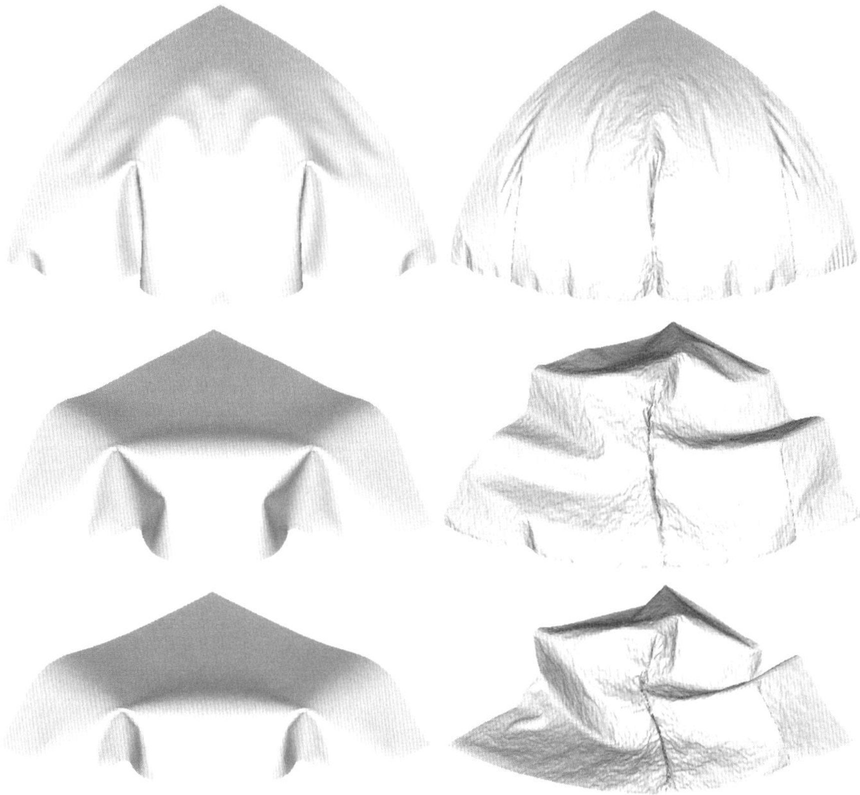

Fig. 7. Deflation of a circular air-bag. Left figure: results obtained with the full bending formulation. Right figure: results obtained with a pure membrane solution

The air bag geometry is initially square with an undeformed diagonal of 1.20. The constitutive material is a linear isotropic elastic one with modulus of elasticity $E = 5.88 \times 10^8$ and Poisson's ratio $\nu = 0.4$. Only one quarter of the geometry has been modelled due to symmetry. Only the normal to the original plane is constrained along the boundaries. The thickness considered is $h = 0.001$ and the inflation pressure is 5000. Using a density $\delta = 1000$, pressure is linearly increased from 0 to the final value in $t = 0.1$.

Two analyses have been performed, a purely membrane one and another one including bending effects. Figure 8 shows the final deformed configurations for three meshes with 289, 1089 and 4225 nodes. The top row corresponds to a full analysis including bending and the central row is a pure membrane analysis. The bottom row is also an analysis including bending where the mesh orientation has been changed.

The top and bottom lines show the final shapes change according to the degree of discretization and mesh orientation due to instabilities and non uniqueness of the solution. The central row shows the pure membrane solution with a wrinkling pattern where the width of the wrinkle is the length of the element.

Fig. 8. Inflation of a square air-bag. Deformed configurations for three different meshes with 800, 3136 and 12416 degrees of freedom

4.3 Inflation of a Square Airbag Against a Spherical Object

The last example of this kind is the inflation of a square airbag supporting a spherical object. This example resembles a problem studied (numerically and experimentally) in [18], where fluid-structure interaction is the main subject. Here fluid is not modelled, and a uniform pressure is applied over all the internal surfaces. The lower surface part of the airbag is limited by a rigid plane and on the upper part a spherical dummy object is set to study the interaction between the airbag and the object.

The airbag geometry is initially square with an undeformed side length of 0.643. The constitutive material used is a linear isotropic elastic one with modulus of elasticity $E = 5.88 \times 10^8$ and Poisson's ratio $\nu = 0.4$. Only one quarter of the geometry has been modelled due to symmetry. The thickness considered is $h = 0.00075$ and the inflation pressure is 250000. Using a density $\delta = 1000$, pressure is linearly increased from 0 to the final value in $t = 0.15$. The spherical object has a radius $r = 0.08$ and a mass of 4.8 (one quarter), and is subjected to gravity load during all the process.

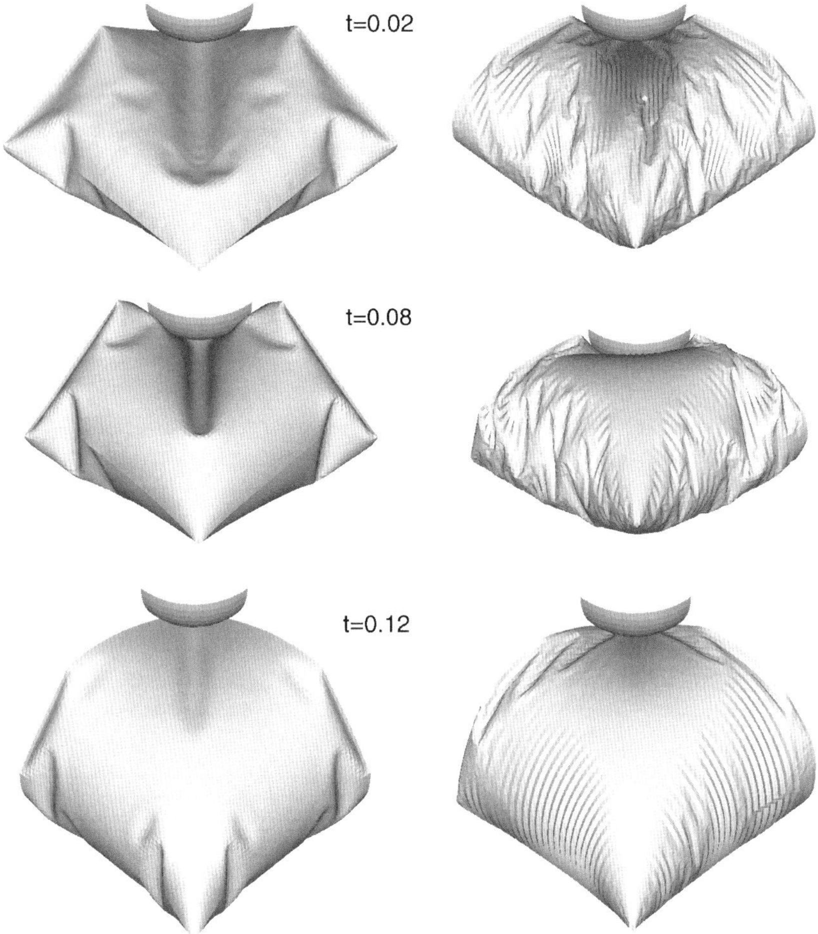

Fig. 9. Inflation of a square airbag against an spherical object. Deformed configurations for different times. Left figure: results obtained with the full bending formulation. Right figure: results obtained with a pure membrane solution

The mesh includes 8192 EBST1 elements and 4225 nodes on each surface of the airbag. Figure 9 shows the deformed configurations for three different times. The sequence on the left of the figure corresponds to an analysis including full bending effects and the sequence on the right is the result of a pure membrane analysis.

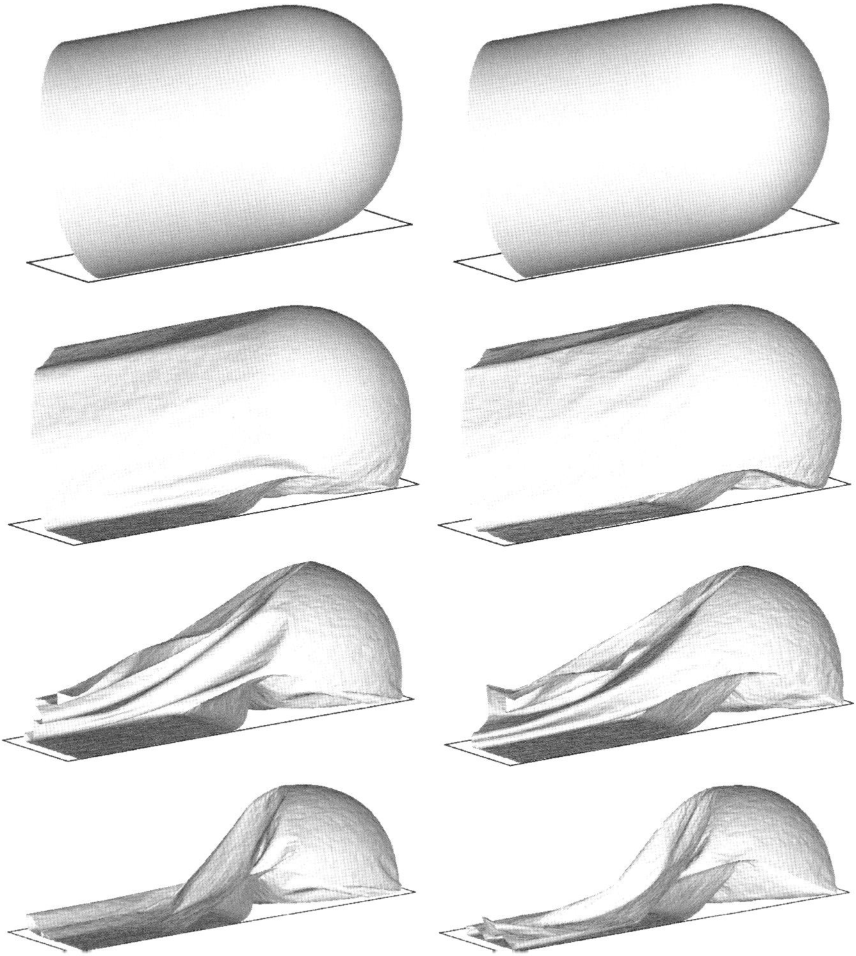

Fig. 10. Inflation and deflation of a closed tube. $L = 6$, $D = 2$, $h = 5 \times 10^{-4}$. Left figure: results obtained with the full bending formulation. Right figure: results obtained with a pure membrane solution

4.4 Inflation/Deflation of a Closed Tube

The last problem is the study of the inflating and de-inflating of a tube with a semi-spherical end cap. The tube diameter is $D = 2$, its total length is $L = 6$ and the thickness $h = 5 \times 10^{-4}$. The material has the following properties $E = 4 \times 10^8$, $\nu = 0.35$, $\varrho = 2 \times 10^3$. The tube is inflated fast until a pressure of 10^4 and when pressure is released the tube de-inflates and falls under self weight. The analysis is performed with a mesh of 16704 EBST1 elements and 8501 nodes modelling a quarter of the geometry. A rigid frictionless base is supposed below. Self contact is also included to avoid penetrations. The evolution of the tube walls during the de-inflating process can be seen in Fig. 10. Note that the central part collapses as expected, while a great part of the semi-spherical cap remains unaltered. For this very thin shell, the differences between a full bending solution and a pure membrane solution are less marked. It must be noted that present element does not presents problems in the very thin limit as the formulation is based on classical thin shell theory and the rotational variables have been elliminates. So the time increment is independent of the thickness.

4.5 Inflation of a Tubular Arch

The last example is the analysis of a tubular arch. This kind of archs are joined together to form large inflatable structures to be used for different purposes. The analized tubular arch has a internal diameter of 0.9; is total length is 11.0 and the heigth is 4.5. The tube thickness is 3×10^{-4}, the constitutive material is polyamid with Young modulus $E = 2.45 \times 10^{8}$ and Poisson ratio $\nu = 0.35$. Due to geometric symmetrys one quarter of the tube was discretized with 33600 triangular elements (17061 nodes). The simulation includes two stages. First the tube is left fall down under gravity action. Second an internal pressure of $p = 883$ is applied in a short time and kept constant afterwards until the full inflation of the tube is reached.

Figure 11 shows deformed configurations for different instants of the process.

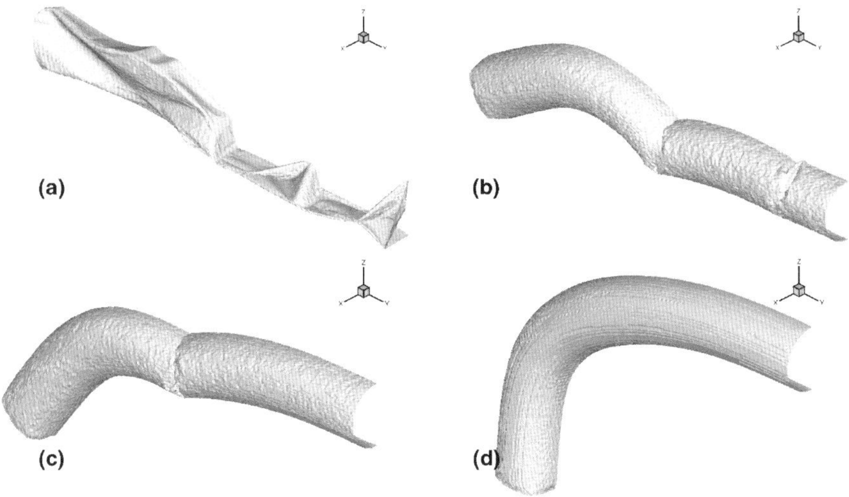

Fig. 11. Inflation of a tubular arch. (a) Deflated tube. (b),(c) Deformed configuration during the inflation process. (d) Final inflated configuration

5 Concluding Remarks

We have presented in the paper the formulation of a rotation-free enhanced basic shell triangle (EBST) using an assumed strain approach. The element is based on an assumed constant curvature field expressed in terms of the nodal deflections of a patch of four elements and an assumed linear membrane strain field for the in-plane behaviour. A simple and economic version of the element using a single integration point has been presented. The EBST1 element has proven to be an excellent candidate for solving practical engineering shell and membrane problems involving complex geometry, dynamics, material non linearity and frictional contact conditions. In the simulation of membranes, bending have been included to avoid massive wrinkling in the compressed zones. The inclusion of bending energy (two integration points through the thickness instead of one) represents an increase of 40% of CPU time.

Acknowledgments

The first author is a member of the scientific staff of the Science Research Council of Argentina (CONICET). The support provided by grants of CONICET and Agencia Córdoba Ciencia S.E. and the support of the company QUANTECH (`www.quantech.es`) providing the code STAMPACK are gratefully acknowledged.

References

1. Taylor RL (2001) Finite element analysis of membrane structures. Publication 203, CIMNE, Barcelona
2. Oñate E (1994) A review of some finite element families for thick and thin plate and shell analysis. Publication 53, CIMNE, Barcelona
3. Hampshire JK, Topping BHV, Chan HC (1992) Three node triangular elements with one degree of freedom per node. Engng Comput 9:49–62
4. Phaal R, Calladine CR (1992) A simple class of finite elements for plate and shell problems. II: An element for thin shells with only translational degrees of freedom. Num Meth Engng 35:979–996
5. Oñate E, Cervera M. (1993) Derivation of thin plate bending elements with one degree of freedom per node. Engineering Computations 10:553–561
6. Oñate E, Zárate F (2000) Rotation-free plate and shell triangles. Num Meth Engng 47:557–603
7. Cirak F, Ortiz M (2000) Subdivision surfaces: A new paradigm for thin-shell finite element analysis. Num Meths Engng 47:2039–2072
8. Flores FG, Oñate E (2001) A basic thin shell triangle with only translational DOFs for large strain plasticity. Num Meths Engng 51:57–83
9. Cirak F, Ortiz M (2001) Fully C^1-conforming subdivision elements for finite deformations thin-shell analysis. Num Meths Engng 51:813–833
10. Flores FG, Oñate E. (2003) Improvements in the membrane behaviour of the three node rotation-free BST shell triangle using an assumed strain approach. Comput Meth Appl Mech Engng (in press)
11. Zienkiewicz OC, Taylor RL (2000) The finite element method. Vol II: Solid Mechanics, Butterworth-Heinemann, place?
12. Oñate E, Flores FG (2004) Advances in the formulation of the rotation-free basic shell. Comput Meth Appl Mech Engng (in press)

13. Ogden RW (1972) Large deformation isotropic elasticity: on the correlation of theory and experiments for incompressible rubberlike solids. Proc Royal Society London A326:565–584
14. Cook RD (1974) Improved two-dimensional finite element, ASCE J Struct Div 100(ST6):1851–1863
15. Huang HC (1989) Static and Dynamic Analysis of Plates and Shells. Springer-Verlag, Berlin
16. Needleman A (1977) Inflation of spherical rubber ballons. Solids and Structures 13:409–421
17. STAMPACK (2003) A General Finite Element System for Sheet Stamping and Forming Problems, Quantech ATZ, Barcelona
18. Marklund PO. Nilsson L (2002) Simulation of airbag inflation processes using a coupled fluid structure approach. Computational Mechanics 29:289–297

FE Analysis of Membrane Systems Including Wrinkling and Coupling

Riccardo Rossi[1], Vitaliani Renato[1], and Eugenio Onate[2]

[1] Università di Padova
ricrossi@caronte.dic.unipd.it
[2] CIMNE, UPC, Barcelona
onate@cimne.upc.es

1 Abstract

Current work summarizes the experience of the writer in the modeling of membrane systems. A first subsection describes an efficient membrane model, together with a reliable solution procedure. The following section addresses the simulation of the wrinkling phenomena providing details of a new solution procedure. The last one proposes an efficient technique to obtain the solution of the fluid structural interaction problem.

2 The Membrane Model

A membrane is basically a 2D solid which "lives" in a 3D environment. Given the lack of flexural stiffness, membranes can react to applied load only by using their in–plane resistance "choosing" the spatial disposition that is best suited to resist to the external forces. The consequence is that membrane structures tend naturally to find the optimal shape (compatible with the applied constraints) for any given load. In this "shape research", they typically undergo large displacements and rotations.

From a numerical point of view, this reflects an intrinsic geometrical non–linearity that has to be taken in account in the formulation of the finite element model. In particular, an efficient Membrane Element should be able to represent correctly arbitrary rotations both of the element as a whole and internally to each element. The possibility of unrestricted rigid body motions constitutes a source of ill-conditioning or even of singularity of the tangent stiffness matrix introducing the need of carefully designed solution procedures.

2.1 Finite Element Model

Current section describes a finite element model that meets all of the requirements for the correct simulation of general membrane systems. The derivation

E. Oñate and B. Kröplin (eds.), Textile Composites and Inflatable Structures, 89–108.

makes use exclusively of orthogonal bases simplifying the calculations and allowing to express all the terms directly in Voigt Notation, which eases the implementation.

Einstein's sum on repeated indices is assumed unless specified otherwise

- $\mathbf{x_I} = \{x_I, y_I, z_I\}^T$ is the position vector of the I–th node in the cartesian space (3D space)
- $\xi = \{\xi, \eta\}^T$ describes the position of a point in the local system of coordinates
- Capital letter are used for addressing to the reference configuration
- $N_I(\xi)$ is the value of the shape function centered on node I on the point of local coordinates ξ

The use of the standard iso-parametric approach allows to express the position of any point as $\mathbf{x}(\xi) = N_I(\xi)\mathbf{x}_I$.
In the usual assumptions of the continuum mechanics it is always possible to define the transformation between the local system of coordinates and the cartesian system as

$$\{\xi + d\xi, \eta\}^T - \{\xi, \eta\}^T \rightarrow \frac{\partial \mathbf{x}(\xi, \eta)}{\partial \xi} d\xi = \mathbf{g}_\xi d\xi \tag{1}$$

$$\{\xi, \eta + d\eta\}^T - \{\xi, \eta\}^T \rightarrow \frac{\partial \mathbf{x}(\xi, \eta)}{\partial \eta} d\eta = \mathbf{g}_\eta d\eta \tag{2}$$

in which we introduced the symbols

$$\mathbf{g}_\xi = \frac{\partial N_I(\xi, \eta)}{\partial \xi} \mathbf{x_I} \tag{3}$$

$$\mathbf{g}_\eta = \frac{\partial N_I(\xi, \eta)}{\partial \eta} \mathbf{x_I} \tag{4}$$

the vectors $\mathbf{g}_\xi$ and $\mathbf{g}_\eta$ of the 3D space, can be considered linearly independent (otherwise compenetration or self contact would manifest) follows immediately that they can be used in the construction of a base of the 3D space. In particular an orthogonal base can be defined as

$$\mathbf{v_1} = \frac{\mathbf{g}_\xi}{\|\mathbf{g}_\xi\|} \tag{5}$$

$$\mathbf{n} = \frac{\mathbf{g}_\xi \times \mathbf{g}_\eta}{\|\mathbf{g}_\xi \times \mathbf{g}_\eta\|} \rightarrow \mathbf{v_2} = \mathbf{n} \times \mathbf{v_1} \tag{6}$$

$$\mathbf{v_3} = \mathbf{g}_\xi \times \mathbf{g}_\eta \tag{7}$$

Vectors $\mathbf{v_1}$ and $\mathbf{v_2}$ describe the local tangent plane to the membrane while the third base vector is always orthogonal. Follows the possibility of defining a local transformation rule that links the local coordinates ξ and the coordinates

$\widehat{\mathbf{x}}$ in the local tangent plane base. This can be achieved by considering that an increment $\{d\xi, 0\}$ maps to the new base as

$$\begin{pmatrix} d\xi \\ 0 \end{pmatrix} \to \begin{pmatrix} \mathbf{g}_\xi \bullet \mathbf{v_1} \\ \mathbf{g}_\xi \bullet \mathbf{v_2} \end{pmatrix} d\xi \ ; \ \begin{pmatrix} 0 \\ d\eta \end{pmatrix} \to \begin{pmatrix} \mathbf{g}_\eta \bullet \mathbf{v_1} \\ \mathbf{g}_\eta \bullet \mathbf{v_2} \end{pmatrix} d\eta \tag{8}$$

this is sinthetized by the definition of the linear application

$$\begin{pmatrix} d\widehat{x}_1 \\ d\widehat{x}_2 \end{pmatrix} = \begin{pmatrix} \mathbf{g}_\xi \bullet \mathbf{v_1} \ \mathbf{g}_\eta \bullet \mathbf{v_1} \\ \mathbf{g}_\xi \bullet \mathbf{v_2} \ \mathbf{g}_\eta \bullet \mathbf{v_2} \end{pmatrix} \begin{pmatrix} d\xi \\ d\eta \end{pmatrix} \to d\widehat{\mathbf{x}} = \mathbf{j} d\xi \tag{9}$$

it should be noted that $\mathbf{g}_\xi \bullet \mathbf{v_3}$ and $\mathbf{g}_\eta \bullet \mathbf{v_3}$ are identically zero consequently no components are left apart in representing the membrane in the new coordinates system. Taking into account the definition of the base vectors the tensor $\mathbf{j}$ becomes (after some calculations)

$$\mathbf{j} = \begin{pmatrix} \|\mathbf{g}_\xi\| & \frac{\mathbf{g}_\xi \bullet \mathbf{g}_\eta}{\|\mathbf{g}_\xi\|} \\ 0 & \frac{\|\mathbf{v_3}\|}{\|\mathbf{g}_\xi\|} \end{pmatrix} \tag{10}$$

and its determinant

$$det(\mathbf{j}) = \|\mathbf{v_3}\| = \|\mathbf{g}_\xi \times \mathbf{g}_\eta\| \tag{11}$$

As the interest is focused on the purely membranal behavior, it is not needed to take in account the deformation of the structure over the thickness, as this can be calculated “a posteriori” once known the deformation of the mid–plane. On the base of this observation, the deformation gradient which describes the deformation of the membrane as a 3D body

$$\mathbf{F}_{3\times 3} = \frac{\partial \mathbf{x}}{\partial \mathbf{X}} \tag{12}$$

can be replaced by

$$\widehat{\mathbf{F}}_{2\times 2} = \frac{\partial \dot{\mathbf{x}}}{\partial \widehat{\mathbf{X}}} = \frac{\partial \dot{\mathbf{x}}}{\partial \xi} \frac{\partial \xi}{\partial \widehat{\mathbf{X}}} = \mathbf{j}\mathbf{J}^{-1} \tag{13}$$

taking in account the behavior over the thickness in the definition of the (two dimensional) constitutive model to be used (for example making the assumption of plane stress). The symbol $\mathbf{J}$ is used here and in the following to indicate $\mathbf{j}$ calculated in the reference position.

under this considerations the subsequent development of the finite element follows closely the standard procedure for a non–linear 2D finite element, the only difference being that the local base changes on all the domain, which makes the linearization slightly more involved.

To proceed further we need therefore to write the Right Cauchy Green Strain tensor $\mathbf{C} = \mathbf{F}^T\mathbf{F}$ which takes the form

$$\mathbf{C} = \left(\mathbf{J}^{-T}\mathbf{j}^T\mathbf{j}\mathbf{J}^{-1}\right) = \left(\mathbf{G}^T\mathbf{g}\mathbf{G}\right) \tag{14}$$

where we introduced the symbols $\mathbf{g} = \mathbf{j}^T\mathbf{j}$ and $\mathbf{G} = \mathbf{J}^{-1}$. Operator $\mathbf{g}$ takes, after some calculations, the simple form

$$\mathbf{g} = \begin{pmatrix} \mathbf{g}_\xi \bullet \mathbf{g}_\xi & \mathbf{g}_\eta \bullet \mathbf{g}_\xi \\ \mathbf{g}_\xi \bullet \mathbf{g}_\eta & \mathbf{g}_\eta \bullet \mathbf{g}_\eta \end{pmatrix} \tag{15}$$

From the definition of he Green Lagrange strain tensor $\mathbf{E} = \frac{1}{2}(\mathbf{C} - \mathbf{I})$ we obtain immediately $\delta\mathbf{E} = \frac{1}{2}\delta\mathbf{C}$. This allows to write the equation of virtual works in compact form as (taking in consideration only body forces and pressure forces)

$$\delta W_{int} = \delta W_{ext} + \delta W_{press} \tag{16}$$

$$\frac{h_0}{2}\int_\Omega \delta\mathbf{C} : \mathbf{S} = h_0 \int_\Omega \delta\mathbf{x} \bullet \mathbf{b} + \int_\omega p\delta\mathbf{x} \bullet \mathbf{n} \tag{17}$$

Internal Work

The term $\frac{h_0}{2}\int_\Omega \delta\mathbf{C} : \mathbf{S}$ describes the work of internal forces during the deformation process. Operator $\mathbf{G} = \mathbf{J}^{-1}$ is referred to the reference configuration and is therefore strictly constant, follows immediately that

$$\delta\mathbf{C} = \mathbf{G}^T \delta\mathbf{g}\mathbf{G} \tag{18}$$

The term $\delta\mathbf{C} : \mathbf{S}$ becomes in Einstein's notation

$$\frac{1}{2}\delta\mathbf{C} : \mathbf{S} = \frac{1}{2}\delta C_{IJ} S_{IJ} = \frac{1}{2}\delta g_{ij} G_{iI} G_{jJ} S_{IJ} \tag{19}$$

introducing the symbols

$$\frac{1}{2}\delta g_{ij} \rightarrow \frac{1}{2}\delta\left\{\mathbf{g}\right\} = \frac{1}{2}\begin{pmatrix} \delta g_{11} \\ \delta g_{22} \\ 2\delta g_{12} \end{pmatrix} \;\; ; \;\; S_{IJ} \rightarrow \left\{\mathbf{S}\right\} = \begin{pmatrix} S_{11} \\ S_{22} \\ S_{12} \end{pmatrix} \tag{20}$$

$$G_{iI} G_{jJ} \rightarrow \left[\mathbf{Q}\right]^T = \begin{pmatrix} (G_{11})^2 & (G_{12})^2 & 2G_{11}G_{12} \\ 0 & (G_{22})^2 & 0 \\ 0 & G_{12}G_{22} & G_{11}G_{22} \end{pmatrix} \tag{21}$$

it is possible to express the (19) in Voigt form as

$$\frac{1}{2}\delta\mathbf{C} : \mathbf{S} = \frac{1}{2}\left\{\delta\mathbf{g}\right\}^T \left[\mathbf{Q}\right]^T \left\{\mathbf{S}\right\} = \frac{1}{2}\left\{\delta\mathbf{g}\right\}^T \left\{\mathbf{s}\right\} \;\; ; \;\; \left\{\mathbf{s}\right\} = \left[\mathbf{Q}\right]^T \left\{\mathbf{S}\right\} \tag{22}$$

considering the definition(15), introducing the symbol $\left\{\delta\mathbf{x}\right\}^T = \left(\left\{\delta\mathbf{x_1}\right\}^T \ldots \left\{\delta\mathbf{x_k}\right\}^T\right)$ and taking in account the isoparametric approximation one obtains

$$\frac{1}{2}\delta g_{11} = \frac{\partial N_I}{\partial\xi}\delta\mathbf{x_I} \bullet \mathbf{g}_\xi = \left\{\delta\mathbf{x}\right\}^T \begin{pmatrix} \frac{\partial N_1}{\partial\xi}\left\{\mathbf{g}_\xi\right\} \\ \ldots \\ \frac{\partial N_k}{\partial\xi}\left\{\mathbf{g}_\xi\right\} \end{pmatrix} \tag{23}$$

$$\frac{1}{2}\delta g_{22} = \frac{\partial N_I}{\partial \xi}\delta \mathbf{x_I} \bullet \mathbf{g}_\eta = \{\delta \mathbf{x}\}^T \begin{pmatrix} \frac{\partial N_1}{\partial \eta}\{\mathbf{g}_\eta\} \\ \dots \\ \frac{\partial N_k}{\partial \eta}\{\mathbf{g}_\eta\} \end{pmatrix} \tag{24}$$

$$\frac{1}{2}\delta 2g_{12} = \frac{\partial N_I}{\partial \eta}\delta \mathbf{x_I} \bullet \mathbf{g}_\xi + \frac{\partial N_I}{\partial \xi}\delta \mathbf{x_I} \bullet \mathbf{g}_\eta = \{\delta \mathbf{x}\}^T \begin{pmatrix} \frac{\partial N_1}{\partial \xi}\{\mathbf{g}_\eta\} + \frac{\partial N_1}{\partial \eta}\{\mathbf{g}_\xi\} \\ \dots \\ \frac{\partial N_k}{\partial \xi}\{\mathbf{g}_\eta\} + \frac{\partial N_k}{\partial \eta}\{\mathbf{g}_\xi\} \end{pmatrix} \tag{25}$$

by defining the matrix

$$[\mathbf{b}]^T = \begin{pmatrix} \frac{\partial N_1}{\partial \xi}\{\mathbf{g}_\xi\} & \frac{\partial N_1}{\partial \eta}\{\mathbf{g}_\eta\} & \frac{\partial N_1}{\partial \xi}\{\mathbf{g}_\eta\} + \frac{\partial N_1}{\partial \eta}\{\mathbf{g}_\xi\} \\ \dots & \dots & \dots \\ \frac{\partial N_k}{\partial \xi}\{\mathbf{g}_\xi\} & \frac{\partial N_k}{\partial \eta}\{\mathbf{g}_\eta\} & \frac{\partial N_k}{\partial \xi}\{\mathbf{g}_\eta\} + \frac{\partial N_k}{\partial \eta}\{\mathbf{g}_\xi\} \end{pmatrix} \tag{26}$$

it is then possible to write

$$\frac{1}{2}\{\delta \mathbf{C}\}^T\{\mathbf{S}\} = \{\delta \mathbf{E}\}^T\{\mathbf{S}\} = \{\delta \mathbf{x}\}^T[\mathbf{b}]^T[\mathbf{Q}]^T\{\mathbf{S}\} \tag{27}$$

Defining the symbol $[\mathbf{B}]$

$$[\mathbf{B}] = [\mathbf{Q}][\mathbf{b}] \tag{28}$$

we finally obtain

$$\{\mathbf{f_{int}}\} = \int_\Omega h_0 [\mathbf{B}]^T\{\mathbf{S}\}\, d\Omega \tag{29}$$

$$\delta W_{int} = \{\delta \mathbf{x}\}^T\{\mathbf{f_{int}}\} \tag{30}$$

External Work

Derivation of the work of external conservative forces follows the standard procedure and can be found on any book on nonlinear finite elements. The expression of the work of follower forces (body forces) is on the other hand a little more involved. In the following the pressure is considered constant, the non linearity being introduced by the change of direction of the normal. For the derivation of the pressure contributions it is much easier to perform the integration over the actual domain then over the reference one.

$$\delta W_{pr} = \int_\omega p\delta \mathbf{x} \bullet \mathbf{n} d\omega = \int_{\xi,\eta} p\delta \mathbf{x} \bullet \mathbf{n} det(j) d\xi d\eta \tag{31}$$

taking in account the definition of the base vectors (6)(7), and considering (11) we obtain immediately

$$\{\mathbf{f_I}\} = \int_{\xi,\eta} N_I(\xi,\eta) p(\xi,\eta) \mathbf{v_3}(\xi,\eta) d\xi d\eta \tag{32}$$

$$\{\mathbf{f_{pr}}\} = \left(\{\mathbf{f_1}\}^T \dots \{\mathbf{f_k}\}^T\right)^T \tag{33}$$

$$\delta W_{pr} = \{\delta \mathbf{x}\}^T\{\mathbf{f_{pr}}\} \tag{34}$$

Linearization

Equation (16) is nonlinear, its practical use needs therefore its linearization. The best rate of convergence is theoretically given by Newton-Raphson technique which guarantees quadratical convergence to the solution. Defining $\Psi = \delta W_{int} - \delta W_{ext} - \delta W_{pr}$ each Newton–Raphson step takes the form

$$d\Psi + \Psi = 0 \tag{35}$$

The term Ψ can be explicitated using expression (30)(34) we therefore miss only the differential $d\Psi$ that can be evaluated from the linearization of the different contributions

Linearization of internal work

The term connected to the internal works can be linearized as follows

$$d\left(W_{int}\right) = d\left(\frac{h_0}{2}\int_{\Omega} \delta\mathbf{C}:\mathbf{S}\right) == \frac{h_0}{2}\int_{\Omega} d\left(\delta\mathbf{C}\right):\mathbf{S} + \frac{h_0}{2}\int_{\Omega} \delta\mathbf{C}:d\left(\mathbf{S}\right) \tag{36}$$

the first terms gives, by using (22)

$$\begin{aligned}\frac{h_0}{2}\int_{\Omega} d\left(\delta\mathbf{C}\right):\mathbf{S} &= \frac{h_0}{2}\int_{\Omega} d\left(\{\delta\mathbf{g}\}^T\right)\{\mathbf{s}\} = \\ &= \frac{h_0}{2}\int_{\Omega} d\left(\frac{\partial\{\delta\mathbf{g}\}^T}{\partial\{\mathbf{x}\}} d\{\mathbf{x}\}\right)\{\mathbf{s}\}\end{aligned} \tag{37}$$

now it can be seen that

$$\begin{aligned}d\left(\frac{1}{2}\{\delta\mathbf{g}\}^T\right) &= \left(\delta\mathbf{g}_\xi \bullet \mathbf{dg}_\xi \; \delta\mathbf{g}_\eta \bullet \mathbf{dg}_\eta \; \delta\mathbf{g}_\eta \bullet \mathbf{dg}_\xi + \delta\mathbf{g}_\xi \bullet \mathbf{dg}_\eta\right)\{\mathbf{s}\} = \\ &= \left(s_{11}\delta\mathbf{g}_\xi \bullet \mathbf{dg}_\xi \; s_{22}\delta\mathbf{g}_\eta \bullet \mathbf{dg}_\eta \; s_{12}\left(\delta\mathbf{g}_\eta \bullet \mathbf{dg}_\xi + \delta\mathbf{g}_\xi \bullet \mathbf{dg}_\eta\right)\right)\end{aligned} \tag{38}$$

substitution of the shape functions gives immediately a set of equalities in the form

$$s_{11}\delta\mathbf{g}_\xi \bullet \mathbf{dg}_\xi = s_{11}\frac{\partial N_I}{\partial\xi}\frac{\partial N_J}{\partial\xi}\delta_{ij}\delta\mathbf{x_I} \bullet \mathbf{dx_{jJ}} = s_{11}\frac{\partial N_I}{\partial\xi}\frac{\partial N_J}{\partial\xi}\delta_{ij}\delta x_{iI} dx_{jJ} \tag{39}$$

which makes possible to write

$$\begin{aligned}d\left(\frac{1}{2}\{\delta\mathbf{g}\}^T\right) = \Bigg(s_{11}\frac{\partial N_I}{\partial\xi}\frac{\partial N_J}{\partial\xi} + s_{22}\frac{\partial N_I}{\partial\eta}\frac{\partial N_J}{\partial\eta} + \\ s_{12}\left(\frac{\partial N_I}{\partial\eta}\frac{\partial N_J}{\partial\xi} + \frac{\partial N_I}{\partial\xi}\frac{\partial N_J}{\partial\eta}\right)\Bigg)\delta_{ij}\delta x_{iI} dx_{jJ}\end{aligned} \tag{40}$$

introducing the vectors $\mathbf{a} = \left(\frac{\partial N_1}{\partial \xi} \ldots \frac{\partial N_k}{\partial \xi}\right)$ and $\mathbf{b} = \left(\frac{\partial N_1}{\partial \eta} \ldots \frac{\partial N_k}{\partial \eta}\right)$ together with the new tensor

$$\mathbf{A} = \left(s_{11}\mathbf{aa} + s_{22}\mathbf{bb} + s_{12}\left(\mathbf{ba} + \mathbf{ab}\right)\right) \quad ; \quad \mathbf{aa} = \mathbf{a} \otimes \mathbf{a} \tag{41}$$

we can simplify greatly the expression in the form

$$d\left(\frac{1}{2}\left\{\delta\mathbf{g}\right\}^T\right) = A_{IJ}\delta ij \delta\mathbf{x_{iI}} d\mathbf{x_{jJ}} \tag{42}$$

or, in Voigt form

$$d\left(\frac{1}{2}\left\{\delta\mathbf{g}\right\}^T\right) = \left(\left\{\delta\mathbf{x_1}\right\}^T \ldots \left\{\delta\mathbf{x_k}\right\}^T\right)\begin{pmatrix} A_{11}\left[\mathbf{I}\right] \ldots & A_{1k}\left[\mathbf{I}\right] \ldots \ldots \ldots \\ A_{k1}\left[\mathbf{I}\right] \ldots & A_{kk}\left[\mathbf{I}\right] \end{pmatrix} \begin{pmatrix} \left\{\mathbf{dx_1}\right\} \\ \ldots \\ \left\{\mathbf{dx_k}\right\} \end{pmatrix} = \left\{\delta\mathbf{x}\right\}^T \left[\mathbf{K_{geo}}\right]\left\{\mathbf{dx}\right\} \tag{43}$$

the derivation of the "material" contribution to the stiffness matrix follows the standard path. Assuming as normally

$$d\mathbf{S} = \frac{\partial\mathbf{S}}{\partial\mathbf{E}} : d\mathbf{E} \rightarrow \left\{\mathbf{dS}\right\} = \left[\mathbf{D_{tan}}\right]\left\{\mathbf{dE}\right\} \tag{44}$$

we obtain immediately

$$\int_\Omega \frac{h_0}{2}\delta\mathbf{C} : d\left(\mathbf{S}\right) = \left(\int_\Omega h_0 \delta\left\{\mathbf{x}\right\}^T\left[\mathbf{B}\right]^T\left[\mathbf{D_{tan}}\right]\left[\mathbf{B}\right]d\Omega\right)\left\{\mathbf{dx}\right\} = \delta\left\{\mathbf{x}\right\}^T\left[\mathbf{K_{mat}}\right]\left\{\mathbf{dx}\right\} \tag{45}$$

$$\left[\mathbf{K_{mat}}\right] = \int_\Omega h_0\left[\mathbf{B}\right]^T\left[\mathbf{D_{tan}}\right]\left[\mathbf{B}\right]d\Omega \tag{46}$$

Linearization of external forces

The linearization of the work W_{ext} is not needed as it describes the work of constant forces. The only term missing is therefore the one relative to the work of the follower forces

$$d\left(\int_\omega p\delta\mathbf{x} \bullet \mathbf{n} d\omega\right) \rightarrow \delta\left\{\mathbf{x_I}\right\}^T d\mathbf{f_I} = \delta\left\{\mathbf{x_I}\right\}^T d\left(\int_\omega pN_I\left\{\mathbf{g}_\xi \times \mathbf{g}_\eta\right\}d\omega\right) \tag{47}$$

differentiating $pN_I\left\{\mathbf{g}_\xi \times \mathbf{g}_\eta\right\}$we obtain

$$d\left(pN_I\left\{\mathbf{g}_\xi \times \mathbf{g}_\eta\right\}\right) = pN_I\left\{\mathbf{dg}_\xi \times \mathbf{g}_\eta\right\} + pN_I\left\{\mathbf{g}_\xi \times \mathbf{dg}_\eta\right\} = pN_I\left\{\mathbf{g}_\xi \times \mathbf{dg}_\eta\right\} - pN_I\left\{\mathbf{g}_\eta \times \mathbf{dg}_\xi\right\} \tag{48}$$

Considering that it is possible to write the cross product of two vectors in Voigt format as

$$\mathbf{c} = \mathbf{a} \times \mathbf{b} \rightarrow \begin{pmatrix} c_1 \\ c_2 \\ c_3 \end{pmatrix} = \begin{pmatrix} 0 & -a_3 & a_2 \\ a_3 & 0 & -a_1 \\ -a_2 & a_1 & 0 \end{pmatrix} \begin{pmatrix} b_1 \\ b_2 \\ b_3 \end{pmatrix} \rightarrow \{\mathbf{c}\} = [\mathbf{a}\times] \{\mathbf{b}\} \tag{49}$$

and taking in account (3) and (4) we obtain

$$d\left(pN_I \left\{\mathbf{g}_\xi \times \mathbf{g}_\eta\right\}\right) = \left(pN_I \frac{\partial N_J}{\partial \eta} \left[\mathbf{g}_\xi \times\right] - pN_I \frac{\partial N_J}{\partial \xi} \left[\mathbf{g}_\xi \times\right]\right) \{\mathbf{dx_J}\} \tag{50}$$

$$[\mathbf{K_{pr}}] = \begin{pmatrix} K_{11} & \dots & K_{1k} \\ \dots & \dots & \dots \\ K_{k1} & \dots & K_{kk} \end{pmatrix} \quad ; \quad [\mathbf{K_{IJ}}] = \left(pN_I \frac{\partial N_J}{\partial \eta} \left[\mathbf{g}_\xi \times\right] - pN_I \frac{\partial N_J}{\partial \xi} \left[\mathbf{g}_\xi \times\right]\right) \tag{51}$$

$$d(\delta W_{pr}) = \{\delta \mathbf{x}\}^T [\mathbf{K_{pr}}] \{\mathbf{dx}\} \tag{52}$$

Linearized formulation

The only step missing is to merge all the terms in (35) to find the final expression. The result of this operation is

$$\{\delta \mathbf{x}\}^T ([\mathbf{K_{geo}}] + [\mathbf{K_{mat}}] - [\mathbf{K_{pr}}]) \{\mathbf{dx}\} = \{\delta \mathbf{x}\}^T (\{\mathbf{f_{ext}}\} - \{\mathbf{f_{int}}\}) \tag{53}$$

invoking the arbitrariety of $\{\delta \mathbf{x}\}$ and introducing the definitions

$$[\mathbf{K_{tan}}] = [\mathbf{K_{geo}}] + [\mathbf{K_{mat}}] - [\mathbf{K_{pr}}] \tag{54}$$

$$\{\mathbf{R}\} = \{\mathbf{f_{ext}}\} - \{\mathbf{f_{int}}\} \tag{55}$$

the principle of virtual works gives for each element

$$[\mathbf{K_{tan}}] \{\mathbf{dx}\} = \{\mathbf{R}\} \tag{56}$$

2.2 Solution Procedure

As briefly outlined at the beginning of the section, membrane systems are possibly subjected to large rigid body motions which reflects in singular or ill-conditioned “static” stiffness matrices. In addition, convergence of the Newton–Raphson algorithm is often difficult as the final solution can be very “far” from the initial guess even for little variations of the applied loads.

Dynamic solution techniques on the other hand are not affected by such problems. Mass and damping contributions remove the singularities from the system and generally provide a better conditioning to the problem. The introduction of dynamic terms provides as well an excellent source of stabilization for the solution (physically the solution can’t change much in a small time), ending up with better convergence properties inside each solution step.

Any standard (non–linear) time integration technique can be theoretically used in conjunction with the proposed FE model for the study of dynamic response of the systems of interest. Some care should be however taken in the choice because the high geometric non–linearities tend to challenge the stability of the time–integration scheme chosen.

Generally speaking, “statics” can be seen as the limit to which a dynamic process tends (under a given constant load). Dynamic systems show a “transient” phase that vanishes in time to reach the so called “steady state”; the presence of damping in the system reduces gradually the oscillations making the system tend to a constant configuration that is the “static” solution. The time needed for the system to reach this final configuration is controlled by the amount of damping. For values of system’s damping exceeding a critical value, the transient phase disappears and the systems reaches directly the final solution without any oscillation.

In many situations the main engineering interest is focused on “static” solutions rather than on the complete dynamic analysis of the system. The previews considerations suggests immediately that “statics” could be obtained efficiently by studying the dynamics of over damped systems. This could be obtained by simply adding a fictitious damping source to the “standard” dynamic problem. The “only” problem is therefore the choice of an idoneous form for such damping. Unfortunately this choice is not trivial, however it possible to observe [10],[8] that the “steady state” solution of the system

$$\mathbf{M}\ddot{\mathbf{x}} + \mathbf{D}\dot{\mathbf{x}} + \mathbf{K}\mathbf{x} = \mathbf{f}(\mathbf{x}) \tag{57}$$

is (statically) equivalent to that of the system

$$\mathbf{D}\dot{\mathbf{x}} + \mathbf{K}\mathbf{x} = \mathbf{f}(\mathbf{x}) \tag{58}$$

which can be seen as the previews for the case of zero density. The advantage of this equivalent system is that the inertia terms are always zero, consequently the system converges smoothly in time to its solution. This final solution is not affected by the particular choice of the damping, however in the author’s experience, an effective choice is $\mathbf{D} = \beta\mathbf{M}$ as proposed by [10].

Table (1) gives the details of the proposed solution procedure, making use of Newmark’s integration scheme. The procedure described differs from a “real” dynamics simulation only on the choice of the damping and of the mass matrix. Any other choice is possible for the time integration scheme to be used. It is of interest to observe that the system described is highly dissipative, energy stability of the time integration scheme is therefore not crucial.

3 Wrinkling Simulation

Given the lack of flexural stiffness, membrane systems are easily subjected to buckling in presence of any compressive load. The idea is that when a

Table 1. Pseudo–Static solution procedure

- for pseudo–static strategy: calculate the constant matrices
$$D = \beta \mathbf{M}$$
set $\mathbf{M} = 0$ after initializing the damping matrix. (if $\mathbf{M}$ is not set to 0, "real" dynamic simulation can be performed)
- choose Newmark constants: a classical choice is
$$\delta = \frac{1}{2} \;;\; \alpha = \frac{1}{4}$$
- evaluate the constants
$$a_0 = \frac{1}{\alpha \Delta t^2} \;;\; a_1 = \frac{\delta}{\alpha \Delta t} \;;\; a_2 = \frac{1}{\alpha \Delta t}$$
$$a_3 = \frac{1}{2\alpha} - 1 \;;\; a_4 = \frac{\delta}{\alpha} - 1 \;;\; a_5 = \frac{\Delta t}{2}\left(\frac{\delta}{\alpha} - 2\right)$$
- predict the solution at time $t + \Delta t$ using for example
$$\mathbf{x}^0_{t+\Delta t} = \mathbf{x}_t + \dot{\mathbf{x}}_t \Delta t$$
$$\dot{\mathbf{x}}_{t+\Delta t} = \dot{\mathbf{x}}_t$$
$$\ddot{\mathbf{x}}_{t+\Delta t} = 0$$
- iterate until convergence
  - calculate the system's contributions
$$\left[\mathbf{K}^{\mathbf{dyn}}_{\mathbf{tan}}\right] = [\mathbf{K_{tan}}] + a_0 [\mathbf{M}] + a_1 [\mathbf{D}]$$
$$\left\{\mathbf{R}^{\mathbf{dyn}}\right\} = \{\mathbf{R}\} - [\mathbf{M}] \left\{\ddot{\mathbf{x}}^{\mathbf{i}}_{\mathbf{t+\Delta t}}\right\} - [\mathbf{D}] \left\{\dot{\mathbf{x}}^{\mathbf{i}}_{\mathbf{t+\Delta t}}\right\}$$
  - solve the system for the correction $\mathbf{dx}$
  - update the results as
$$\mathbf{x}^{i+1}_{t+\Delta t} = \mathbf{x}^{i}_{t+\Delta t} + \mathbf{dx}$$
$$\Delta \mathbf{x} = \mathbf{x}^{i+1}_{t+\Delta t} - \mathbf{x}_t$$
$$\dot{\mathbf{x}}_{t+\Delta t} = a_1 \Delta \mathbf{x} - a_4 \dot{\mathbf{x}}_t - a_5 \ddot{\mathbf{x}}_t$$
$$\ddot{\mathbf{x}}_{t+\Delta t} = a_0 \Delta \mathbf{x} - a_2 \dot{\mathbf{x}}_t - a_3 \ddot{\mathbf{x}}_t$$
- go to next time step

compressive stress tends to appear on a part of a structure, it is immediately removed by local instability phenomena, that manifest with the formation of little "waves" of direction perpendicular to the direction of stresses. Prediction of the size of those "waves" commonly called "wrinkles" is not generally possible as their disposition is somehow random and connected to initial imperfections. However their average size is strictly connected to the bending stiffness meaning in particular that for the problems of interest the wrinkle tend to become quite little in comparison with the total size of the structure. It has been proved to be feasible [2] and [1] to describe correctly the formation of the wrinkles using extensively mesh refinement procedures together with low order thin-shell elements. An analogous approach using higher order shells and a fixed reference mesh, joint with some comparison with experimental data can be found in [11]. A key point to be taken in account is that this procedures need a mesh density inversely proportional to the expected size of the wrinkles. In other words the smaller are the wrinkles, the more elements are needed to correctly describe the phenomena. As in our structures, the thickness is very low compared to the other dimensions, the referenced approaches would become soon too expensive.

An alternative procedure is based on the "enrichment" of the elements involved in the simulation. The idea is to renounce to a description of the single wrinkle and to focus the analysis of the average stress and displacement field. This allows to consider in the analysis elements of size bigger then the expected wrinkle size introducing the effect of the local instability in the calculation of the stress or strain field at integration point level. We would like to stress that this approach is not necessarily less "precise" than the former. Indeed no information on the wrinkling size is provided, however the global stress field is correctly described. It is as well important to highlight how the position of the wrinkles is never known given its strong dependence on the initial imperfections, therefore the only reliable result is the individuation of the "wrinkled zone" that can be correctly described by both methods.

3.1 Enriched Material Model

Over the years many different proposals to perform the element enrichment were presented. Mainly two different approaches survived, one based on manipulations of the gradient of deformations, the second connected with a redefinition of the constitutive model.

The former, proposed by Roddeman in [6] and [7], is based on the definition of an effective deformation gradient obtained superimposing to the normal displacement field a term connected with the formation of wrinkles. This modification allows to describe correctly the shortening of the average plane of the membrane in presence of compressive stresses.

The latter is based on a modification of the stress-strain relationship, meaning that the constitutive law is modified not to allow compressive stresses. The main advantage of this second techniques, is to make the implementation completely independent from the element used, characteristic that makes them very attractive for the practical implementation.

A "new" material model, based on the modification of a standard linear material is introduced in current section. This formulation is based on the penalization of the elastic characteristics of the material in the system of the principal stresses. In simple words, the material is softened in the direction of the compressive stresses and keeps its characteristics in the other direction.

This is achieved by a two step procedure, based on a phase of assessment of the state of the membrane and on a phase of modification of the material tangent matrix.

Many different choices are theoretically possible in combining the two different phases, however in the writer experience, iterative application of the wrinkling correction inside the same time step leads generally to a very slow or unstable convergence behavior. The proposed solution procedure is therefore based on a "explicit" approach in the form

- standard pseudo–static solution step
- check state of each element
- modify material
- go to next "time" step

This procedure is very efficient as it takes full advantage of the pseudo–static solution procedure the only additional cost being linked to the evaluation of the state and to the penalization of the constitutive matrix. As during each time step the material is "constant", no additional source of non-linearity is introduced therefore the element retains its convergence properties. The stabilization of the stress-field is guaranteed by the dynamic process that, together with the stabilization introduced in the material model effectively damps out the oscillations.

The reader should note that the aim of the proposed technique is to get a reliable static solution. There is absolutely no guarantee that "on the way" to the static solution the wrinkling procedure converges inside each time step, however, when all the movement is dissipated so the structure reached the final configuration, wrinkling arrived to a constant solution.

Assessment of the state of the membrane

One of the crucial steps in the procedure is the evaluation of the state of the membrane. In particular it is necessary to "decide" if the membrane is (or rather should be) in biaxial tension, in uniaxial tension or completely unstressed because of the formation of wrinkles. The assessment procedure, is based on the introduction of the fictitious stress σ^* that represents the stress that would exist on the membrane if formation of the wrinkles was not allowed. This is related to the total stress from the relation

$$[\sigma^*] = [\mathbf{D_{original}}] : \{\mathbf{E}\} \tag{59}$$

the principal direction of σ^* can be calculated as

$$c_1 = \sigma^*_{11} + \sigma^*_{22} \ ; \ c_2 = \sigma^*_{11} - \sigma^*_{22} \ ; \ c_3 = \sqrt{(\frac{c_2}{2})^2 + (\sigma^*_{12})^2}$$

$$\sigma^*_I = c_1 + c_2 \ ; \ \sigma^*_{II} = c_1 - c_2 \ ; \ \alpha^* = tan^{-1}\left(\frac{2\sigma^*_{12}}{c_2}\right) \ ; \tag{60}$$

by introducing the tensors

$$[\mathbf{E}] = \begin{pmatrix} \epsilon_{11} & \frac{\gamma_{12}}{2} \\ \frac{\gamma_{12}}{2} & \epsilon_{22} \end{pmatrix} \; ; \; \{\mathbf{n}_{\sigma^*}\} = \begin{pmatrix} cos(\alpha^*) \\ sin(\alpha^*) \end{pmatrix} \tag{61}$$

it is possible to express the strain corresponding to the principal stresses as

$$\{\epsilon^*\} = \{\mathbf{n}_{\sigma^*}\}^T [\mathbf{E}] \{\mathbf{n}_{\sigma^*}\} \tag{62}$$

this strains can be used together with the corresponding principal stresses in assessing the state of the membrane, using the so called "mixed criteria". The "decision" proceeds as follows:

- ($\sigma^*_{II} > 0$) biaxial tension → "**taut state**"
- ($\sigma^*_{II} < 0$) and ($\epsilon_{sigma_I} > 0$) uniaxial tension → "**wrinkled state**"
- otherwise (all compressed) → "**slack state**"

Modification of the material

Once known the state, the material has to be modified to remove undesired compression. This is obtained by modifying the directions in which direction appears, basically removing the stiffness contribution in that direction. The procedure is distinguished for the various cases:

- **"taut case":** the $[\mathbf{D_{tan}}]$ is the original matrix as the whole membrane is in tension and acts with its whole stiffness
- **"wrinkled state"** in this case one direction has to be penalized leaving the other unchanged. Introducing the matrix

$$\begin{matrix} c = cos(\alpha^*) \\ s = sin(\alpha^*) \end{matrix} \; ; \; [\mathbf{R}(\alpha^*)] = \begin{pmatrix} c^2 & s^2 & -2sc \\ s^2 & c^2 & 2sc \\ sc & -sc & c^2 - s^2 \end{pmatrix} \tag{63}$$

  the penalization can be applied following the steps

  - $[\mathbf{D_{rotated}}] = [\mathbf{R}[-\alpha]] [\mathbf{D_{original}}] \left[\mathbf{R}[-\alpha]^{\mathbf{T}}\right]$
  - $[\mathbf{D_{modified}}] = \begin{pmatrix} D_{rot_{11}} & PD_{rot_{12}} & D_{rot_{13}} \\ PD_{rot_{21}} & PD_{rot_{22}} & PD_{rot_{23}} \\ D_{rot_{31}} & PD_{rot_{32}} & D_{rot_{33}} \end{pmatrix}$
  - $[\mathbf{D_{modified}}] = [\mathbf{R}[\alpha]] [\mathbf{D_{modified}}] [\mathbf{R}[\alpha]]^T$

- **"slack state"** the membrane is compressed in all directions. No contribution to the stiffness should be provided, consequently the whole constitutive matrix can be penalized as $[\mathbf{D_{modified}}] - P [\mathbf{D_{modified}}]$

This modification procedure guarantees that the stress $\{\mathbf{S}\} = [\mathbf{D_{modified}}] \{\mathbf{E}\}$ presents no compression.

The penalty factor "P" plays a central role in the stability of the wrinkling procedure. The problem is that when some parts of the structure are softened in some direction the stress redistributes, often causing a cyclic change of state of other parts of the structure. The use of a constant penalty factor as proposed for example in [5], causes some parts of the structure to be basically switched on and off when they change of state. An improvement can be obtained through the definition of a variable penalty factor, which makes the transition smoother helping the convergence. Introducing the parameter σ_{max} that indicates the maximum tolerable compression, P_{max} as the max penalty factor and defining $P_\sigma = \frac{\sigma_{max}}{\sigma}$ a suitable formulation for the penalty parameter can be obtained as

$$P_\sigma < P_{max} \rightarrow P = P_{sigma} \;; \tag{64}$$

$$P_\sigma < 0 or P_\sigma > 1 \rightarrow P = 1.0 \tag{65}$$

stability can be further improved by taking in account the loading history of each element. This should be considered as a purely numerical artifice to minimize oscillations of the stress field and can be expressed as:

$$if(State \equiv OldState) \rightarrow \text{leave P unchanged}$$

$$otherwise \rightarrow P = P_{old} * coeff$$

it should be however checked that the modified value for P is allowable.

3.2 Validation

A few examples are proposed in current subsection to validate the procedure presented. Given the nature of the problem, it is very difficult to obtain an analytical or experimental prove of the effectiveness of the procedure, validation is therefore based on a set of numerical experiments.

It has already been highlighted that a realistic representation of the wrinkling behavior can be obtained using a sufficiently high number of elements; simulation can therefore be performed on dense meshes, introducing initial imperfections to initialize the formation of the wrinkles. This way the compressive stresses are correctly removed, and the results obtained can be used in validating the proposed wrinkling procedure.

A few test examples are proposed here showing the results obtained with the proposed approach.

CIRCULAR AIRBAG: The circular airbag is probably one of the best examples to be used in testing the efficacy of the wrinkling procedure. The simulation proposed was carried using

$$\rho = 2700 \left[\frac{Kg}{m^3}\right] \;;\; E = 7000 \left[\frac{N}{mm^3}\right] \;;\; \nu = 0.3$$

$$Thickness = 0.001\,[mm] \;;\; Radius = 1000\,[mm]$$

coarse mesh – no wrinkling correction:

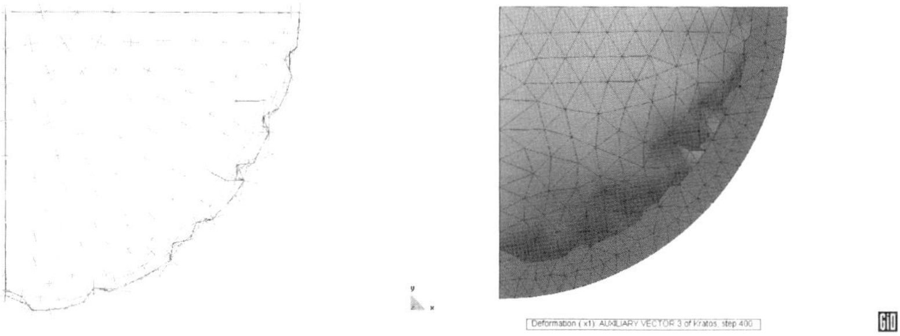

(a) Plot of Principal PK2 stresses (b) Deformed VS reference configuration

dense mesh – no wrinkling correction:

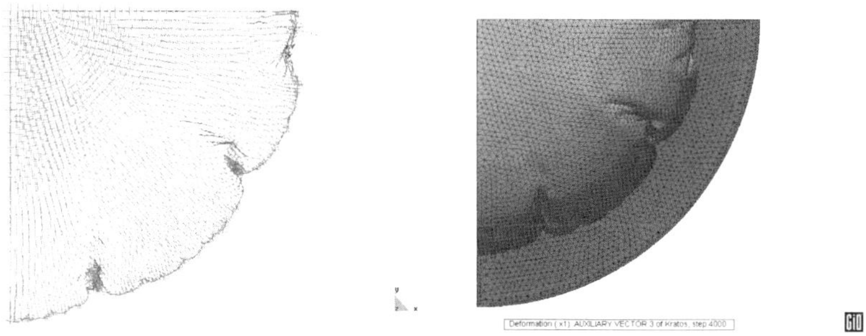

(a) Plot of Principal PK2 stresses (b) Deformed VS reference configuration

coarse mesh – wrinkling correction applied:

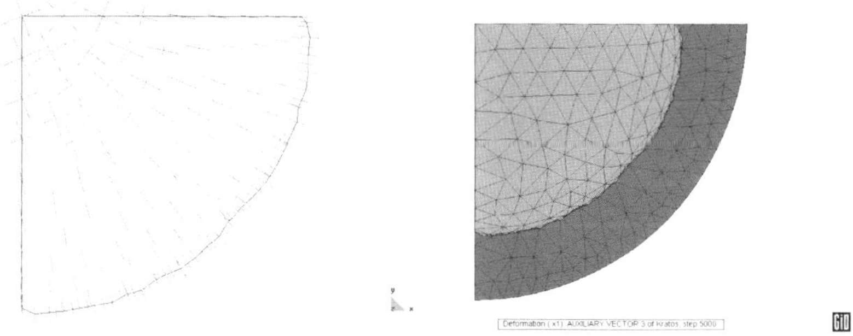

(a) Plot of Principal PK2 stresses (b) Deformed VS reference configuration

Fig. 1. Inflated Circular Airbag

Symmetry boundary conditions were applied and the problem was evaluated with and without wrinkling algorithm. The same airbag was simulated using different meshes increasing progressively the mesh density. The results reported here refer to a coarse mesh of 236 elements and to a denser one of 4802 elements. For this example, a very dense mesh is needed to capture the formation of folds and wrinkles that eliminate the compression. Fig. 1(c) and 1(d) suggests immediately how the formation of wrinkles and deep folds (larger wrinkles) correctly removes the compressive stresses. It is very relevant to highlight how the location of the folds changes in different simulations but their "distance" tends to be the same.

It can be easily checked how even different runs of the same structure with the same mesh can lead to different wrinkling patterns. The only realistic result is therefore the extension of the wrinkled zone.

The solution obtained on the coarse mesh without any improvement (see Fig. 1(a), 1(b)) is poor both in terms of stresses and displacements The introduction of the wrinkling correction allows to catch the correct behavior using a much coarser mesh. Considering the results on the dense mesh as the reference solution, Fig. 1(e), 1(f) clearly shows how a remarkable improvement is obtained both in terms of stresses and displacements using the wrinkling correction. Table 2 in particular highlights how the results of the analysis on the coarse mesh with the wrinkling correction are practically coincident to the reference solution confirming the efficacy of the approach.

Table 2. Circular Airbag: Data sampled at center point

	Dense	**Coarse** No Correction	**Coarse** Corrected
Displacement: [m]	0.465	0.37	0.47
S_I: $\left[\frac{N}{mm^2}\right]$	9.55E7	11.2E7	9.51E7
S_{II}: $\left[\frac{N}{mm^2}\right]$	9.05E7	7.21E7	9.09E7

SHEAR TEST: A simple shear test is performed by imposing displacements on one side of a square membrane. The parameters used are the same as for the previews example. The dimension of the side is $1000[mm]$ and the imposed displacement is $200[mm]$. Two cases are considered the first (Fig. 2(c), 2(d) using the standard approach on a dense mesh, the second (Fig. 2(a), 2(b)) applying the proposed correction on a coarser mesh. Local buckling is correctly reproduced by the first approach that is considered a representation of the "true" behavior of the membrane; this result is achieved imposing an initial imperfection in the form of a very small out-of-plane load. The formation of the tensile diagonal is correctly reproduced in the second simulation using the enriched material model. The improved procedure allows as well to describe correctly the deformed shape of the square (it can be easily checked that the "normal" solution has straight sides).

ANNULUS UNDER SHEAR: The last proposed example is an annulus under shear constituted by a thin membrane blocked by two rigid disks on the inner and outer boundaries. The inner disk is rotated by 10 counterclockwise causing the membrane to wrinkle. Fig. 3(b) shows the results of the wrinkling procedure applied to a coarse mesh. Comparison with the reference results (Fig. 3(d)) shows excellent agreement in terms of principle PK2 stresses.

coarse mesh – wrinkling correction applied:

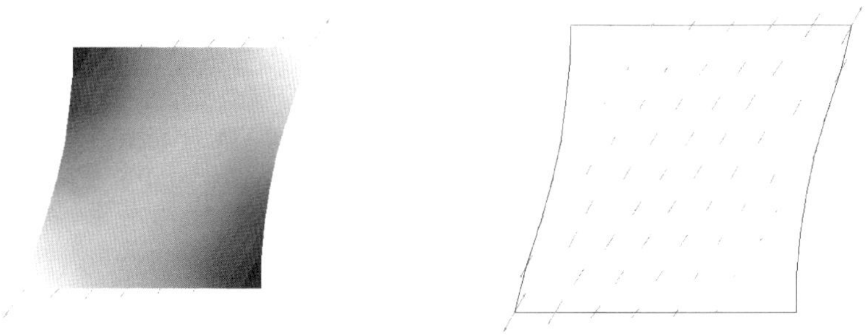

(a) Values of Principal PK2 Stresses (b) Plot of Principal PK2 stresses

dense mesh – no wrinkling correction:

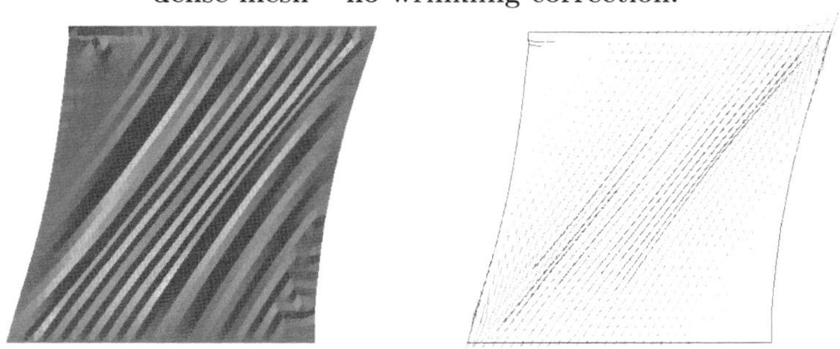

(a) Wrinkled configuration (b) Plot of Principal PK2 stresses

Fig. 2. Quadrilatera under shear

4 Coupling Issues - The Case of the Sails

Coupled fluid–membrane analysis is a challenging problem involving high non–linearities both on the side of the structure and of that of the fluid. The physical problem is however pretty clear: the membrane, is immersed in a fluid field. The presence of the structure influences the flow of the fluid, which exerts a force on the membrane. This force causes a deformation, changing the boundary conditions for the fluid flow and consequently the force exerted. Given the high flexibility of the structure, the coupling becomes strong.
This section addresses the coupled simulation of membrane systems, with particular reference to the "static" simulation of boat sails.

Before proceeding to the description of "our" method we should observe that sails are aerodynamic bodies which tend to a "stable" configuration with the fluid flow reaching a sort of steady state condition. The main engineering interest is therefore connected to the determination of this "final" configuration which represents a sort of "static solution" of the problem.

It is theoretically possible to deal with the coupled process using different strategies, including in particular "implicit (coupling) procedures" as pro-

coarse mesh – wrinkling correction applied:

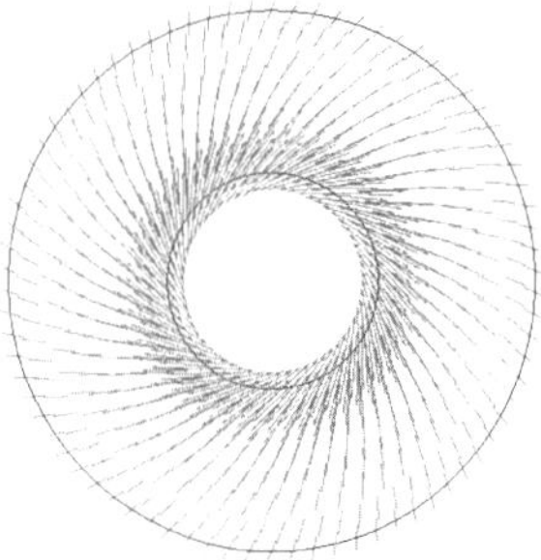

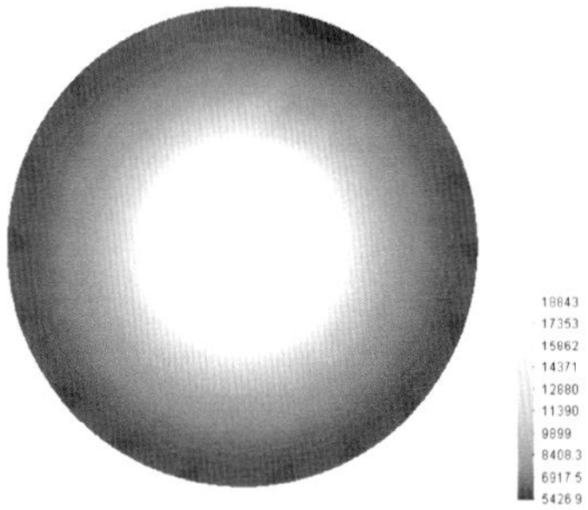

Principal PK2 Stresses (b) Values of First Principal PK2

dense mesh – no wrinkling correction:

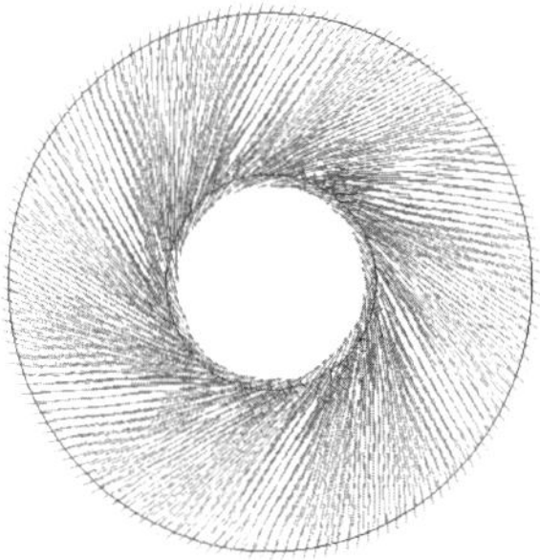

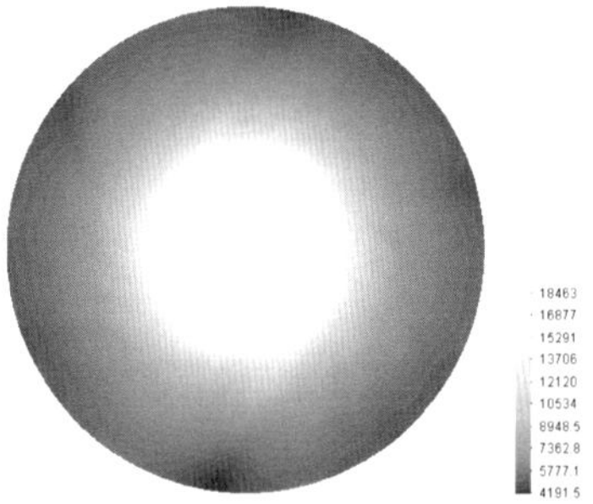

Principal PK2 Stresses Values of First Principal PK2

Fig. 3. Annulus subjected to torsion

posed for example in [3] or "explicit" ones as described in [9]. Classical arguments in favor of one or of the other are connected to the numerical stability and computational efficiency of the different techniques. The traditional objection to the use of "explicit" ones is linked to the stringent requirements on the time step. Time step constraint for the stability of the coupling procedure is in fact normally more stringent than the one for the single-field solution. This strong requirement is connected to the lack of energy conservation at the interface between the various fields, which tends to introduce spurious energy contributions in the system.

It is however possible to observe that the pseudo–dynamic solution procedure presented in section 2 has very high dissipative properties and is perfectly suited for the research of coupled "static" solutions

The objection to the use of "explicit" coupling schemes is therefore no longer applicable as our artificial damping can easily remove any spurious energy contribution introduced by the couping process. Given this observation "explicit" procedures are much more efficient than the corresponding "im-

Fig. 4. Interaction of a genoa and a main sail - pressures at the end of the coupled analysis

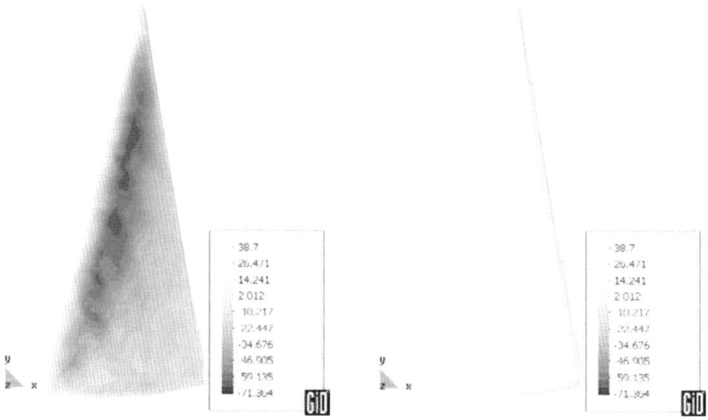

(a) genoa - leeward face (b) genoa - windward face

Fig. 5. Pressure field on the genoa at the end of the coupled analysis

plicit" ones as the single step is much cheaper. Box (3) proposes an efficient coupled solution strategy.

Fig. (4) presents the results of the coupled analysis of a genoa and main sail; a genoa alone is presented in Fig (4) The results obtained by this approach are presented in Fig. (4) in application to the simulation of a genoa and main sail.

- Predict Structural Solution
- Deform the mesh of the fluid domain According the predicted shape of the structure (variables needed for ALE formulation of the fluid should be calculated). The mesh–movement should keep the quality of the mesh, minimizing the deformation of the elements close to the structure, see [4]
- Advance in time the fluid on the deformed mesh
- transfer stresses FROM the fluid boundary TO the structural Boundary (stresses can be transferred as calculated)
- Advance the structure in using the pseudo–static solution technique
- proceed to next time step

Table 3. "Pseudo–static" coupling procedure

References

1. Ortiz M. Cirak F. Fully c1 conforming subdivision elements for finite deformation thin shell analysis. *IJNME*, 51:813–833, 2001.
2. Schroeder P Cirak F., Ortiz M. Subdivision surfaces: a new paradigm for thin shell finite element analysis. *IJNME*, 47:2039–2072, 2000.
3. W.A. Wall D.P. Mok. Partitioned analysis schemes for the transient interaction of incompressible flows and nonlinear flexible structures. In *trends in computational structural mechanics*, Barcelona, 2001.
4. G.Bugeda S.R. Idelsohn E. Onate, J. Garcia. A general stabilized formulation for incompressible fluid flow using finite calculus and the finite element method. In *Towards a New Fluid Dynamics with its challenges in Aeronautics*, Barcelona, 2002.
5. Jenkins C. Schur W. Liu X. Large deflecion analysis of pneumatic envelopes using a penalty parametr modified material model. *Finite Elements in Analysis and Design*, 37:233–251, 2001.
6. Drukker J. et al. Roddeman D.G. The wrinkling of thin membranes: Part 1 - theory. *Journal of Applied Mechanics*, 54:884–887, 1987.
7. Drukker J. et al. Roddeman D.G. The wrinkling of thin membranes: Part 2 - numerical analysis. *Journal of Applied Mechanics*, 54:888–892, 1987.
8. R. Rossi. A finite element formulation for 3d membrane structures including a wrinkling modified material model. Technical Report 226, CIMNE, 2003.
9. B. Larroutorou S.Piperno, C.Farhat. Partitioned procedures for the transient solution of coupled aeroelastic problems - part2 - energy transfer analysis and three dimensional applications. *Computer Methods in Applied Mechanics and Engineering*, 124:79–112, 1995.
10. R.L. Taylor. Finite element analysis of membrane structures. Technical report, CIMNE, 2001.
11. Pellegrino S. Wong Y.W. Computation of wrinkling amplitudes in thin membranes. In *43rd AIAA/ASME/ASCE/AHS/ASC conference*, Denver, 2002.

Wrinkles in Square Membranes

Y.W. Wong[1] and S. Pellegrino[2]

[1] SKM Consultants (M) Sdn Bhd, Suite E-15-01, Plaza Mont' Kiara, No. 2, Jalan Kiara, Mont' Kiara, 50480 Kuala Lumpur, MALAYSIA
WYWong@skmconsulting.com.my
[2] Department of Engineering, University of Cambridge, Trumpington Street, Cambridge, CB2 1PZ, U.K. pellegrino@eng.cam.ac.uk

This paper investigates the wrinkling of square membranes of isotropic material, subject to coplanar pairs of equal and opposite corner forces. These membranes are initially stress free and perfectly flat. Two wrinkling regimes are observed experimentally and are also reproduced by means of finite-element simulations. A general methodology for making preliminary analytical estimates of wrinkle patterns and average wrinkle amplitudes and wavelengths, while also gaining physical insight into the wrinkling of membranes, is presented.

1 Introduction

Thin, prestressed membranes will be required for the next generation of spacecraft, to provide deployable mirror surfaces, solar collectors, sunshields, solar sails, etc. Some applications require membranes that are perfectly smooth in their operational configuration, but many other applications can tolerate membranes that are wrinkled; in such cases the deviation from the nominal shape has to be known. The design of membrane structures with biaxial pretension, which would have a smooth surface, significantly increases the overall complexity of the structure and hence, for those applications in which small wrinkles are acceptable, engineers need to be able to estimate the extent, wavelength and amplitude of the wrinkles.

The wrinkling of membranes has attracted much interest in the past, starting from the development of the tension field theory [1]. Simpler formulations and extensions of this theory were later proposed [2-7]. All of these formulations, with accompanying numerical solutions [8,9], model the membrane as a *no-compression, two-dimensional continuum with negligible bending stiffness.* Many studies of membrane wrinkling have been carried out during the past three years, and have been presented at the 42nd, 43rd, and 44th AIAA SDM Conferences [10-12].

E. Oñate and B. Kröplin (eds.), Textile Composites and Inflatable Structures, 109–122.

This paper considers a uniform elastic square membrane (which is a simple model of a square solar sail) of side length $L + 2r_1$ and thickness t that is prestressed by two pairs of equal and opposite concentrated forces, T_1 and T_2, uniformly distributed over a small length d at the corners, as shown in Fig. 1. This membrane is isotropic with Young's Modulus E and Poisson's ratio ν; it is also initially stress free and perfectly flat (before the application of the corner forces).

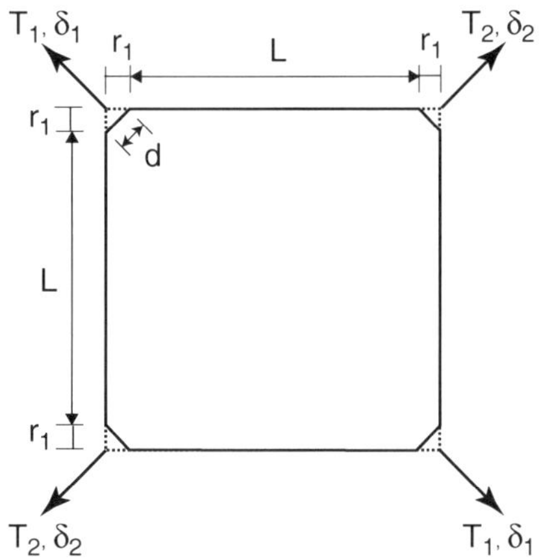

Fig. 1. Membrane subjected to corner forces.

We use this problem to present a general and yet simple analytical method for making preliminary estimates of wrinkle patterns and average wrinkle amplitudes and wavelengths in membrane structures. We also present a finite element simulation method for making more accurate estimates. The results from both our analytical approach and finite element simulations are compared with experimental measurements.

The layout of the paper is as follows. Section 2 describes two regimes of wrinkling that were observed experimentally. Section 3 presents our methodology for tackling wrinkling problems analytically, and hence derives solutions for the square membrane problem. Section 4 presents a finite-element simulation technique, whose results are compared with measurements and analytical results in Section 5. Section 6 concludes the paper.

2 Experimental Observations

Fig. 2 shows photographs of the wrinkle patterns in a Kapton membrane with $L = 500$ mm, $t = 0.025$ mm, and $d = 25$ mm. For symmetric loading ($T_1 = T_2$) the wrinkle pattern is fairly symmetric, as shown in Fig. 2(a), with wrinkles radiating from each corner; the central region is free of wrinkles. For a load ratio of $T_1/T_2 = 2$ the wrinkles grow in amplitude but remain concentrated at the corners. Then, for $T_1/T_2 = 3$ a large diagonal wrinkle becomes visible, whose amplitude grows further for $T_1/T_2 = 4$.

(a) (b) (c) (d)

Fig. 2. Wrinkled shapes for T_1 equal to (a) 5 N, (b) 10 N, (c) 15 N, and (d) 20 N; $T_2 = 5$ N in all cases.

3 Analytical Approach

Our analytical approach is in four parts, as follows.

First, we identify a two-dimensional stress field that involves no compression anywhere in the membrane; the regions where the minor principal stress is zero are then assumed to be wrinkled and the wrinkles are assumed to be along the major principal stress directions. Ideally, both equilibrium and compatibility should be satisfied everywhere by the selected stress field, but analytical solutions in closed-form —obtained by tension field theory— exist only for simple boundary conditions. We have recently shown [13] that a carefully chosen, simple stress field that satisfies only equilibrium can provide quick solutions that are useful for preliminary design. More accurate stress fields can be obtained from a two-dimensional stress analysis with membrane finite elements, as briefly discussed in Section 4.

Second, we note that the bending stiffness of the membrane is finite, although small, and hence a compressive stress will exist in the direction perpendicular to the wrinkles. Because of its small magnitude, this stress was neglected in our previous analysis of the stress field. We assume that this compressive stress varies only with the wavelength of the wrinkles and set it equal to the critical buckling stress of a thin plate in uniaxial compression. Thus, the stress across the wrinkles is a known function of the wrinkle wavelength.

Third, we enforce equilibrium in the out-of-plane direction. Since the stress distribution is known, except for the wrinkle wavelength, a single equation of equilibrium will determine the wrinkle wavelength.

Fourth, the wrinkle amplitudes are estimated by considering the total strain in the membrane as the sum of two components, a material strain and a wrinkling strain.

3.1 Stress Field

Fig. 3 shows three equilibrium stress fields that can be used to analyse membranes under (a) a symmetric loading, (b) an asymmetric loading with $T_1/T_2 \leq 1/(\sqrt{2}-1)$, and (c) an asymmetric loading with $T_1/T_2 \geq 1/(\sqrt{2}-1)$. In each case the membrane is divided into regions which are subject to simple stress states.

The stress field in Fig. 3(a) is purely radial in the corner regions, with

$$\sigma_r = \frac{T}{\sqrt{2}rt} \tag{1}$$

where $r < r_1 + L/2$ is the radial distance measured from the apex. Hence, σ_r is uniform on any circular arc and all other stress components are zero. The central region, defined by circular arcs of radius $R = r_1 + L/2$, is subject to uniform biaxial stress of magnitude $T/\sqrt{2}Rt$.

Note that near the point of application of each corner load a small, biaxially stressed region bounded by the radius $r_1 = d/\sqrt{2}$ has been defined. In these regions both normal stress components are T/dt.

For moderately asymmetric loading, see Fig. 3(b), we consider corner stress fields similar to those given by Eq. (1), hence

$$\sigma_r = \frac{T_i}{\sqrt{2}rt} \tag{2}$$

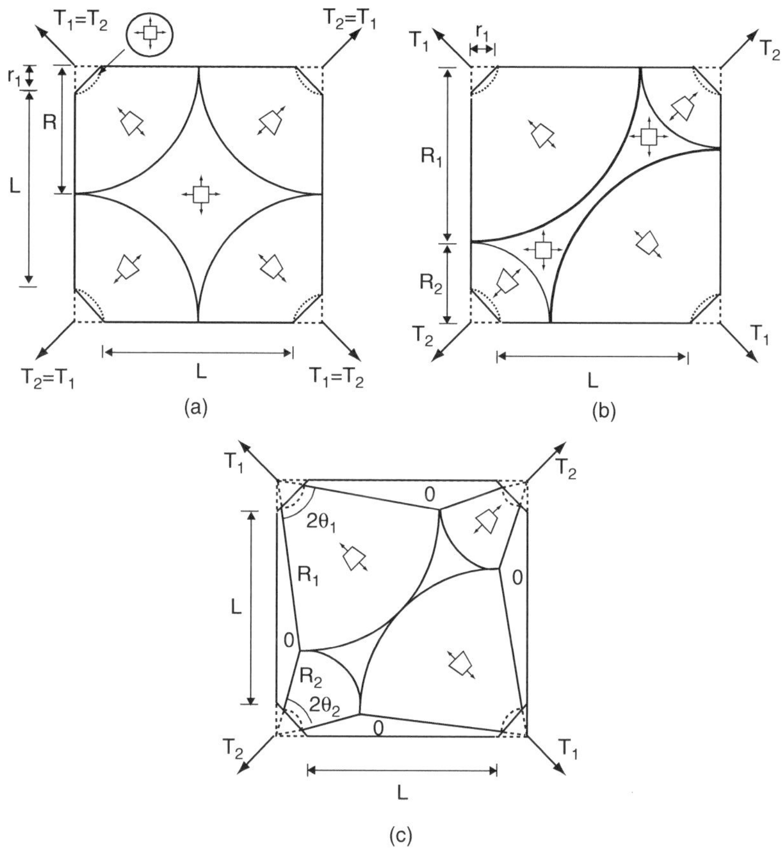

Fig. 3. Stress fields.

but vary the outer radii of these stress fields, in such a way that the radial stress is still uniform on the four arcs bounding the central region. Hence, we need to choose R_1 and R_2 such that $R_1/R_2 = T_1/T_2$ and $R_1 + R_2 = L + 2r_1$. This approach is valid until the two larger arcs reach the centre of the membrane, which happens for

$$\frac{R_1}{R_2} = \frac{T_1}{T_2} = \frac{1}{\sqrt{2}-1} \tag{3}$$

For larger values of T_1/T_2 we consider the stress field shown in Fig. 3(c); note that the diagonal region between the two most heavily loaded corners of the membrane is subject to zero transverse stress, and hence a single diagonal wrinkle can form. Also note that the edges of the membrane are now unstressed. The stress in each corner region is now given by

$$\sigma_r = \frac{T_i}{2rt\sin\theta_i} \tag{4}$$

and hence for the central region to be biaxially stressed the condition

$$\sigma_r = \frac{T_1}{2R_1 t \sin\theta_1} = \frac{T_2}{2R_2 t \sin\theta_2} \tag{5}$$

has to be satisfied. Given T_1 and T_2, one can find –by geometry together with Eq. (5)– a unique set of values for the half-angles defining the corner regions, θ_1, θ_2, and for the radii, R_1, R_2, thus fully defining the stress field.

The values of θ_1 and θ_2 remain constant for any particular value of T_1/T_2, and so the only variable in Eq. (4) is r. Hence, the slack regions will grow as the load ratio is increased. Finally, it should be noted that both of our earlier stress fields can be obtained as special cases of the last one.

3.2 Wrinkle Details

A critical compressive stress, $\sigma_{\rm cr}$, must exist in the direction transverse to the wrinkles. We will assume that this stress is given by the buckling stress of an infinitely long plate of width λ

$$\sigma_{\rm cr} = -\frac{\pi^2}{\lambda^2}\frac{Et^2}{12(1-\nu^2)} \tag{6}$$

In the case of fan-shaped wrinkles we will set λ equal to the half-wavelength mid way between the corner and the edge of the fan.

To estimate the wrinkle details, we begin by considering a simple analytical expression for the shape of the wrinkled surface. For example, in the case of a symmetrically loaded membrane we assume that at each corner there is a set of uniform, radial wrinkles whose out-of-plane shape can be described in the polar coordinate system of Fig. 4 by

$$w = A \sin\frac{\pi(r-r_1)}{R_{\rm wrin} - r_1} \sin 2n\theta \tag{7}$$

where A is the wrinkle amplitude, n the total number of wrinkles at the corner —each subtending an angle of $\pi/2n$— and θ is an angular coordinate measured from the edge of the membrane.

Since the stress in the corner regions is uniaxial there is the possibility of wrinkles forming there. The radial strain is

$$\epsilon_r = \sigma_r/E \tag{8}$$

where σ_r is given by Eq. (1). The corresponding radial displacement, $u(r)$ (positive outwards), can be obtained from

$$u = \int \epsilon_r dr + c \tag{9}$$

where the constant of integration c can be obtained by noting that $u \approx 0$ at $r = R$, i.e. at

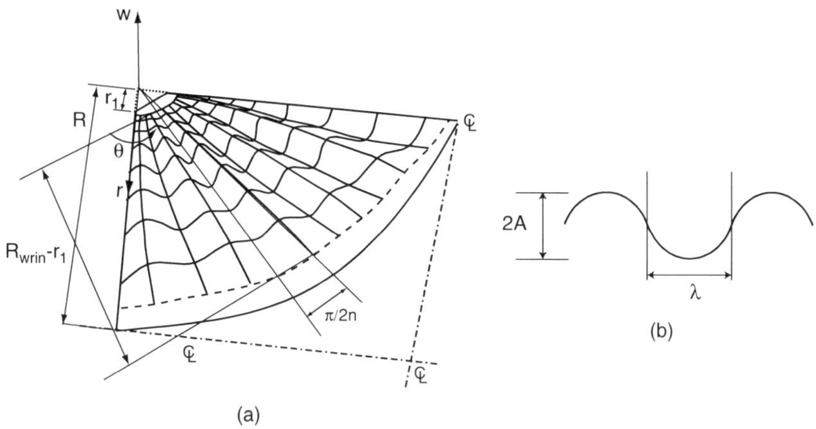

Fig. 4. Corner wrinkles: (a) overall shape; (b) central cross section and definition of half-wavelength.

the edge of the biaxially stressed region. Therefore,

$$u = \frac{T}{\sqrt{2}Et} \ln \frac{r}{R} \tag{10}$$

The hoop strain required for geometric compatibility is

$$\epsilon_{\theta_g} = \frac{u}{r} \tag{11}$$

and the hoop material strain is

$$\epsilon_{\theta_m} = -\nu \frac{\sigma_r}{E} \tag{12}$$

Wrinkles will form when ϵ_{θ_g} is larger in magnitude than ϵ_{θ_m} (note that both strains are negative), hence combining Eqs. (1), (10)–(12), we obtain

$$\ln \frac{R}{r} \geq \nu \tag{13}$$

The radius of the wrinkled region, R_{wrin}, is the largest r for which Eq. (13) is satisfied. Within the wrinkled region, i.e. for $r < R_{\text{wrin}}$, an additional "wrinkling" strain is required

$$\epsilon_{\theta_g} = \epsilon_{\theta_m} + \epsilon_{\theta_{\text{wrin}}} \tag{14}$$

The wrinkling strain is related to the wrinkle amplitude, and for the wrinkle shape defined by Eq. (7) it can be shown that at $r = (R_{\text{wrin}} - r_1)/2$

$$\epsilon_{\theta_{\text{wrin}}} = -\frac{\pi^2 A^2}{4\lambda^2} \tag{15}$$

Substituting Eq. (10) into Eq. (11) and Eq. (1) into Eq. (12), and then both into Eq. (14) we find that A has to satisfy

$$\frac{\sqrt{2}T}{Et(R_{\text{wrin}} - r_1)} \ln \frac{R_{\text{wrin}} - r_1}{2R} = -\frac{\sqrt{2}\nu T}{Et(R_{\text{wrin}} - r_1)} - \frac{\pi^2 A^2}{4\lambda^2} \tag{16}$$

Next, we work out the number of wrinkles by considering out-of-plane equilibrium of the wrinkled membrane at a point of maximum out-of-plane displacement, e.g. at $r = (R_{\text{wrin}} - r_1)/2, \theta = \pi/4n$. The equilibrium equation is

$$\sigma_r \kappa_r + \sigma_\theta \kappa_\theta = 0 \tag{17}$$

where κ_r and κ_θ are the curvatures in the radial and hoop directions, respectively, which can be obtained by differentiating Eq. (7). Hence,

$$\kappa_r = -\frac{A\pi^2}{(R_{\text{wrin}} - r_1)^2} \quad \text{and} \quad \kappa_\theta = -\frac{16An^2}{(R_{\text{wrin}} - r_1)^2} \tag{18}$$

The transverse stress component σ_θ is set equal to σ_{cr}. Substituting Eqs. (1), (6) and (18) into Eq. (17) gives

$$\frac{\sqrt{2}T}{(R_{\text{wrin}} - r_1)t} - \frac{4Et^2n^2}{3(1-\nu^2)\lambda^2} = 0 \tag{19}$$

Since λ is related to the number of wrinkles by

$$\lambda = \frac{R_{\text{wrin}} - r_1}{2} \frac{\pi}{2n} \tag{20}$$

we can substitute for λ into Eq. (19) and solve for n to obtain

$$n = \sqrt[4]{\frac{3\sqrt{2}\pi^2 T(R_{\text{wrin}} - r_1)(1-\nu^2)}{64Et^3}} \tag{21}$$

Given Eqs. (20) and (21) we can predict the wrinkle amplitude A by solving Eq. (16), which gives

$$A = \frac{2\lambda}{\pi} \sqrt{\frac{\sqrt{2}T}{Et(R_{\text{wrin}} - r_1)} \left(\ln \frac{2R}{R_{\text{wrin}} - r_1} - \nu \right)} \tag{22}$$

In the case $T_1 \neq T_2$ it is straightforward to generalize Eqs. (19) and (22) to find the wavelength and amplitude of the wrinkles in each corner region. However, for $T_1/T_2 \geq 1/(\sqrt{2}-1)$ the two larger corner stress fields come into contact, see Fig. 3(c), and hence a single diagonal wrinkle can form between the two most heavily loaded corners. This much larger wrinkle can be analysed following a similar approach [13], to obtain the following expressions for half-wavelength and amplitude

$$\lambda = \sqrt[4]{\frac{2\pi^2 R_1 (R_1 - r_1)^2 E t^3 \sin\theta_1}{3(1-\nu^2)T_1}} \tag{23}$$

and

$$A = \frac{2\sqrt{\lambda(\delta_1 + \delta_2)}}{\pi} \tag{24}$$

Here δ_1 and δ_2 are the radial displacements of the corners loaded by T_1 and T_2, respectively; these corner displacements can be estimated from

$$\delta_i \approx \frac{T_i}{2Et\sin^2\theta_i}\left[\theta_i \ln\frac{R_i}{r_1} + (1-\nu)\left(\frac{A_i}{R_i^2} + \theta_i - \frac{1}{2}\tan\theta_i\right)\right] \quad i = 1, 2 \tag{25}$$

Here, A_i is the area of a part of the central, biaxially stressed region that is associated with the loads T_i. More refined estimates can be obtained from a two-dimensional finite-element analysis, see Section 5.

4 Finite-Element Simulations

We have recently shown [14] that wrinkling of a thin membrane can be accurately modelled using the thin shell elements available in the commercial finite-element package ABAQUS [15]. The analysis is carried out by introducing initial geometrical imperfections, obtained from an initial eigenvalue analysis, followed by a geometrically non-linear post-buckling analysis using the pseudo-dynamic *STABILIZE solution scheme. This approach, although expensive in computational terms, is so far the only method that can reveal full wrinkle details and can be relied upon as an almost exact replication of physical experimentation. An alternative approach is the Iterative Modified Properties (IMP) method [9] which uses a combined stress-strain wrinkling criterion in a two-dimensional membrane model. The IMP method has been recently implemented as an ABAQUS user subroutine and has been shown to accurately predict the extent of the wrinkled regions and the two-dimensional stress distribution —but of course not the details of the wrinkles.

Both of these modelling techniques were used to simulate a 0.025 mm thick, 500×500 mm^2 square Kapton membrane (E=3530 N/mm^2 and $\nu = 0.3$). The membrane was loaded at each corner through a spreader beam by a 0.1 mm thick, 25 mm × 20 mm Kapton tabs—as in the experiment of Section 2.

A uniform mesh of 200 by 200 square elements was used to model the whole structure, in order to capture the fine wrinkle details in the corners.

In the shell model, the Kapton membrane and the corner tabs were modelled using S4R5 thin shell elements of different thickness. At the corners, the shell elements were connected to "Circ" beam elements through the *MPC,

TIE function. The central node was constrained against translation in the x- and y-direction; all side edges were left free. Both the out-of-plane rotations of the membrane and all in-plane bending degrees of freedom of the corner beams were restrained. The corner loads were applied as distributed loads along the truncated corners of the membrane.

In the IMP model, the membrane was modelled using M3D4 membrane elements, whose constitutive behaviour was modelled through a UMAT subroutine. The corner tabs were modelled with S4 shell elements and the same beam elements as for the shell model were used.

4.1 Simulation Details

Two load steps were applied, first a symmetric loading of $T_1 = T_2 = 5$ N. Second, $T_2 = 5$ N was maintained while T_1 was increased up to 20 N.

The analysis procedure was essentially identical for all of the simulations. First, a uniform prestress of 0.5 N/mm^2 was applied, to provide initial out-of-plane stiffness to the membrane. This was achieved by means of *INITIAL CONDITION, TYPE=STRESS. Next, a non-linear-geometry analysis was carried out, with *NLGEOM, to check the equilibrium of the prestressed system. Then, a linear eigenvalue analysis step was carried out (for the thin shell model only) in order to extract possible wrinkling mode-shapes of the membrane under a symmetrical loading. Four such mode-shapes were selected, based on their resemblance to the expected final wrinkled shape, and were introduced as initial geometrical imperfections. Finally, an automatically stabilised post-wrinkling analysis was performed, with *STATIC, STABILIZE. This analysis is very sensitive because the magnitude of the wrinkles is very small, hence an increment of 0.001 of the total load had to be selected. The default stabilize factor was reduced to 10^{-12} to minimise the amount of fictitious damping; this was the smallest amount required to stabilise the solution.

Fig. 5(a) shows the symmetrically wrinkled shape obtained for $T_1 = T_2 =$ 5 N. The wrinkle amplitudes are very small and have been enlarged 100 times for clarity. Fig. 5(b) shows the corresponding shape for $T_1/T_2 = 4$. Here the distinguishing feature is a large diagonal wrinkle between the two more heavily loaded corners, a number of smaller wrinkles can also be seen near the corners with smaller loads.

Fig. 6 shows the wrinkle profiles measured at three different cross sections, for $T_1/T_2 = 4$, plotted against those obtained from the simulations. Note that experiments and simulations match closely in the central region; the wrinkle wavelengths, in particular, are predicted quite accurately. But ABAQUS predicts smaller displacements of the edges of the membrane, due to the fact that the initial shape of the physical model has not been captured with sufficient accuracy (the edges of a Kapton sheet are naturally curled).

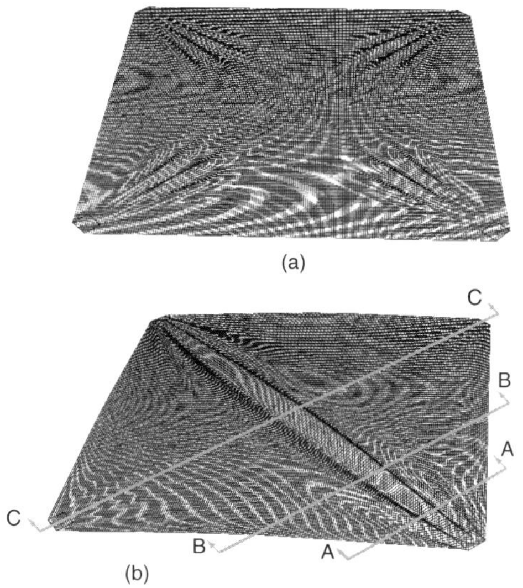

Fig. 5. Wrinkled shapes for (a) $T_1/T_2 = 1$ and (b) $T_1/T_2 = 4$.

Table 1. Wrinkles under symmetric loading (all values in mm)

	n		λ at $r = 70$ mm		A	
T (N)	Eq. 21	Experiment	Eq. 20	Experiment	Eq. 22	Experiment
5	9.6	8	11.6	11.0	0.16	0.12
20	13.6	11	8.3	9.7	0.22	0.14

5 Validation of Analytical and Finite-Element Models

Table 1 compares the predicted wrinkle half-wavelengths and amplitudes from Eqs. (20)–(22) with experimental measurements for a symmetrically loaded membrane, i.e. for $T_1/T_2 = 1$.

For the load case $T_1/T_2 = 4$, Table 2 compares the analytical predictions for the large diagonal wrinkle at the centre of the membrane, for two different membrane thicknesses, with predictions from ABAQUS and experimental measurements. The wavelength predictions are very close. The wrinkle amplitudes have been estimated first using only analytical solutions, Eq. (24) with δ_1 and δ_2 from Eq. (25), and second using Eq. (24) but with corner deflections from a 2-D IMP analysis. We can see that the fully analytical estimates are up to 88% higher than those measured experimentally, but if we use the corner displacements from the 2D finite-element analysis we obtain estimates that are only 48% and 16% higher than the measurements.

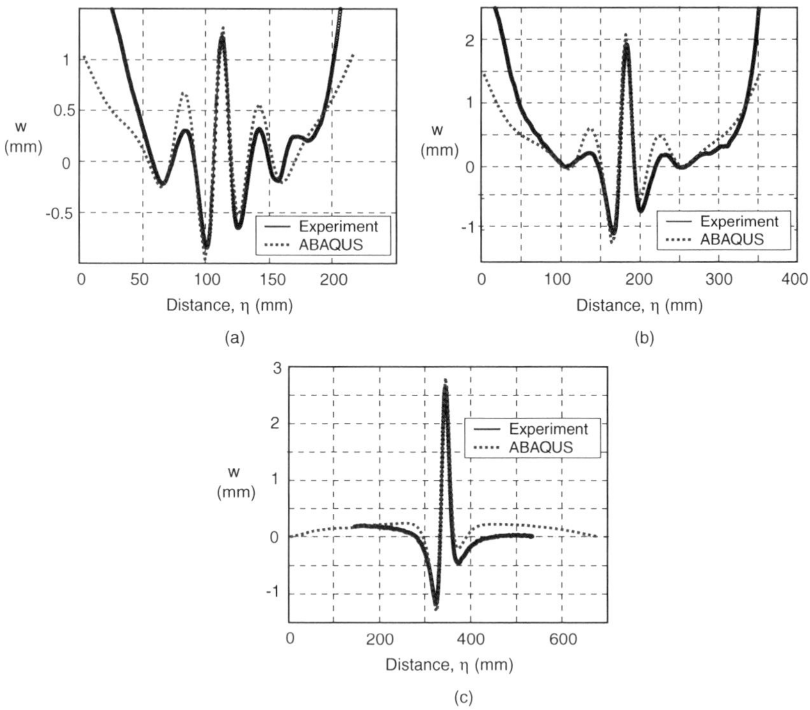

Fig. 6. Comparison of experimental measurements with ABAQUS results for $T_1/T_2 = 4$ and for cross sections at distance of (a) 105 mm, (b) 177 mm, (c) 346 mm from corner.

Table 2. Wrinkle half-wavelengths and amplitudes for $T_1/T_2 = 4$ at $r = 346$ (all values in mm)

	λ			A			
t	Eq. 23	Experiment	FE	Eqs 24, 25	Eq. 24+FE[1]	Experiment	FE[2]
0.025	24.6	23.8	22.3	3.5	2.8	1.9	2.0
0.050	41.3	33.9	35.6	3.2	2.1	1.8	1.6

6 Discussion and Conclusions

The wrinkling of square membranes subject to coplanar pairs of equal and opposite corner forces has been investigated. It has been shown that two wrinkling regimes exist. The first regime occurs for symmetric loading and asymmetric loading up to approximately $T_1/T_2 = 2.41$, and it is characterised

[1] Membrane model with IMP.

[2] Shell model.

by relatively small, radial corner wrinkles. The second regime occurs for asymmetric loading with approximately $T_1/T_2 > 2.41$, and is characterised by a single, large diagonal wrinkle, plus small radial wrinkles at all four corners. Each regime has been observed experimentally and also reproduced to great accuracy in a finite-element simulation.

A general methodology for gaining insight into the wrinkling of membranes, which was first proposed for uniformly wrinkled membranes in shear [16], has been presented. For the first wrinkling regime ($T_1/T_2 < 2.41$) it is reasonable to assume uniform, fan-shaped wrinkles expressed in terms of sine functions. With these assumptions, analytical expressions have been derived for the number of corner wrinkles and their average maximum amplitude. For the second wrinkling regime ($T_1/T_2 > 2.41$) an analogous derivation leads to an expression for the amplitude of the central wrinkle. These expressions have been found to overestimate wrinkle amplitudes by up to 50%, and hence they appear to be sufficiently accurate for preliminary design.

Finite element analysis using thin shell elements has been shown to be able to replicate real physical experimentation with an accuracy typically better than 10% on amplitude. However, a very fine mesh had to be used to resolve the small corner wrinkles; hence a complete simulation takes up to several days on a 2GHz Pentium 4 PC.

Acknowledgments

We thank Professors C.R. Calladine and K.C. Park for helpful suggestions. Partial support from NASA Langley Research Center, research grant NAG-1-02009, Integrated membranous-microelement space structures technology (technical monitor Dr K. Belvin) is gratefully acknowledged.

References

1. Wagner H (1929) Flat sheet metal girder with very thin metal web, Zeitschrift für Flugtechnik Motorlurftschiffahrt 20: 200–207, 227–233, 256–262, 279-284
2. Reissner E (1938) On tension field theory. Proc. 5th Int. Cong. Appl. Mech., 88–92
3. Stein M, Hedgepeth JM (1961) Analysis of Partly Wrinkled Membranes. NASA Langley Research Center, NASA TN D-813
4. Mansfield EH (1969) Tension field theory a new approach which shows its duality with inextensional theory. Proc 12th Int Cong Appl Mech, 305–320
5. Mansfield EH (1989) The Bending and Stretching of Plates, second edition. Cambridge University Press, Cambridge
6. Pipkin AC (1986) The relaxed energy density for isotropic elastic membranes. IMA J Appl Math, 36: 85–99
7. Epstein M, Forcinito MA (2001) Anisotropic membrane wrinkling: theory and analysis. Int J Solids Structures 38: 5253–5272

8. Jenkins C, Leonard JW (1991) Nonlinear dynamic response of membranes: State of the art. ASME Appl Mech Reviews 44: 319–328
9. Adler A (2000) Finite Element Approaches for Static and Dynamic Analysis of Partially Wrinkled Membrane Structures. PhD Dissertation, University of Colorado at Boulder
10. Proc. 42nd AIAA/ASME/ASCE/AHS/ASC Structures, Structural Dynamics and Materials Conference, 16-19 April 2001, Seattle, WA, AIAA, Reston
11. Proc 43rd AIAA/ASME/ASCE/AHS/ASC Structures, Structural Dynamics, and Materials Conference and Exhibit, Denver, CO, 22-25 April 2002, AIAA, Reston
12. Proc. 44th AIAA/ASME/ASCE/AHS/ASC Structures, Structural Dynamics, and Materials Conference and Exhibit, Norfolk, VA, 7-10 April 2003, AIAA, Reston
13. Wong YW, Pellegrino S, Park KC (2003) Prediction of wrinkle amplitudes in square solar sails. Proc 44th AIAA/ASME/ASCE/AHS/ASC Structures, Structural Dynamics, and Materials Conference and Exhibit, Norfolk, VA, 7-10 April 2003, AIAA-2003-1982
14. Wong YW, Pellegrino S (2002) Computation of wrinkle amplitudes in thin membranes. Proc 43rd AIAA/ASME/ASCE/AHS/ASC Structures, Structural Dynamics, and Materials Conference and Exhibit, Denver, CO, 22-25 April 2002, AIAA-2002-1369
15. Hibbit, Karlsson and Sorensen, Inc. (2001) ABAQUS Theory and Standard User's Manual, Version 6.2, Pawtucket, RI, USA
16. Wong YW, Pellegrino S (2002) Amplitude of wrinkles in thin membranes. In: Drew HR, Pellegrino S (eds) New Approaches to Structural Mechanics, Shells and Biological Structures. Kluwer Academic Publishers, Dordrecht, 257-270

F.E.M. for Prestressed Saint Venant-Kirchhoff Hyperelastic Membranes

Antonio J. Gil

Civil and Computational Engineering Centre, School of Engineering, University of Swansea, Singleton Park, SA2 8PP, United Kingdom a.j.gil@swansea.ac.uk

Summary. *This chapter presents a complete numerical formulation for the nonlinear structural analysis of prestressed membranes with applications in Civil Engineering. These sort of membranes can be considered to undergo large deformations but moderate strains, consequently nonlinear continuum mechanics principles for large deformation of prestressed bodies will be employed in order to proceed with the analysis. The constitutive law adopted for the material will be the one corresponding to a prestressed hyperelastic Saint Venant-Kirchhoff model. To carry out the computational resolution of the structural problem, the Finite Element Method (FEM) will be implemented according to a Total Lagrangian Formulation (TLF), by means of the Direct Core Congruential Formulation (DCCF). Eventually, some numerical examples will be introduced to verify the accuracy and robustness of the aforementioned formulation.*

Key words: Tension membrane structures, Total Lagrangian Formulation, Direct Core Congruential Formulation, Hyperelastic Saint Venant-Kirchhoff material, Newton-Raphson method

1 Introduction

Tension structures constitute a structural form providing remarkable opportunities in the fields of architecture and civil engineering. Nowadays, numerous practical examples can be located throughout the entire world because of the acceptance by designers and their upward trend in popularity -see [1], [2], [3], [4]-. The increasing necessity of creating large enclosed areas, unobstructed by intermediate supports, has allowed the introduction of limitless possibilities for doubly curved surface forms -see [5]-. Although there are several categories that fall into the general term of tension structures, this paper will focus on the well known prestressed membranes, from which prestressed cables can be analyzed by extension. As an instance of these models, Fig. 1 reflects several views of a prestressed cable reinforced membrane.

We will focus on those particular membranes where strains can be assumed moderate, despite having large deformations. Two different and successive loading cases may be distinguished according to their effects on the stabilization of the

E. Oñate and B. Kröplin (eds.), Textile Composites and Inflatable Structures, 123–142.

prestressed membrane. The first one or prestressed loading is developed to provide the necessary in-surface rigidity to the membrane in order to support the second loading step. The latter, also named in service loading step is comprised of a wide group of loads: snow, wind or live loads among others.

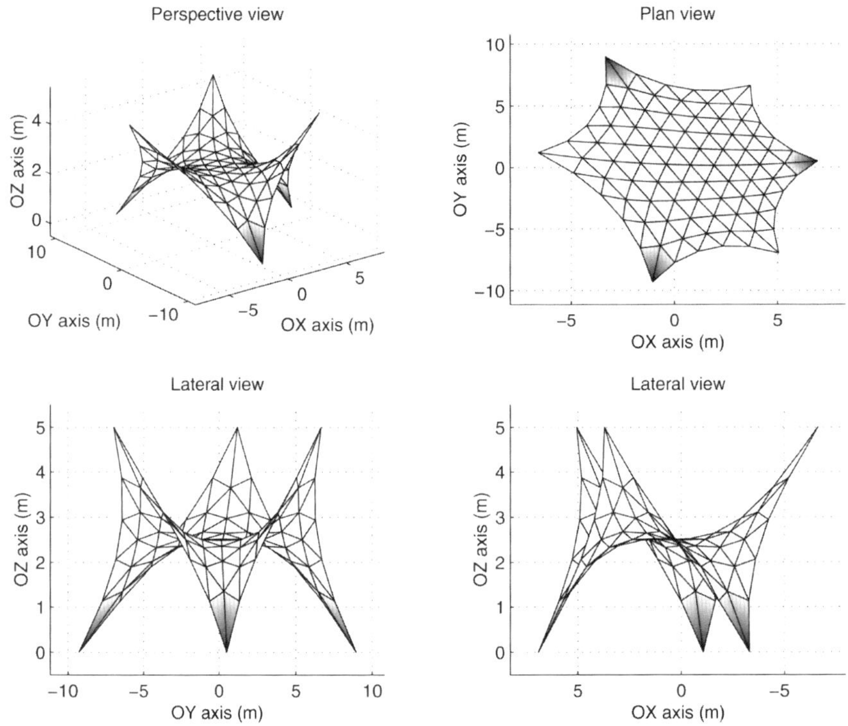

Fig. 1. Prestressed cable reinforced membrane

The theory of hyperelastic membranes, as for instance, propounded by [6], [7] and [8] treats the problem from an analytical point of view. Some simplicity may be accomplished if the Von Karman compatibility equations are used -see [9] and [10]-, whereby rotations are considered to be moderate. Regardless of the important implications of this approach for the theoretical understanding of these structures, a main disadvantage is that it results in a nonlinear partial differential system of equations with impossible analytical resolution.

Because of this lack of numerical results, variational approaches ought to be taken into consideration as the best means to provide feasible solutions. Some authors have treated the problem of finite hyperelasticity set on rubberlike membrane materials by means of the Finite Element Method (FEM). By following this approach, interesting papers are those due to [11], [12], [13], [14], [15] and [16]. For these cases, the Updated Lagrangian Formulation (ULF) is considered to be the most suitable for the derivation of the tangent stiffness matrix. This matrix is required by the Newton-Raphson iterative scheme for the solution of the nonlinear equilibrium equations.

The discussion to follow is divided into six parts. The second section reviews the classical nonlinear strong form equations. The consideration of the Saint Venant-Kirchhoff material as the adopted model will very conveniently provide a linear constitutive relationship of easy implementation. The third section entails a comprehensive explanation of the Finite Element semidiscretization of the previously obtained strong form. After the weak form is derived in a straightforward manner, the displacement field is interpolated by means of shape functions based on a Lagrangian mesh geometry. The resulting formulation will be the so called Total Lagrangian Formulation (TLF). Afterwards, the exact linearization of the Total Lagrangian weak form of the momentum balance is carried out in detail. For the sake of further computing implementation reasons, the Direct Core Congruential Formulation (DCCF) is reviewed as the most appropriate formulation.

Eventually, based on the aforementioned formulation, two numerical examples for both a cable network and for a prestressed membrane, are presented. These cases will show adequate performance as the required quadratically convergence of the Newton-Raphson method is obtained. Some conclusions are presented at the end.

2 Strong Formulation: General Structural Principles

Before establishing the formulation in terms of particular finite elements, that is, cable or membrane elements, we will develop in this section the general equations that govern the behaviour of prestressed membrane structures. For a complete understanding, it is necessary to consider three successive configurations of the material body: an initial nominally stressed state $\Re_0$, a primary state $\Re_t$ and a secondary state $\Re_{t^*}$, for the time instants t and t^*, respectively. It is important to point out that the term nominally stressed state is employed to describe a self-equilibrated configuration where the internal stresses are as small as required by the designer. Usually, $\Re_0$ represents the nominally stressed state found at a form finding state. $\Re_t$ symbolizes the actual in service prestressed state prior to the live loading, which may be different to $\Re_0$ and due to constructions prestresses. Finally, $\Re_{t^*}$ stands for the live loading in service state.

Between these latter two stages, a displacement field $\mathbf{u} = (u_1, u_2, u_3)$ may be defined in $\mathbf{R}^3$. To differentiate the coordinates of a body particle along the deformation path, the following convention will be employed throughout the remainder of this paper:

- X_A, $(A = 1, 2, 3)$ for the initial nominally stressed configuration $\Re_0$.
- X_j^{pret}, $(j = 1, 2, 3)$ for the initial prestressed configuration $\Re_t$ or primary state.
- x_i, $(i = 1, 2, 3)$ for the current spatial configuration $\Re_{t^*}$ or secondary state.

Henceforth, we will consider as incremental those quantities which proceed from the movement from the primary to the secondary state. The spatial coordinates for the time t^* of a particle can be related to its material coordinates in the initial nominally stressed configuration $\Re_0$ according to the classical mapping equation $x_i = x_i(X_A, t^*)$. By recalling the chain rule, relations among deformation gradient tensors

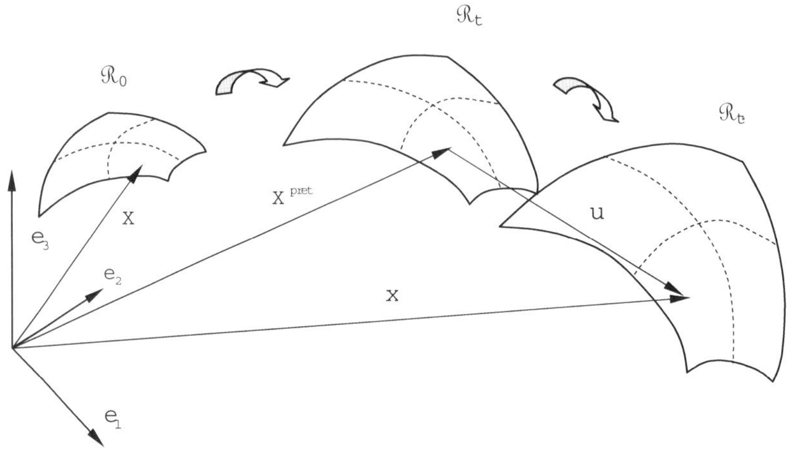

Fig. 2. Deformation path

and their respective jacobians are summarized as:

$$x_{i,A} = x_{i,j} X_{j,A}^{pret} \quad \Rightarrow \quad J^* = J' J \tag{1}$$

where the implied summation convention for repeated indices as well as the comma differentiation symbol $x_{i,j} = \partial x_i / \partial X_j^{pret}$ have been introduced to simplify the algebra. The term J above represents the jacobian at the primary state, J^* stands for the jacobian at the end of the secondary state and $J^{'}$ symbolizes the jacobian as a consequence of the incremental deformation.

The description of the deformation and the measure of strain are essential parts of nonlinear continuum mechanics. From a kinematically point of view, a material particle position in the primary and secondary states may be expressed in terms of the incremental displacement field. Analogously, the deformation gradient tensor may be introduced to characterize adequately the deformation path as:

$$x_i = X_i^{pret} + u_i \quad \Rightarrow \quad x_{i,A} = X_{i,A}^{pret} + u_{i,A} = X_{i,A}^{pret} + u_{i,j} X_{j,A}^{pret} \tag{2}$$

where:

$$u_{i,j} = \frac{\partial u_i}{\partial X_j^{pret}} \tag{3}$$

In contrast to linear elasticity, many different measures of strain may be used in nonlinear continuum mechanics. Nevertheless, the Green-Lagrange strain tensor is considered to be the most appropriate measure specially when dealing with moderate strains. The Green-Lagrange strain tensor for the secondary state with respect to the initial nominally stressed configuration is defined in terms of the deformation gradient tensor and the Delta-Kronecker tensor as:

$$E_{AB}^* = \frac{1}{2}(x_{i,A} x_{i,B} - \delta_{AB}) \tag{4}$$

By substituting equation (2) into equation (4):

$$E_{AB}^* = \frac{1}{2}(X_{i,A}^{pret} X_{i,B}^{pret} - \delta_{AB}) + \frac{1}{2}(2 X_{i,A}^{pret} X_{j,B}^{pret} e_{ij} + u_{i,A} u_{i,B}) \tag{5}$$

with:

$$e_{ij} = \frac{1}{2}(u_{i,j} + u_{j,i}) \tag{6}$$

The difference between the Green-Lagrange strain tensor for the primary and secondary states can be carried out in a straightforward manner as:

$$E^*_{AB} - E_{AB} = X^{pret}_{i,A} X^{pret}_{j,B} e_{ij} + \frac{1}{2} u_{i,A} u_{i,B} \tag{7}$$

By applying the chain rule:

$$E^*_{AB} - E_{AB} = X^{pret}_{i,A} X^{pret}_{j,B} (e_{ij} + \frac{1}{2} u_{s,i} u_{s,j}) = X^{pret}_{i,A} X^{pret}_{j,B} E^{*relat}_{ij} \tag{8}$$

where the tensor E^{*relat}_{ij} has been introduced for the sake of convenience and it represents a relative measure of the strain at the secondary state by taking the primary one as an adequate reference.

In nonlinear problems, various stress measures can be defined. In this paper, in addition to the Cauchy or real stress tensor, two tensorial entities referred to as the second Piola-Kirchhoff and the nominal stress tensors are to be used. The latter is known as well as the transpose of the first Piola-Kirchhoff stress tensor. By considering as initial configuration the initial nominally stressed one, the Cauchy stress tensor σ^*_{ij} may be related to the nominal stress tensor P^*_{Aj} and the second Piola-Kirchhoff stress tensor S^*_{AB} as:

$$J^* \sigma^*_{ij} = x_{i,A} P^*_{Aj} = x_{i,A} x_{j,B} S^*_{AB} \qquad x_{i,A} = \frac{\partial x_i}{\partial X_A} \tag{9}$$

The same relationship may be developed when the initial prestressed configuration is adopted to be the reference state. The super index *relat* is added to distinguish the new nonlinear stress tensors with respect to those shown in the above formula:

$$J' \sigma^*_{ij} = x_{i,s} P^{*relat}_{sj} = x_{i,s} x_{j,t} S^{*relat}_{st} \qquad x_{i,s} = \frac{\partial x_i}{\partial X^{pret}_s} \tag{10}$$

Formulae (9) and (10) can be modified to set up expressions (11) which summarize the relationship among the nominal and second Piola-Kirchhoff stress tensors obtained in both the initial undeformed configuration $\Re_0$ and the primary state $\Re_t$.

$$\begin{aligned} P^{*relat}_{sj} &= J^{-1} X^{pret}_{s,A} X^{pret}_{t,B} x_{j,t} S^*_{AB} \\ S^{*relat}_{st} &= J^{-1} X^{pret}_{s,A} X^{pret}_{t,B} S^*_{AB} \end{aligned} \tag{11}$$

The local equilibrium equations in the secondary state may be expressed with respect to three possible descriptions: $\Re_0$, $\Re_t$ and $\Re_{t^*}$, these being a Lagrangian formulation for the first two configurations and an Eulerian formulation for the final configuration. These expressions may be gathered as follows:

$$\sigma^*_{ji,j} + \rho^* b_i = 0 \ \text{ in } \ \Re_{t^*}, \ \text{ with } f^*_i = \sigma^*_{ji} n^*_j d\Gamma^* \tag{12}$$

$$P^*_{Ai,A} + \rho_0 b_i = 0 \ \text{ in } \ \Re_0, \ \text{ with } f^*_i = P^*_{Ai} n_A d\Gamma_0 \tag{13}$$

$$P_{ji,j}^{*relat} + \rho b_i = 0 \text{ in } \Re_t, \text{ with } f_i^* = P_{ji}^{*relat} n_j d\Gamma \tag{14}$$

The formula (14) along with the boundary conditions and continuity conditions -see [17]-, represents the strong formulation of the structural problem according to a Lagrangian description with respect to a reference stressed configuration. This is the equation that will be used from now on.

Many engineering applications, particularly the one which concerns us, involve moderate strains and large rotations. Therefore, in these kind of problems the effects of large deformation are primarily due to rotations. The response of the material may then be modeled as an extension of the well known linear elastic law by replacing the Cauchy stress tensor by the second Piola-Kirchhoff stress one and the small strain tensor by the Green-Lagrange strain one.

This material behaviour is named Saint Venant-Kirchhoff hyperelastic or simply Kirchhoff material. By accounting for the hyperelastic pattern of this constitutive model, the second Piola-Kirchhof stress tensor may be formulated in an elegant way by means of the Helmholtz free energy -also known as internal strain energy-. Thus, the second Piola-Kirchhoff stress tensor in the current configuration may be formulated by means of a Taylor series expansion truncated after the first order as follows:

$$S_{AB}^* = \frac{\partial w_{int}}{\partial E_{AB}^*} = \frac{\partial w_{int}}{\partial E_{AB}} + \frac{\partial^2 w_{int}}{\partial E_{AB} \partial E_{CD}} (E_{CD}^* - E_{CD}) \tag{15}$$

The accuracy of this Taylor series depends directly on the smallness of the step $E_{CD}^* - E_{CD}$. For tension membrane structures, as it was previously mentioned, this is a valid assumption. Thus:

$$S_{AB}^* = S_{AB} + C_{ABCD} X_{i,C}^{pret} X_{j,D}^{pret} E_{ij}^{*relat} \tag{16}$$

By recalling (11) and (16):

$$S_{st}^{*relat} = J^{-1} X_{s,A}^{pret} X_{t,B}^{pret} S_{AB} + J^{-1} X_{s,A}^{pret} X_{t,B}^{pret} C_{ABCD} X_{i,C}^{pret} X_{j,D}^{pret} E_{ij}^{*relat} \tag{17}$$

The fourth order tensor of elastic moduli can be referred to the prestressed configuration as follows:

$$C_{ijkl} = J^{-1} X_{i,A}^{pret} X_{j,B}^{pret} C_{ABCD} X_{k,C}^{pret} X_{l,D}^{pret} \tag{18}$$

Eventually, equation (17) may be reformulated to give the final expression:

$$S_{ij}^{*relat} = \sigma_{ij} + C_{ijkl} E_{kl}^{*relat} \tag{19}$$

This final formula is set up to show the constitutive law for a prestressed Saint Venant-Kirchhoff hyperelastic material. The second Piola-Kirchhoff stress tensor is expressed in terms of an easy linear relationship which depends on three tensorial entities: Cauchy stress tensor in the primary state, fourth order tensor of elastic moduli and the Green-Lagrange strain tensor of the secondary state referred to the primary state.

Recall -see [18]- that in a Saint Venant-Kirchhoff hyperelastic material, the constitutive tensor can be formulated as follows:

$$C_{ijkl} = \lambda \delta_{ij} \delta_{kl} + 2\mu \delta_{ik} \delta_{jl} \tag{20}$$

where λ and μ are known as the Lamé constants. These two constants can be related to the classical Young modulus E and Poisson's ratio ν as follows:

$$\lambda = \frac{\nu E}{(1+\nu)(1-2\nu)} \qquad \mu = \frac{E}{2(1+\nu)} \tag{21}$$

Another important feature which needs to be obtained is the incremental strain energy accumulated into the structure along the deformation path from the primary to the secondary states. By performing again a Taylor series expansion truncated after the second order, the internal strain energy functional per unit of nominally stressed volume may be developed as:

$$w_{int}^{*} = w_{int} + \frac{\partial w_{int}}{\partial E_{AB}}(E_{AB}^{*} - E_{AB}) + \frac{1}{2}\frac{\partial^2 w_{int}}{\partial E_{AB}\partial E_{CD}}(E_{AB}^{*} - E_{AB})(E_{CD}^{*} - E_{CD}) \tag{22}$$

This above formula can be rewritten as:

$$w_{int}^{*} = w_{int} + \Omega + \Upsilon \tag{23}$$

where the terms Ω and Υ can be depicted as:

$$\Omega = S_{AB} X_{i,A}^{pret} X_{j,B}^{pret} E_{ij}^{*relat} = J\sigma_{ij} E_{ij}^{*relat} \tag{24}$$

$$\Upsilon = \frac{1}{2} C_{ABCD} X_{i,A}^{pret} X_{j,B}^{pret} E_{ij}^{*relat} X_{k,C}^{pret} X_{l,D}^{pret} E_{kl}^{*relat} = \frac{1}{2} J C_{ijkl} E_{ij}^{*relat} E_{kl}^{*relat} \tag{25}$$

By substituting (24) and (25) back into (22):

$$w_{int}^{*} - w_{int} = J[\sigma_{ij} E_{ij}^{*relat} + \frac{1}{2} C_{ijkl} E_{ij}^{*relat} E_{kl}^{*relat}] = J\Psi \tag{26}$$

By integrating over the initial undeformed volume V^0 corresponding to the configuration $\Re_0$, and by applying the mass conservation principle from this volume V^0 to the prestressed volume V^{pret} corresponding to the configuration $\Re_t$, we obtain the incremental Helmholtz's free energy functional, which is given as:

$$\Delta W_{int} - \int_{V^0} (w_{int}^{*} \quad w_{int}) dV = \int_{V^0} J\Psi dV - \int_{V^{pret}} \Psi dV = \int_{V^{pret}} w_{int}^{*relat} dV \tag{27}$$

Therefore, the internal strain energy functional per unit of volume of the primary state takes the final form:

$$w_{int}^{*relat} = \sigma_{ij} E_{ij}^{*relat} + \frac{1}{2} C_{ijkl} E_{ij}^{*relat} E_{kl}^{*relat} \tag{28}$$

3 Finite Element Discretization

3.1 From the Strong to the Weak Formulation

The above mentioned primary and secondary states can be understood as an initial prestressed state $\Re^{pret}$ and a final in service loading state $\Re$ due to the consideration of live and dead load. Henceforth, the coordinates of any body's particle, in both prestressed and final loaded states, are related by means of the incremental displacement field $\mathbf{u}$ as follows:

$$\mathbf{x} = \mathbf{X}^{pret} + \mathbf{u}, \quad x_i = X_i^{pret} + u_i \tag{29}$$

According to this nomenclature, the strong formulation of the problem in a Lagrangian description with respect to the prestressed configuration is summarized in Fig. 3. The configurations $\Re^{pret}$ represents a material body of domain V^{pret} with frontier Γ^{pret}. As can be observed, the super index ^{*relat} has been suppressed for the sake of simplicity.

1. Balance of linear momentum.

$$\frac{\partial P_{ji}}{\partial X_j^{pret}} + \rho^{pret} b_i = 0 \;\; in \;\; V^{pret} \tag{30}$$

2. Transformation of stress tensors.

$$J\sigma_{ij} = \frac{\partial x_i}{\partial X_s^{pret}} P_{sj} = \frac{\partial x_i}{\partial X_s^{pret}} \frac{\partial x_j}{\partial X_t^{pret}} S_{st} \text{ with } J = det(\frac{\partial x_i}{\partial X_j^{pret}}) \tag{31}$$

3. Green-Lagrange strain tensor.

$$E_{ij} = \frac{1}{2}(\frac{\partial x_k}{\partial X_i^{pret}} \frac{\partial x_k}{\partial X_j^{pret}} - \delta_{ij}) \tag{32}$$

4. Constitutive law.

$$S_{ij} = \sigma_{ij}^{pret} + C_{ijkl} E_{kl} \tag{33}$$

5. Internal strain energy functional per unit volume of prestressed configuration.

$$w_{int} = \sigma_{ij}^{pret} E_{ij} + \frac{1}{2} C_{ijkl} E_{ij} E_{kl} \tag{34}$$

6. Boundary conditions.

$$t_i = P_{ji} n_j^{pret} = \bar{t}_i \;\; on \;\; \Gamma_{t_i}^{pret} \qquad u_i = \bar{u}_i \;\; on \;\; \Gamma_{u_i}^{pret} \tag{35}$$

Fig. 3. Strong formulation for a Lagrangian description.

Thus, the weak form may be developed in a Total Lagrangian Format (TLF) by means of the so called Principle of Virtual Work. By neglecting inertia forces:

$$\delta W_{int}(\delta u_i, u_i) = \delta W_{ext}(\delta u_i, u_i) \tag{36}$$

$$\delta W_{int} = \int_{V^{pret}} \delta F_{ij} P_{ji} dV = \int_{V^{pret}} \delta \mathbf{F}^T : \mathbf{P} dV = \int_{V^{pret}} \delta E_{ij} S_{ij} dV = \int_{V^{pret}} \delta \mathbf{E} : \mathbf{S} dV \tag{37}$$

$$\delta W_{ext} = \int_{V^{pret}} \delta u_i b_i dV + \int_{\Gamma^{pret}} \delta u_i \bar{t}_i d\Gamma = \int_{V^{pret}} \delta \mathbf{u}^T \cdot \mathbf{b} dV + \int_{\Gamma^{pret}} \delta \mathbf{u}^T \cdot \bar{\mathbf{t}} d\Gamma \tag{38}$$

where the work conjugacy property of the tensors $\mathbf{S}$ and $\mathbf{P}$ with $\mathbf{E}$ and $\mathbf{F}^T$ respectively, has been employed for the equalities in (37); $\mathbf{b}$ is the body force vector and $\bar{\mathbf{t}}$ are the surface tractions.

3.2 Semidiscretization of the Weak Form

The weak form equations obtained formerly may be combined with a finite element discretization of the displacement field in terms of the nodal values $\mathbf{u}^I$ and shape functions N^I as:

$$u_i = {u_i}^I N^I \quad i = 1,2,3; \quad I = 1 \dots N \tag{39}$$

enables the nodal equivalent internal and external vector forces, $\mathbf{f}_{int}$ and $\mathbf{f}_{ext}$, respectively, to be obtained in a straightforward manner for a given node I as:

$$f^I_{int,i} = \int_{V^{pret}} P_{ji} \frac{\partial N^I}{\partial X_j^{pret}} dV = \int_{V^{pret}} F_{ik} S_{kj} \frac{\partial N^I}{\partial X_j^{pret}} dV \tag{40}$$

$$f^I_{ext,i} = \int_{V^{pret}} b_i N^I dV + \int_{\Gamma^{pret}} \bar{t}_i N^I d\Gamma \tag{41}$$

Assembling these forces for all the nodes of the Lagrangian mesh gives the global equilibrium equations:

$$\mathbf{f}_{int} = \mathbf{f}_{ext} \implies \mathbf{f}_{res} = \mathbf{f}_{int} - \mathbf{f}_{ext} = \mathbf{0} \tag{42}$$

where $\mathbf{f}_{int}$ is the global vector of internal forces, $\mathbf{f}_{ext}$ is the global vector of external forces and $\mathbf{f}_{res}$ is the global vector of residual forces. This last vector represents clearly the out of balance forces as a result of the strong nonlinearity contained into the structural problem.

3.3 Linearization of the Global Equilibrium Equations

The set of equations depicted by (42) presents a geometrically nonlinear feature, so an iterative solution scheme will be required. Among all the available methods, the second-order Newton-Raphson one accomplishes the best convergence properties. The total tangent stiffness matrix required by the later one is formed by linearizing the global equilibrium equations (42) in the direction of the incremental displacement $\mathbf{u}$.

By carrying out the linearization of the global vector of internal forces, it turns out to be:

$$d\mathbf{f}^I_{int} = d\mathbf{f}^{mat^I}_{int} + d\mathbf{f}^{geo^I}_{int} = (\mathbf{K}^{mat^{IJ}} + \mathbf{K}^{geo^{IJ}}) d\mathbf{u}^J \tag{43}$$

$$K^{mat^{IJ}}_{ij} = \int_{V^{pret}} F_{ik} \frac{\partial N^I}{\partial X_p} C_{pklm} F_{jl} \frac{\partial N^J}{\partial X_m} dV \tag{44}$$

$$K^{geo^{IJ}}_{ij} = \delta_{ij} \int_{V^{pret}} \frac{\partial N^I}{\partial X_l} S_{lk} \frac{\partial N^J}{\partial X_k} dV \tag{45}$$

where $\mathbf{K}^{mat^{IJ}}$ and $\mathbf{K}^{geo^{IJ}}$ stand for the elemental material or constitutive stiffness matrix and the elemental geometrical or initial stress stiffness matrix, respectively.

By assuming that the body forces $\mathbf{b}$ and external surface tractions $\bar{\mathbf{t}}$ not associated to pressure forces remain constant and by taking into account that the pressure component is dependent upon the geometry due to changing orientation and surface area of the structure, the linearization of the global vector of external forces is given through the following derivation:

$$\mathbf{f}^I_{ext} = \int_\Gamma \mathbf{t}N^I d\Gamma = -p\int_\Gamma \mathbf{n}N^I d\Gamma \tag{46}$$

By applying the Nanson rule for the unit normal $\mathbf{n}$ -see for instance [17] or [19] for details- and particularizing for an isoparametric three-node linear finite element:

$$\mathbf{f}^I_{ext} = -p\int_{\Gamma^{pret}} J\mathbf{F}^{-T}\mathbf{n}^{pret}N^I d\Gamma = \frac{-p\Gamma^{pret}}{3}\left(\frac{\partial \mathbf{x}}{\partial \xi_1^{pret}} \times \frac{\partial \mathbf{x}}{\partial \xi_2^{pret}}\right) \tag{47}$$

$$K^{p^{IJ}}_{ij} = \frac{-p\Gamma^{pret}\epsilon_{ilm}}{3}\left(\delta_{lj}\frac{\partial N^J}{\partial \xi_1^{pret}}F_{m2} + F_{l1}\delta_{mj}\frac{\partial N^J}{\partial \xi_2^{pret}}\right) \tag{48}$$

where p is the pressure scalar acting on the considered finite element, ξ_1^{pret} and ξ_2^{pret} are the local plane coordinates and $\boldsymbol{\epsilon}$ is the so called alternating third order tensor.

3.4 Direct Core Congruential Formulation (DCCF)

From the computational viewpoint, a very elegant procedure termed the Direct Core Congruential Formulation (DCCF) may be applied to perform the implementation stage of the formulation developed above. This methodology, which is hardly used in the existing literature is due to pioneer studies in [20] and [21]. The main ideas behind this formulation can be discovered in the notable paper due to [22].

The scope of the DCCF is establishing the set of global equilibrium equations whose unknowns are the components of the displacement gradient tensor $\mathbf{G}$ which is given as:

$$G_{ij} = \frac{\partial u_i}{\partial X_j^{pret}} \tag{49}$$

Therefore, this new set of equations is completely independent on the geometry of the structure and on the adopted discretization properties. Afterwards, every single component of the displacement gradient tensor may be easily expressed in terms of the nodal displacements of the Lagrangian mesh. Naturally, it is right then when properties concerning geometry and discretization are brought to light. The consideration of only traslational degrees of freedom for the nodes of the Lagrangian mesh makes the DCCF specially simple and easy of being implemented. The Fig. 4 shows a summary of this formulation:

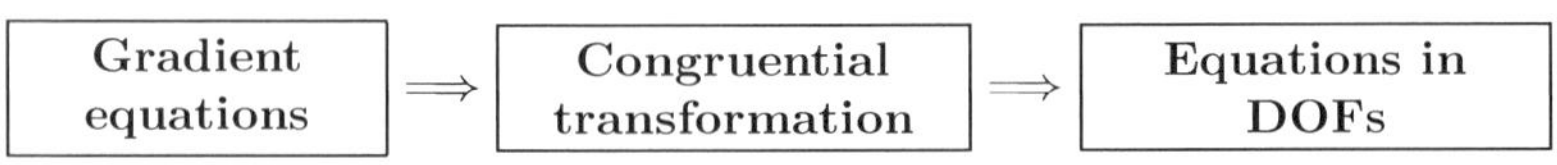

Fig. 4. DCCF scheme.

By employing a classical vectorization of the displacement gradient tensor -see for instance [17] and [19]- into a column vector $\mathbf{g}$ and by expressing this in terms of the nodal displacements, it turns:

$$\mathbf{g} = \mathbf{B}\mathbf{u} \tag{50}$$

where $\mathbf{B}$ is the matrix of the gradients of the shape functions for the chosen finite element and $\mathbf{u}$ is the column vector that gathers the nodal displacements for the N nodes of a single finite element:

$$\mathbf{u}^T = \begin{pmatrix} u_1^1 \ u_2^1 \ u_3^1 \dots u_1^N \ u_2^N \ u_3^N \end{pmatrix} \tag{51}$$

Analogously, the 2x2 submatrix of in-plane components of the Green-Lagrange strain tensor $\mathbf{E}$ may be vectorized into a column vector $\mathbf{e}$ by following the kinematic Voigt rule as:

$$\mathbf{e}^T = \begin{pmatrix} E_{11} \ E_{22} \ 2E_{12} \end{pmatrix} \tag{52}$$

Every component of the new vector $\mathbf{e}$ may be expressed in terms of the vector $\mathbf{g}$ as:

$$e_i = \mathbf{h}_i^T \mathbf{g} + \frac{1}{2}\mathbf{g}^T \mathbf{H}_i \mathbf{g} \tag{53}$$

where $\mathbf{h}_i$ and $\mathbf{H}_i$ are a vector and a symmetric matrix of numerical values comprising $(1,0)$. Eventually, the in-plane components of the second Piola-Kirchhoff stress tensor may be transformed into a vector by means of the kinetic Voigt rule as:

$$\mathbf{s}^T = \begin{pmatrix} S_{11} \ S_{22} \ S_{12} \end{pmatrix} \tag{54}$$

By substituting equations (53) and (54) back into (37) the vector of global internal forces may be rewritten in an easier way as:

$$\mathbf{f}_{int} = \int_{V^{pret}} \mathbf{B}^T \boldsymbol{\phi}_{int} dV \qquad \boldsymbol{\phi}_{int} = s_i \mathbf{h}_i + s_i \mathbf{H}_i \mathbf{g} \tag{55}$$

where summation is implied for repeated indices according to Einstein's notation. By proceeding in the same manner, the contributions to the generic total tangent stiffness matrix $\mathbf{K}^{mat}$ and $\mathbf{K}^{geo}$ may be obtained as:

$$\mathbf{K}^{mat} = \int_{V^{pret}} \mathbf{B}^T \mathbf{S}^{mat} \mathbf{B} dV \qquad \mathbf{K}^{geo} = \int_{V^{pret}} \mathbf{B}^T \mathbf{S}^{geo} \mathbf{B} dV \tag{56}$$

$$\mathbf{S}^{mat} = (\mathbf{h}_i + \mathbf{H}_i \mathbf{g}) C_{ij} (\mathbf{h}_j^T + \mathbf{g}^T \mathbf{H}_j) \qquad \mathbf{S}^{geo} = s_i \mathbf{H}_i \tag{57}$$

where the fourth order tensor of elastic moduli, C_{ij}, has been transformed into a 3x3 matrix by applying the Voigt rule vectorization procedure to equation (33) to come out with:

$$\mathbf{s} = \boldsymbol{\sigma}^{pret} + \mathbf{C}\mathbf{e} \tag{58}$$

In the next sections, two numerical examples will be presented with the purpose of demonstrating the capabilities of the described technique. The Finite Element Method was applied to study both a prestressed cable network and a prestressed membrane. As a consequence, the DCCF was particularized on two different finite elements: a two-noded and a three-noded linear isoparametric finite elements. The latter is described in some detail in the appendix located at the end of the chapter.

4 Numerical Example: Cable Network

This example, extracted from [23], represents a squared hyperbolic paraboloid with a 12 foot horizontal side. The difference between the highest and lowest points is measured as 1.5 feet. The sides of the structural model are composed of rigid beams which prevent any displacement.

The structure is modelled by means of a cable network depicted at Fig. 5, where a representation of nodes and elements can be observed. The external nodes are adopted to be fixed and the only allowed displacement is for the inner nodes.

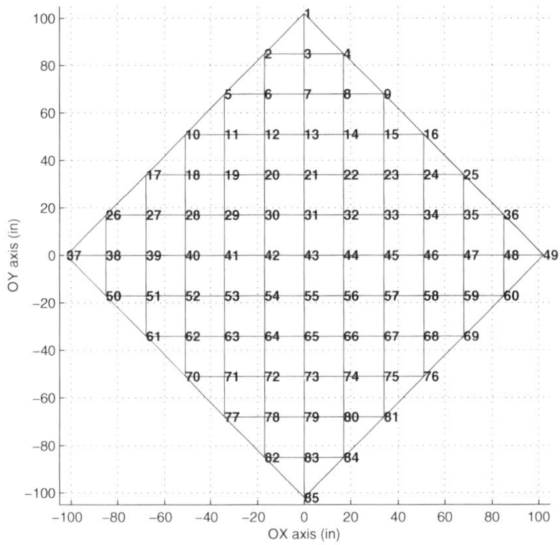

Fig. 5. Broughton: cable discretization.

All the cables are subjected to a previous prestressed state of 1.2 Kip. In addition to that, the cables will be represented by a unique material of 0.618e3 Kip for the mechanical factor EA.

The structural model undergoes under service conditions a downward point load of value 0.22 Kip applied at node 79 -see figure (5)-. The analysis has been solved by using a Newton-Raphson numerical scheme in one incremental step.

Figs. 6 and 7 represent both original and displaced shapes. Analogously, Table 1 reflects a survey of displacements according to the different spatial directions for some remarkable internal nodes. Results according to reference [23] and present work are displayed.

Fig. 8 gathers the convergence curves for the numerical method which has been employed. Eventually, a representation of the Cauchy axial forces is shown in Fig. 9.

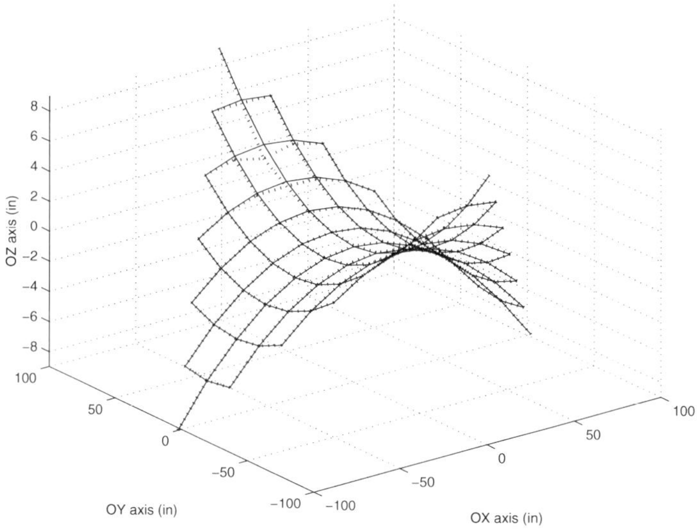

Fig. 6. Structural model: cable isometric view.

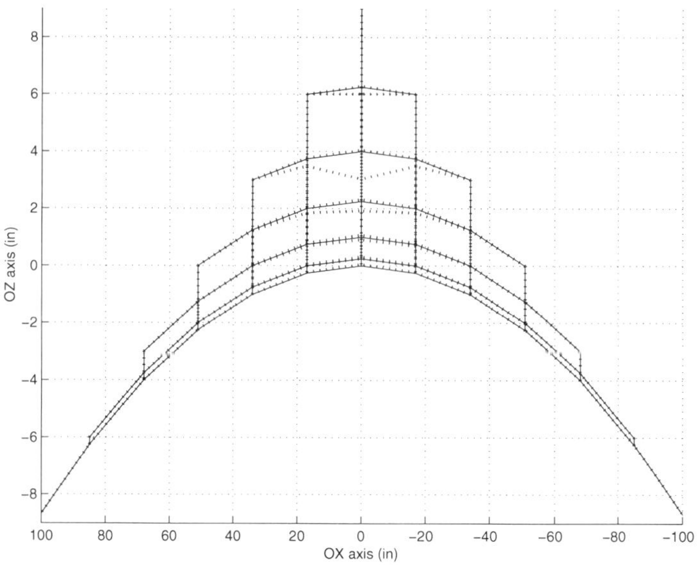

Fig. 7. Structural model: cable lateral view.

5 Numcrical Example: Membrane

This example is considered in [24]. It is a square plane membrane initially prestressed. The edges of the membrane are completely fixed. The side length is 240 in and the thickness is measured as 0.004167 in. Mechanical properties for the ma-

	Broughton			Present work		
Node	u	v	w	u	v	w
83	0.0000	-0.0334	-0.2410	0.0000	-0.0340	-0.2415
79	0.0000	-0.1390	-0.9650	0.0000	-0.1388	-0.9667
43	0.0000	-0.0410	0.0510	0.0000	-0.0414	0.0516
23	0.0000	-0.0020	0.0160	-0.0002	-0.0016	0.0155

Table 1. Displacements (in).

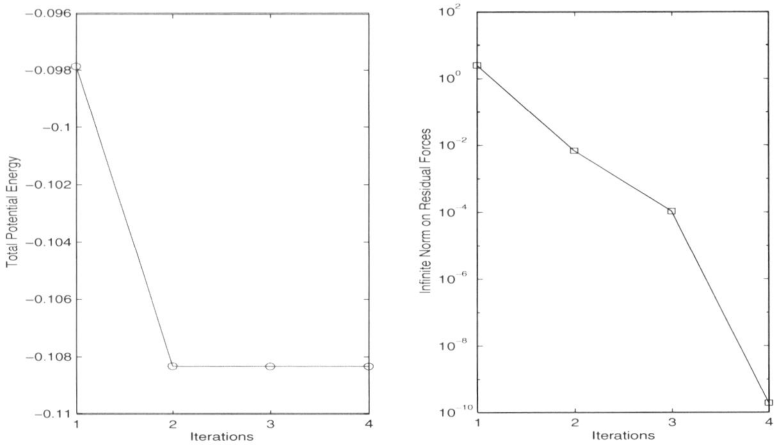

Fig. 8. Convergence curves.

terial are adopted to have the following values: 30,000 Ksi for the Young modulus and 0.3 for the Poisson ratio. The prestressing effect is considered to be 80,000 psi isotropically distributed. The mesh adopted is the same as the one employed in the original reference for the sake of comparison purposes. More accurate results would have been achieved with a finer mesh.

In service loading conditions are taken to be a point transverse load applied in middle. The load which takes a value of 10 Kip is taken downwards. According to the Fig. 10, the Lagrangian mesh is comprised of 32 isoparametric three-node linear elements and 25 nodes.

To accomplish the final solution, the employed method was the Newton-Raphson method for a single load increment. Two convergence curves are gathered in Fig. 11. The first shows the evolution of the total potential energy along the iterations path, whilst the second represents the infinite norm on residual forces vector with respect to the number of iterations as well. The second curve shows perfectly the required quadratic convergence of the Newton-Raphson algorithm.

Fig. 12 shows the displacement field along the cartesian axis OX and OZ, respectively. Both representations reveal in a clear manner the axisymmetry of the membrane. Table 2 details the displacement values for three different nodes of the mesh. The accuracy of the numerical example can be observed by checking the results with those obtained in [24].

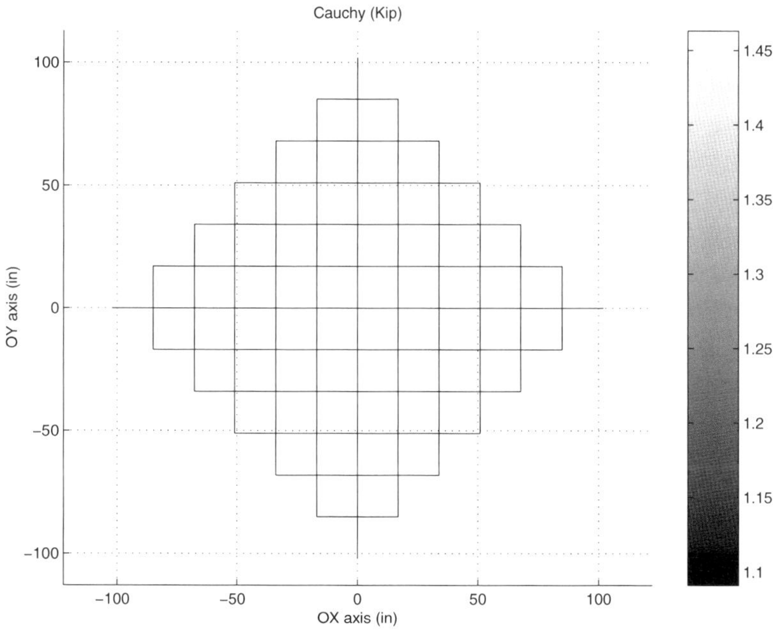

Fig. 9. Axial forces(Kip).

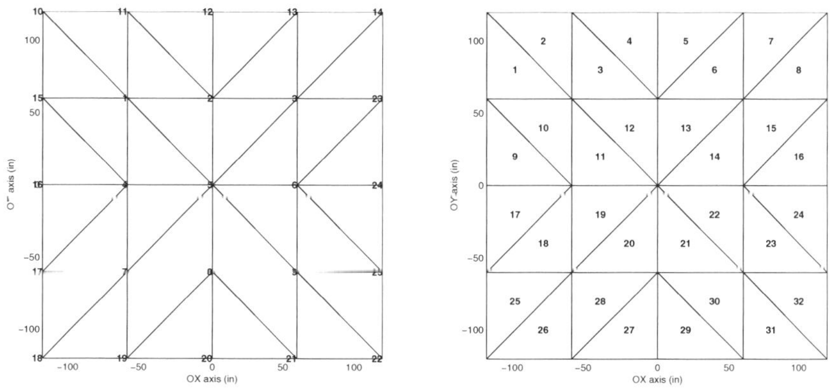

Fig. 10. Numerical example: Discretization.

Analogously, principal Cauchy stresses σ_I and σ_{II} can be viewed in Fig. 13. Table 3 presents the numerical values for three different elements of the membrane as well as its comparison with those of [24]. Perfect agreement can be deduced.

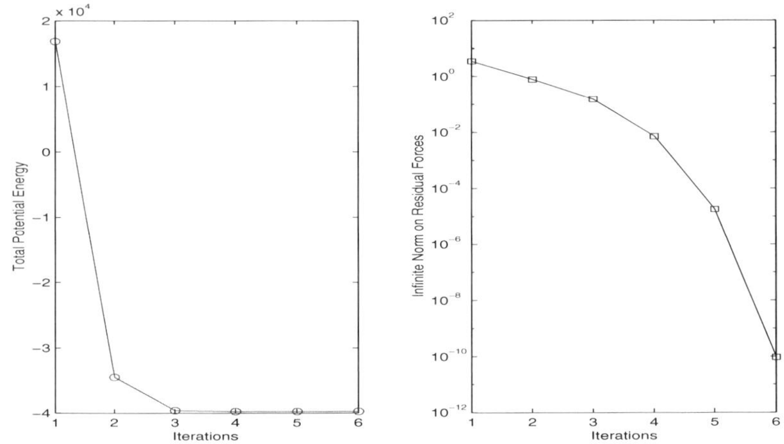

Fig. 11. Numerical example: Convergence curves.

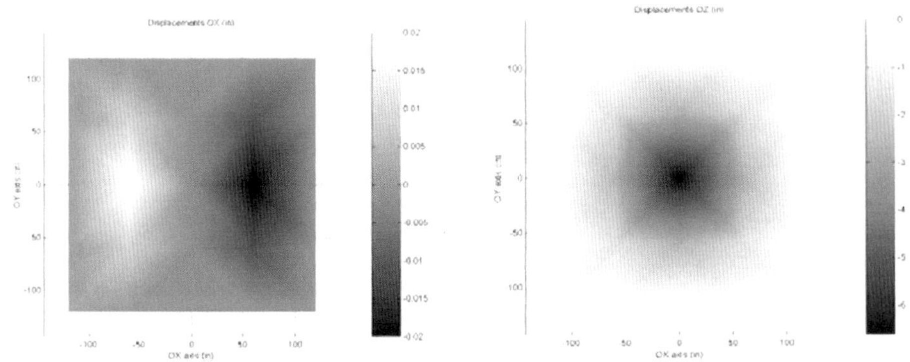

Fig. 12. Numerical example: Displacements OX & OZ.

	Levy & Spillers			Present work		
Node	u	v	w	u	v	w
1	0.015	-0.015	-1.431	0.014	-0.014	-1.423
2	0.000	-0.017	-2.605	0.000	-0.017	-2.600
5	0.000	0.000	-6.642	0.000	0.000	-6.626

Table 2. Numerical example: Displacements (in).

6 Concluding Remarks

This chapter has offered a complete consistent numerical formulation for the structural analysis of prestressed hyperelastic Saint Venant-Kirchoff membranes. Two structural loading steps can be considered successively: in the first, the prestressing loads are transferred to the structure whereas in the second, in service loads such as live loads or wind load arise. By taking into account this approach, an initial

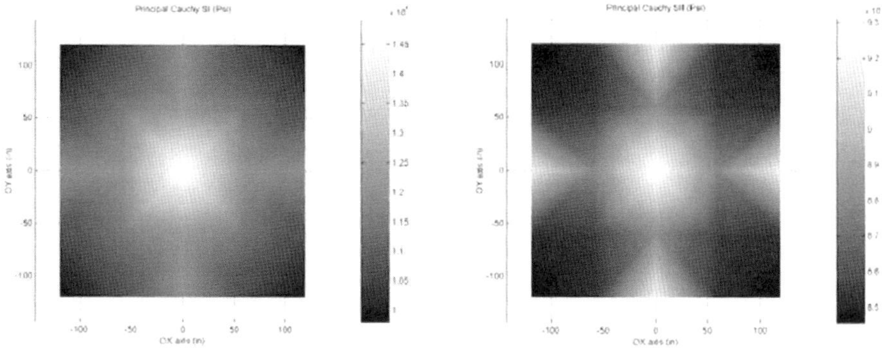

Fig. 13. Numerical example: Cauchy stresses σ_I & σ_{II}.

	Levy & Spillers			Present work		
Element	σ_{xx}	σ_{yy}	σ_{xy}	σ_{xx}	σ_{yy}	σ_{xy}
1	97377.6	85212.4	-2801.5	97300.1	85163.9	-2796.7
3	83510.2	96859.1	-8657.1	83501.5	96830.3	-8630.7
11	144691.0	97830.7	-15615.6	144470.8	97849.2	-15582.4

Table 3. Numerical example 1: Cauchy stresses (Psi).

prestressed configuration can be used to refer all the in service conditions back to it. A Total lagrangian Formulation was developed because of its implicit features. Afterwards, a finite element spatial discretization scheme along with a linearization or perturbation technique were employed over the equivalent internal nodal forces to derive the total tangent stiffness matrix. A straightforward computational implementation is allowed by means of the Direct Core Congruential Formulation. Eventually, the robustness of the whole formulation has been demonstrated by the convergence achievement of some examples.

7 Acknowledgments

I would like to emphasize my gratitude to Professor J. Bonet and Dr. Richard D. Wood, at the Civil and Computational Engineering Centre, University of Swansea, for many helpful comments and discussions.

A.1 Appendix: Three-Node Isoparametric Linear Finite Element

To reduce the continuously distributed structural system to a discrete one, the membrane is approximated by a discrete model consisting of a finite number *elem* of flat three-node isoparametric elements with linear shape functions. The geometry of each element in the initial prestressed state is thus defined by a plane of uniform thickness t bounded by straight lines which intersects at three points called nodes.

A fixed local coordinate system $O\xi_1^{pret}\xi_2^{pret}\xi_3^{pret}$, in addition to the global reference frame, is established for each element. The behaviour of each element is to be first described independently in terms of its local coordinates and is then to be transformed into global coordinates. For the sake of simplicity, it is assumed that each element lies in the $O\xi_1^{pret}\xi_2^{pret}$ plane of its local coordinate system. The shape functions are evaluated to obtain:

$$N^I(\xi_1^{pret},\xi_2^{pret}) = \frac{1}{2\Gamma^{pret}}(a^I + b^I\xi_1^{pret} + c^I\xi_2^{pret})$$

$$a^I = \xi_1^{pret^J}\xi_2^{pret^K} - \xi_1^{pret^K}\xi_2^{pret^J} \quad b^I = \xi_2^{pret^J} - \xi_2^{pret^K} \quad c^I = \xi_1^{pret^K} - \xi_1^{pret^J}$$

$$\Gamma^{pret} = \tfrac{1}{2}(c^K b^J - c^J b^K) \qquad I,J,K = 1,2,3 \tag{A.1}$$

where Γ^{pret} is the area of the initial prestressed triangle. Note that the quantities a^I, b^I and c^I -named Zienkiewicz's coefficients, see for instance [24]- are independent of the deformation of the membrane and are computed directly from the geometry of the initially prestressed shape. According to the Direct Core Congruential Formulation, the vectorized displacement gradient tensor is given in its transpose form as:

$$\mathbf{g}^T = \begin{pmatrix} g_1 & g_2 & g_3 & g_4 & g_5 & g_6 & g_7 & g_8 & g_9 \end{pmatrix}$$
$$\mathbf{g}^T = \begin{pmatrix} \frac{b^I u_1^I}{2\Gamma^{pret}} & \frac{b^I u_2^I}{2\Gamma^{pret}} & \frac{b^I u_3^I}{2\Gamma^{pret}} & \frac{c^I u_1^I}{2\Gamma^{pret}} & \frac{c^I u_2^I}{2\Gamma^{pret}} & \frac{c^I u_3^I}{2\Gamma^{pret}} & 0 & 0 & 0 \end{pmatrix} \tag{A.2}$$

The vector of global internal forces is particularized for the triangular flat element by using the DCCF as:

$$\mathbf{f}_{int} = t\Gamma^{pret}\mathbf{B}^T\phi_{int} = \frac{t}{2}\begin{pmatrix} b^1\mathbf{I}_3 & c^1\mathbf{I}_3 \\ b^2\mathbf{I}_3 & c^2\mathbf{I}_3 \\ b^3\mathbf{I}_3 & c^3\mathbf{I}_3 \end{pmatrix}\phi_{int} \tag{A.3}$$

where:

$$\phi_{int} = s_i\mathbf{h}_i + s_i\mathbf{H}_i\mathbf{g} = \begin{pmatrix} s_1(1+g_1) + s_3 g_4 \\ s_1 g_2 + s_3(1+g_5) \\ s_1 g_3 + s_3 g_6 \\ s_2 g_4 + s_3(1+g_1) \\ s_2(1+g_5) + s_3 g_2 \\ s_2 g_6 + s_3 g_3 \end{pmatrix} \tag{A.4}$$

The total tangent stiffness matrix may be computed by means of the congruential transformation and thus the submatrix due to the contribution of the nodes I and J is depicted as:

$$\mathbf{K}^{IJ} = \frac{t}{4\Gamma^{pret}}(b^I b^J\mathbf{S}_{11} + c^I c^J\mathbf{S}_{22} + b^I c^J\mathbf{S}_{12} + c^I b^J\mathbf{S}_{21} \tag{A.5}$$

where:

$$\mathbf{S}_{ij} = \mathbf{S}_{ij}^{geo} + \mathbf{S}_{ij}^{mat} \qquad i,j = 1,2 \tag{A.6}$$

The geometrical and material components of the stiffness matrix $\mathbf{S}$ can be developed as follows:

$$\mathbf{S}^{geo} = \begin{pmatrix} \mathbf{S}_{11}^{geo} & \mathbf{S}_{12}^{geo} \\ \mathbf{S}_{21}^{geo} & \mathbf{S}_{22}^{geo} \end{pmatrix} \qquad \mathbf{S}^{mat} = \begin{pmatrix} \mathbf{S}_{11}^{mat} & \mathbf{S}_{12}^{mat} \\ \mathbf{S}_{21}^{mat} & \mathbf{S}_{22}^{mat} \end{pmatrix} \tag{A.7}$$

$$\mathbf{S}_{11}^{geo} = s_1 \mathbf{I}_3 \qquad \mathbf{S}_{11}^{mat} = C_{11}\mathbf{f}_1\mathbf{f}_1^T + C_{33}\mathbf{f}_2\mathbf{f}_2^T + C_{13}\mathbf{f}_1\mathbf{f}_2^T + C_{31}\mathbf{f}_2\mathbf{f}_1^T \tag{A.8}$$

$$\mathbf{S}_{12}^{geo} = s_3 \mathbf{I}_3 \qquad \mathbf{S}_{12}^{mat} = C_{13}\mathbf{f}_1\mathbf{f}_1^T + C_{32}\mathbf{f}_2\mathbf{f}_2^T + C_{12}\mathbf{f}_1\mathbf{f}_2^T + C_{33}\mathbf{f}_2\mathbf{f}_1^T \tag{A.9}$$

$$\mathbf{S}_{21}^{geo} = s_3 \mathbf{I}_3 \qquad \mathbf{S}_{21}^{mat} = C_{31}\mathbf{f}_1\mathbf{f}_1^T + C_{23}\mathbf{f}_2\mathbf{f}_2^T + C_{33}\mathbf{f}_1\mathbf{f}_2^T + C_{21}\mathbf{f}_2\mathbf{f}_1^T \tag{A.10}$$

$$\mathbf{S}_{22}^{geo} = s_2 \mathbf{I}_3 \qquad \mathbf{S}_{22}^{mat} = C_{33}\mathbf{f}_1\mathbf{f}_1^T + C_{22}\mathbf{f}_2\mathbf{f}_2^T + C_{32}\mathbf{f}_1\mathbf{f}_2^T + C_{23}\mathbf{f}_2\mathbf{f}_1^T \tag{A.11}$$

where s_i is the ith-component of the second Piola-Kirchfoff stress tensor in Voigt notation, $\mathbf{I}_3$ is the identity 3x3 matrix, C_{ij} is a component of the fourth order tensor of elastic moduli in Voigt notation and the vectors $\mathbf{f}_1$ and $\mathbf{f}_2$ constitute the first and second column of the deformation gradient tensor which can be expressed in their transpose forms as:

$$\mathbf{f}_1^T = \begin{pmatrix} 1+g_1 & g_2 & g_3 \end{pmatrix} \qquad \mathbf{f}_2^T = \begin{pmatrix} g_4 & 1+g_5 & g_6 \end{pmatrix} \tag{A.12}$$

References

1. J. Schlaich. On some recent lightweight structures. *Journal of the IASS*, 43(139):69–79, 2002.
2. K. Ishii. *Membrane Structures in Japan.* SPS Publishing Company, Tokio, 1995.
3. H. Berger. Form and function of tensile structures for permanent buildings. *Engineering Structures*, 21:669–679, 1999.
4. H. Berger. *Light Structures. Structures of light.* Birkhuser, 1996.
5. W. J. Lewis. Lightweight tension membranes: an overview. *Civil Engineering*, 126:171–181, 1998.
6. A. E. Green and W. Zerna. *Theoretical Elasticity.* Oxford University Press, 2nd edition, 1968.
7. J. W. Leonard. *Tension structures.* McGraw-Hill, New York, 1988.
8. F. Otto. *Tensile structures.* MIT Press, Cambridge, 1967.
9. S. Timoshenko and S. Woinowsky-Krieger. *Theory of plates and shells.* McGraw-Hill, New York, 1959.
10. M. A. Crisfield. *Non-linear finite element analysis of solids and structures*, volume 1: Essentials. John Wiley and Sons, 1991.
11. J. T. Oden and T. Sato. Finite strains and displacements of elastic membranes by the finite element method. *International Journal of Solids and Structures*, 3:471–488, 1967.
12. F. Grutmann and R. L. Taylor. Theory and finite element formulation of rubberlike membrane shells using principal stretches. *International Journal for Numerical Methods in Engineering*, 35:1111–1126, 1992.
13. E. A. Souza, D. Peric, and D. R. J. Owen. Finite elasticity in spatial description linearization aspects with 3-d membrane applications. *International Journal for Numerical Methods in Engineering*, 38:3365–3381, 1995.

14. B. Wu, X. Du, and H. Tan. A three-dimensional finite element analysis of membranes. *Computers and Structures*, 59(4):601–605, 1996.
15. R. L. Taylor. Finite element analysis of membrane structures. *Publication CIMNE*, 203:1–34, 2001.
16. J. Bonet, R. D. Wood, J. Mahaney, and P. Heywood. Finite element analysis of air supported membrane structures. *Computer methods in applied mechanics and engineering*, 190:579–595, 2000.
17. T. Belytschko, W. K. Liu, and B. Moran. *Nonlinear finite elements for continua and structures.* John Wiley and Sons, 2000.
18. G. A. Holzapfel. *Nonlinear solid mechanics: a continuum approach for Engineering.* John Wiley and Sons, 2000.
19. J. Bonet and R. D. Wood. *Nonlinear continuum mechanics for finite element analysis.* Cambridge University Press, 1997.
20. R. H. Mallett and P. V. Marcal. Finite element analysis of nonlinear structures. *Journal of the Structural Division*, 94(ST9):2081–2105, 1968.
21. S. Rajasekaran and D. W. Murray. Incremental finite elements matrices. *Journal of the Structural Division*, 99(ST12):2423–2438, 1973.
22. L. A. Crivelli and C. A. Felippa. A three-dimensional non-linear Timoshenko beam based on the core-congruential formulation. *Internationl Journal of Numerical Methods in Engineering*, 36:3647–3673, 1993.
23. P. Broughton and P. Ndumbaro. *The analysis of cable and catenary structures.* Thomas Teldford, London, 1994.
24. R. Levy and W. Spillers. *Analysis of geometrically nonlinear structures.* Chapman and Hall, 1995.

Equilibrium Consistent Anisotropic Stress Fields in Membrane Design

Kai-Uwe Bletzinger[1], Roland Wüchner[2] and Fernass Daoud[3]

[1] Lehrstuhl für Statik
Technische Universität München
80290 München, Germany
kub@bv.tum.de

[2] Lehrstuhl für Statik
Technische Universität München
80290 München, Germany
wuechner@bv.tum.de

[3] Lehrstuhl für Statik
Technische Universität München
80290 München, Germany
daoud@bv.tum.de

Key words: form finding, anisotropic surface stress, minimal surface, mesh control.

Abstract. *This paper deals with the control of mesh distortions which may appear during the form finding procedure of membrane design. The reason is the unbalance of surface stresses either due to the interaction of edge and surface or incompatibilities along the sewing lines of adjacent membrane patches. An approach is presented which is based on a rational modification of the surface tension field. The criterion is based on the control of the element distortion and derived from differential geometry. Several examples demonstrate the success of the method.*

1 Introduction

Two different lines of research have developed which deal with the generation of structural shapes: the fields of "form finding" and "structural optimization" , respectively. The methods of form finding are usually restricted to tensile structures (cables and membranes) whereas the methods of structural optimization are far more general and can usually be applied to any kind of structure. So far, The differences of the two approaches are not only the level

E. Oñate and B. Kröplin (eds.), Textile Composites and Inflatable Structures, 143–151.

of specialization but also their aims. Form finding methods are designed to determine structural shapes from an inverse formulation of equilibrium and are derived from the simulation of physical phenomena of soap films and hanging models. In the case of soap films the structural shape is defined by the equilibrium geometry of a prescribed field of tensile surface stresses. It is well known that the shapes related to isotropic surface stresses are minimal surfaces which have minimal surface area content within given edges. Minimal surfaces have the additional property of zero mean curvature, or, with respect to pneumatically loaded surfaces of constant mean curvature.

Using a variational approach for the solution we realize that only those shape variations are meaningful which result in a variation of the area content. In other words, the variation of the position of any point on the surface must have a component normal to the surface. A variation of the position along the surface will not alter the area content. That means, that if a finite element method is used to solve the problem and the surface is discretized by a mesh of elements and nodes the stiffness with respect to a movement of the nodes tangential to the surface vanishes. This problem is well known since long from shape optimal design also where the design parameters must be chosen such that their modification must have an effect on the structural shape. Shape optimal design is controlled by the modification of the boundary.

There exist several remedies. Two techniques have been accepted as state of the art in shape optimal design: (i) linking the movement of internal nodes to key nodes using mapping techniques from CAGD, the so called design element technique, and, (ii) defining move directions for nodes on the boundary to guarantee relevant shape modifications. The positive side effect is that the number of optimization variables can drastically be reduced by this approach which is very attractive for optimal design. On the other hand, however, the space of possible shapes is also reduced. That is unacceptable if a high variability of shape modification is needed, as e.g. for the shape design of tensile structures or for the problem to find shapes of equilibrium of tension fields at the surface of liquids or related fields. Then methods are needed which are able to stabilize the nodal movement such that all three spatial coordinates of any finite element node may be variables in the shape modification process. For the special case of form finding of tensile structures the updated reference strategy (URS) is designed to find the shape of equilibrium of pre-stressed membranes. It is a general approach which can be applied to any kind of special element formulation (membranes or cables). A stabilization term is used which fades out as the procedure converges to the solution. The method is based on the specific relations of Cauchy and 2nd Piola-Kirchhoff stress tensors which appear to be identical at the converged solution [5], [8], [6]. Other alternative stabilization approaches have the same intention but used other methodologies, one may find many references in [4], [7], [3], [1].

All methods, however, will have problems or even fail if they are applied to physically meaningless situations without solution. E.g. it is not sure if a minimal surface exists for a given edge, or, a practical question from tent design, it is practically impossible to a priori satisfy equilibrium along the common edge of membrane strips which are anisotropically pre-stressed in different directions. The unbalance of stresses can be detected by a cumulative distortion of the FE-mesh during iteration. Methods are needed to control the mesh distortion by adapting the stress distribution. The present approach defines a local criterion to modify the pre-stress such that the element distortion is controlled. Put into the context of the updated reference strategy it appears to converge to homogeneous meshes during the regular time of iteration needed by the form finding procedure. The additional numerical effort is negligible. The results are surfaces of balanced shape representing equilibrium as close as possible at the desired distribution of stresses. The approach will be extended to other situations where mesh control might be necessary, e.g. in general shape optimal design or other surface tension problems [2].

2 The Updated Reference Strategy

The basic idea of URS will be briefly shown to understand the following chapters. A detailed description is given in [5]. Suppose one wants to find the equilibrium shape of a given surface tension field $\boldsymbol{\sigma}$. The solution is defined by the stationary condition

$$\delta w = \int_a \boldsymbol{\sigma} : \delta \mathbf{u}_{,\mathbf{x}} \; da = \int_A det\mathbf{F}(\boldsymbol{\sigma} \cdot \mathbf{F}^{-T}) : \delta \mathbf{F} \; dA = 0 \qquad (1)$$

which represents the vanishing virtual work of the Cauchy stresses $\boldsymbol{\sigma}$ and $\delta \mathbf{u}_{,\mathbf{x}}$ is the spatial derivative of the virtual displacements, $\mathbf{F}$ is the deformation gradient, a and A are the surface area of the actual and the reference configuration, respectively. As mentioned in the introduction the direct discretization of (1) w.r.t. all spatial coordinates will give a singular system of equations. The problem is stabilized by blending with the alternative formulation of (1) in terms of 2nd Piola-Kirchhoff stresses $\mathbf{S}$:

$$\delta w = \lambda \int_A det\mathbf{F}(\boldsymbol{\sigma} \cdot \mathbf{F}^{-T}) : \delta \mathbf{F} \; dA + (1-\lambda) \int_A (\mathbf{F} \cdot \mathbf{S}) : \delta \mathbf{F} \; dA = 0 \qquad (2)$$

where the coefficients of $\mathbf{S}$ are assumed to be constant and identical to those of $\boldsymbol{\sigma}$. The continuation parameter λ must be chosen small enough, even zero is possible. As the solution of (2) is used as the reference configuration for a following analysis the procedure converges to the solution of (1). Then $\mathbf{F}$ becomes the identity and both stress tensors are identical. The stabilization fades out.

3 Equilibrium of Surface Stresses

As also mentioned in the introduction it is not always possible to find a shape of equilibrium for each combination of edge geometry and surface stress distribution, isotropic or anisotropic. In these cases the size of some elements will steadily increase during iteration although URS is designed to control the mesh quality. That is because it is not possible to satisfy equilibrium of the given surface stresses at the location of those elements. For example consider a catenoid, Fig. 1. If the height exceeds the critical value the surface will collapse as shown. During iteration that is indicated by the increasing length of the related elements. To avoid the collapse the meridian stresses should be increased.

Now, consider the catenoid with an constant anisotropic stress field with larger meridian stresses, Fig. 2. The surface does not collapse anymore, but still the elements at the top get out of control as indicated by the tremendous increase of size. Again, the physical explanation is that equilibrium cannot be found with a constant distribution of meridian stresses over the height.

Fig. 1. Collapsed catenoid with different end rings.

Fig. 2. Catenoid with anisotropic pre-stress (meridian/ring=3/1).

A similar situation arises in membrane design. Typically those structures are sewed together by several strips, each of them made of anisotropic material and anisotropically pre-stressed. Along the common line there exists also an intrinsic unbalance of equilibrium of stresses which must be handled with during the form finding procedure. All cases demand for automatically adjusted stresses.

4 Element Size Control

The element size is used as indicator to adjust the surface stress field. At each Gauss point of the element discretization an additional constraint is intro-

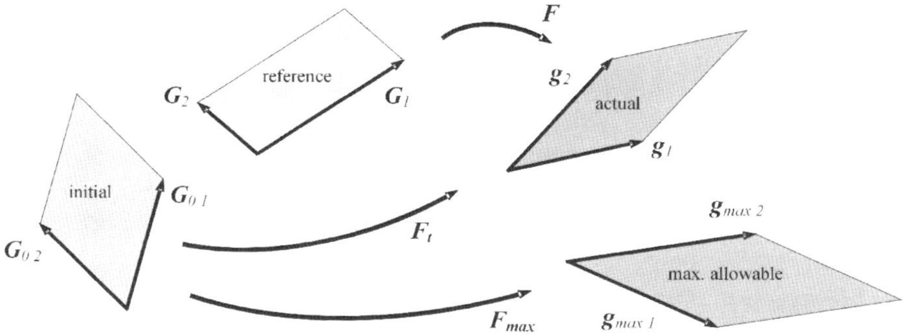

Fig. 3. Configurations during form finding.

duced to check for the length of base vectors against the allowable maximal deformation. The standard optimization approach by the Lagrangian multiplier technique is prohibitive because of the size of the problem. Therefore, a local criterion is defined which can be solved at each Gauss point independently. The approach is analogous to what is done in elastic-plastic analysis. In contrast to that, now, the surface stresses are constant until the critical deformation is reached and will then be adjusted to the further change of deformation.

Consider the following configurations and the related definition of base vectors: the initial configuration, $\mathbf{G_0}$, where the form finding was started, the reference configuration, $\mathbf{G}$, defined by the update procedure of URS, and the actual configuration, $\mathbf{g}$, as the state of equilibrium of the current time step. An additional configuration, $\mathbf{g}_{max}$ is defined which states the maximum allowable element size. There are several deformation gradients $\mathbf{F}$ defined which map differential entities between the configurations, Fig. 3.

The deformation gradient $\mathbf{F}_t^{(k)}$in time step k which describes the total deformation is multiplicatively created by $\mathbf{F}_t^{(k)} = \mathbf{F} \cdot \mathbf{F}_t^{(k-1)}$ where $\mathbf{F} = \mathbf{g}_i \otimes \mathbf{G}^{i(k)}$ and $\mathbf{F}_t^{(k)} = \mathbf{g}_i^{(k)} \otimes \mathbf{G}_0^i$.

If the actual deformation exceeds the allowable limit, i.e. $\| \mathbf{F}_t^{(k)} \| > \| \mathbf{F}_{max} \|$, a modified surface stress tensor $\boldsymbol{\sigma}_{mod}$ is generated by the following rule of nested pull back and push forward operations:

1. apply the surface Cauchy stress $\boldsymbol{\sigma}$ to the "max. allowable" configuration
2. determine the related 2nd Piola-Kirchhoff stress $\mathbf{S}_0$ w.r.t. to the initial configuration by pull back using $\mathbf{F}_{max}$
3. push $\mathbf{S}_0$ forward to the actual configuration applying $\mathbf{F}_t$

The resulting operation is:

$$\boldsymbol{\sigma}_{mod} = \frac{det\,\mathbf{F}_{max}}{det\,\mathbf{F}_t} \mathbf{F}_t \cdot \mathbf{F}_{max}^{-1} \cdot \boldsymbol{\sigma} \cdot \mathbf{F}_{max}^{-T} \cdot \mathbf{F}_t^T \tag{3}$$

and, considering the 2nd Piola-Kirchhoff stress tensor $\mathbf{S}$ (used in the stabilizing term of URS):

$$\boldsymbol{\sigma}_{mod} = \frac{det\,\mathbf{F}_{max}}{det\,\mathbf{F}\ det\,\mathbf{F}_t}\mathbf{F}_t \cdot \mathbf{F}_{max}^{-1} \cdot \mathbf{F} \cdot \mathbf{S} \cdot \mathbf{F}^T \cdot \mathbf{F}_{max}^{-T} \cdot \mathbf{F}_t^T \tag{4}$$

The "max. allowable" configuration is defined by multiples of the base vectors of the actual configuration:

$$\mathbf{g}_{max\ i} = \beta_{(i)}\mathbf{g}_i \tag{5}$$

Then, we can determine the related modified deformation gradient $\mathbf{F}_{mod}$

$$\mathbf{F}_{mod} = \mathbf{F}_t \cdot \mathbf{F}_{max}^{-1} \cdot \mathbf{F} = \frac{1}{\beta_{(i)}}\mathbf{g}_i \otimes \mathbf{G}^i \tag{6}$$

The change of element size is determined by tracing the length change of base vectors during the deformation process. Introducing the ratio $\alpha_i = \| \mathbf{g}_{(i)} \| \,/\, \| \mathbf{G}_{(i)} \|$ (no summation) the total change of geometry in time step k is defined as: $\alpha_{ti}^{(k)} = \alpha_{(i)}\alpha_{t(i)}^{(k-1)}$. If the total change of geometry is larger than the allowable maximum $\alpha_{max\ i}$ then the constraint is active. The factor β_i which is used to define the "max. allowable" configuration can now be determined as:

$$\begin{aligned} \text{if} \quad \alpha_{ti}^{(k)} > \alpha_{max\ (i)} \quad \text{then :} \quad & \alpha_{max\ (i)} = \alpha_{t(i)}^{(k)}\ \beta_i \\ \therefore \quad & \beta_i = \frac{\alpha_{max(i)}}{\alpha_{t(i)}^{(k)}} \end{aligned} \tag{7}$$

Considering both surface base vectors $\mathbf{g}_1$ and $\mathbf{g}_2$ the determinant of $\mathbf{F}_{mod}$ is now

$$det\,\mathbf{F}_{mod} = (\beta_1\beta_2)^{(-1)}\ det\,\mathbf{F} \tag{8}$$

If constraints are active the state of stresses must change with time and the modified Cauchy stress tensor $\boldsymbol{\sigma}_{mod}^{(k+1)}$ for the next time step $k+1$ which is acting in the "max. allowed" configuration follows as:

$$\boldsymbol{\sigma}_{mod}^{(k+1)} = \frac{\beta_1\beta_2}{det\,\mathbf{F}}\mathbf{F}_{mod} \cdot \mathbf{S}^{(k)} \cdot \mathbf{F}_{mod}^T \tag{9}$$

In the context of URS the components of $\boldsymbol{\sigma}_{mod}^{(k+1)}$ will also be used as the components of $S^{(k+1)}$ in the next time step. They are:

$$\begin{aligned} \sigma_{mod}^{11(k+1)} &= \frac{\beta_2}{\beta_1\ det\,\mathbf{F}}S^{11(k)} \\ \sigma_{mod}^{22(k+1)} &= \frac{\beta_1}{\beta_2\ det\,\mathbf{F}}S^{22(k)} \\ \sigma_{mod}^{12(k+1)} &= \frac{1}{det\,\mathbf{F}}S^{12(k)} \end{aligned} \tag{10}$$

The modification rule (10) simply says that the surface stress of an element which became too long must be increased to shorten it and vice versa.

5 Examples

5.1 Catenoid

This example shows the effect of different values of α_{max} on the final shape of a catenoid which was initially isotropically pre-stressed. The choice of α_{max} controls the maximal deformation of the elements. A value of 1.0 means almost no element deformation. Compare with Fig. 1 how collapse could be omitted.

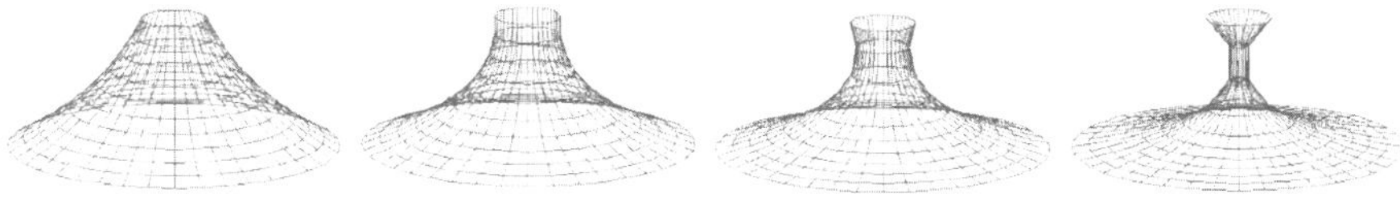

(a) $\alpha_{max} = 1.0$ (b) $\alpha_{max} = 1.2$ (c) $\alpha_{max} = 1.5$ (d) $\alpha_{max} = 2.0$

Fig. 4. Stabilized catenoid of initially isotropic surface stress.

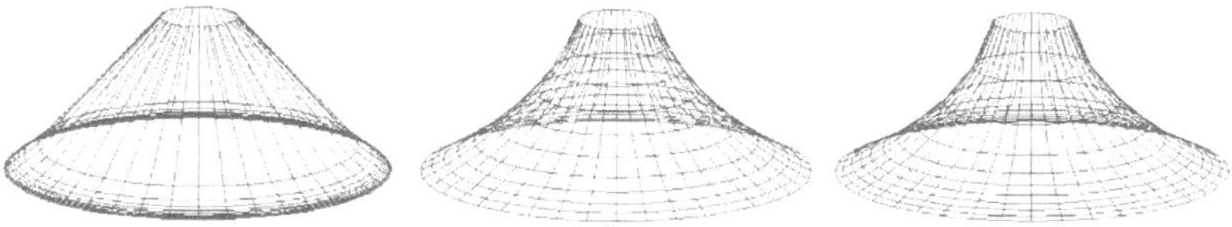

(a) no constraint (b) $\alpha_{max} = 1.2$ (c) $\alpha_{max} = 2.0$

Fig. 5. Initially anisotropically pre-stressed catenoid (meridian/ring=1.2/1).

5.2 Tent Hüfingen

This tent is composed by 5 membrane patches which are anisotropically pre-stressed towards the center of the tent. A top view shows the consequences of unbalanced stresses at the common edges of adjacent patches where the mesh is distorted, Fig. 6, left. The mesh quality is maintained using a value of $\alpha_{max} = 1.0$, Fig. 6, right. Fig. 7 shows the generated shapes using different values for the mesh control factor.

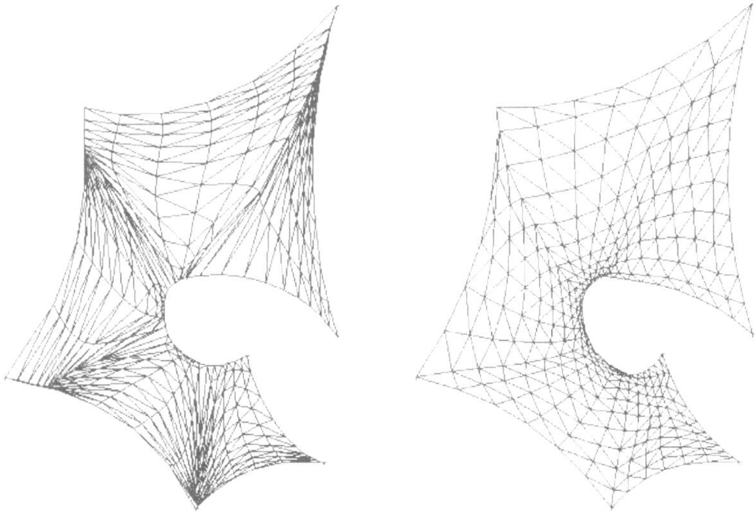

Fig. 6. Top view after form finding, without or with mesh control (left, right).

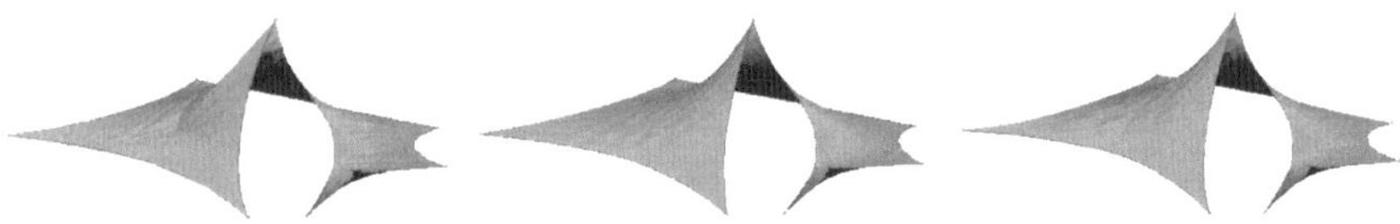

(a) no mesh control (b) $\alpha_{max} = 1.0$ (c) $\alpha_{max} = 1.5$

Fig. 7. Effects of varied mesh control

6 Conclusion

Motivated by the specific problems of form finding an approach was developed to control the mesh quality of the finite element discretization. The idea of the method is to adjust the value of the applied surface stresses by a simple scaling of the stress components which is derived from geometric considerations. The method is very effective without additional effort. Several examples demonstrate the success. Further developments are directed towards the generalization of the approach for general problems of shape optimal design where similar situations occur.

References

1. B. Maurin and R. Motro. Cutting pattern of fabric membranes with the stress composition method. *Int. Journal of space structures*, 14:121–130, 1999.
2. D. Peric and P. H. Saksono. On finite element modeling of surface tension: variational formulation and applications. In W. Wall, K.-U. Bletzinger, and K. Schweizerhof, editors, *Trends in Computational Structural Mechanics*, pages 731–740. CIMNE, Barcelona, 2001.
3. K. Ishii. Form finding analysis in consideration of cutting patterns of membrane structures. *Int. Journal of space structures*, 14:105–120, 1999.
4. K. Linkwitz. Formfinding by the Direct Approach and pertinent strategies for the conceptual design of prepressed and hanging structures. *Int. Journal of space structures*, 14:73–88, 1999.
5. K.-U. Bletzinger and E. Ramm. A general finite element approch to the form finding of tensile structures by the updated reference strategy. *Int. Journal of space structures*, 14:131–146, 1999.
6. K.-U. Bletzinger and R. Wüchner. Formfinding of anisotropic pre-stressed membrane structures. In W. Wall, K.-U. Bletzinger, and K. Schweizerhof, editors, *Trends in Computational Structural Mechanics*, pages 595–603. CIMNE, Barcelona, 2001.
7. M. Barnes. Form finding and analysis of tension structures by dynamic relaxation. *Int. Journal of space structures*, 14:89–104, 1999.
8. R. Wüchner. Formfindung von Membrantragwerken unter Berücksichtigung anisotroper Vorspannung. Master's thesis, Lehrstuhl für Statik, TU München, 2000.

Efficient Finite Element Modelling and Simulation of Gas and Fluid Supported Membrane and Shell Structures

T. Rumpel, K. Schweizerhof and M. Haßler[1]

Institute of Mechanics, University Karlsruhe (TH), Kaiserstr. 12, 76128 Karlsruhe, Germany, Karl.Schweizerhof@bau-verm.uni-karlsruhe.de

Summary. In statics, the large deformation analysis of membrane or shell structures loaded and/or supported by gas or fluid can be based on a finite element description for the structure only. Then in statics the effects in the gas or the fluid have to be considered by using the equations of state for the gas or the fluid, the information about the current volume and the current shape of the structure. The interaction of the gas/fluid with the structure, which can be also otherwise loaded, is then modelled by a pressure resulting from the gas/fluid always acting normal to the current wetted structural part. This description can be also directly used to model slow filling processes without all the difficulties involved with standard discretization procedures. In addition the consistent derivation of the nonlinear formulation and the linearization for a Newton type scheme results in a particular formulation which can be cast into a very efficient solution procedure based on a sequential application of the Sherman-Morrison formula. The numerical examples show the efficiency and the effects of the developed algorithms which are particularly important for structures, where the volume of the gas of fluid has to be considered.

Key words: Pressure loading, hydrostatics, large deformations, finite elements, deformation dependent loading, membranes, shell structures

1 Introduction

The simulation of the inflation resp. filling and the support of thin membrane or shell type structures by gas or fluids can be usually performed in an efficient way by assuming an internal pressure in the structure which acts normal to the inner surface [2], [3], [12] besides any other loading. The restriction of this model is that it does not take into account the change of the volume of the gas or fluid due to the deformation of the structure even if there is no further inflation or filling. Also the pressure may change due to temperature modifications of the gas/fluid. In both cases the volume of the gas resp. the mass conservation of the fluid has to be considered in the model [1], [7]; this

E. Oñate and B. Kröplin (eds.), Textile Composites and Inflatable Structures, 153–172.

is also important for stability considerations of the gas/fluid filled structure under any other external loading. Then in the case of gas filling the internal gas pressure formally provides an additional rank-one update of the FE stiffness matrix which stabilizes from an engineering point of view an almost completely flexible structure. Hydrostatics with a free fluid surface, also leads to an additional rank-one update of the FE stiffness matrix, whereas compressible, heavy fluids lead to a rank-two update [5].

Thus in the case of fluid filling [6], [8], [10] or a mixture of gas and fluid [9] the filling of membrane-like structures can be performed without a separate discretization of the fluid with e.g. FE or Finite Volumes or similar. Also fully fluid filled structures can be analyzed without separate discretization of the fluid [9]. Several cases have to be distinguished, fluid with a free surface [8], [10], structure under overpressure of fluid [9] and fluid with free surface but gas overpressure. The contribution shows the derivation of the variational formulation and the corresponding Finite Element discretization for compressible fluids under gravity loading. A particular focus is on the consistent linearization of the nonlinear equations and the accompanying constraint equations. Also the specific solution of the linearized equation system based on a sequential application of the Sherman-Morrison formula is presented. The numerical examples show large deformation analyses of gas and fluid filled shell structures with rather thin flexible walls under various conditions, such as filling and loading.

2 Governing Equations

The mathematical description of static fluid structure interaction can be based on the principle of stationarity for the total potential energy δW of a fluid in an elastic structure and additional equations describing the physical behavior of different fluids or gases.

2.1 Virtual Work Expression

The variation of the elastic potential of the structure is specified by $\delta^{el} V$, $\delta^i \Pi$ denotes the virtual work of the pressure loading which acts between the fluid i and the structure, $\delta^{ex} \Pi$ is the virtual work of other external forces acting on the structure

$$\delta W = \delta^{el} V + \delta^i \Pi - \delta^{ex} \Pi = 0 \tag{1}$$

The interaction term between fluid and structure is described by a body fixed pressure force ${}^i p \, {}^* \mathbf{n}$, with a non-normalized normal vector ${}^* \mathbf{n} = \mathbf{e}_\xi \times \mathbf{e}_\eta$ and the pressure level ${}^i p$, see equation (2). $\mathbf{e}_\xi = \frac{\partial x}{\partial \xi}$, $\mathbf{e}_\eta = \frac{\partial x}{\partial \eta}$ denote covariant non-normalized vectors on the wetted surface of the structure

$$\delta^i \Pi = \int_\eta \int_\xi {}^i p \, {}^* \mathbf{n} \cdot \delta \mathbf{u} \, d\xi d\eta \tag{2}$$

The pressure acts normal to the surface element $d\xi d\eta$ along the virtual displacement $\delta\mathbf{u}$. Therefore a virtual work expression of a follower force is given. Possible physical properties of the fluid i are summarized in the following paragraphs:

2.2 Compressible Fluids

If the dead weight of a fluid is neglected, we can distinguish between a pneumatic model, see [1], [7] and a hydraulic description. The corresponding constitutive equations are the Poisson's law for a pneumatic ($i = p$) and the Hooke's law ($i = h$) for a hydraulic model.

Pneumatic Model

In realistic physical situations the investigations can be restricted to conservative models, which entails the application of the adiabatic state equation

$$ {}^{p}p\ v^{\kappa} = {}^{p}P\ V^{\kappa} = const. \tag{3} $$

${}^{p}p$, v are the state variables (pressure and volume) of the gas in the deformed state, capital letters denote the initial state and κ the isentropy constant.

Hydraulic Model

For an analysis of hydraulic systems the fluid pressure is given by Hooke's law. ${}^{h}p$ is the mean pressure in the fluid determined by the bulk modulus K and the relative volume change of the fluid with V as initial volume

$$ {}^{h}p(v) = -K\frac{v - V}{V} \tag{4} $$

2.3 Hydrostatic Loading – Incompressible Fluids under Gravity Loading

For partially filled structures the liquid can be treated as incompressible, see [6], [8], [10]. The pressure distribution is given by the hydrostatic pressure law, with ρ as the constant density, $\mathbf{g}$ as the gravity and with the difference of the upper liquid level ${}^{o}\mathbf{x}$ and an arbitrary point $\mathbf{x}$ on the wetted structure. A conservative description is achieved, if the volume conservation of the liquid is taken into account during the deformation of liquid and structure, too

$$ {}^{g}p = \rho\mathbf{g}\cdot({}^{o}\mathbf{x} - \mathbf{x}) \tag{5} $$

$$ \text{and} \quad v = const. \tag{6} $$

2.4 Compressible Hydrostatic Loading – Compressible Fluids Under Gravity Loading

A further important case is the composition of dead weight and compressibility of fluid ($i = hg$), see [5],[9]. The corresponding pressure law for technical applications can be found by combining Hooke's law and mass conservation with the assumption of an uniform density distribution throughout the fluid. The hydrostatic pressure law for compressible fluids can be derived from a variational analysis of the gravity potential and the virtual work expression of the pressure resulting from Hooke's law

$$
\begin{aligned}
{}^{hg}p &= {}^{c}p - {}^{x}p - {}^{h}p &\quad (7)\\
&= \rho(v)\mathbf{g}\cdot(\mathbf{c}-\mathbf{x}) - {}^{h}p &\quad (8)\\
\text{with} \quad \rho(v)v &= const. &\quad (9)
\end{aligned}
$$

${}^{c}p = \rho(v)\mathbf{g}\cdot\mathbf{c}$ is the pressure at the center $\mathbf{c}$ of volume, ${}^{x}p = \rho(v)\mathbf{g}\cdot\mathbf{x}$ denotes the pressure at an arbitrary point $\mathbf{x}$ on the wetted structure. In the view of a mesh-free representation of the fluid, the constitutive equations are dependent on the shape and on the volume of the gas or fluid enclosed by the structure or by parts of the structure. It must be noted that the term ${}^{c}p$ is responsible for the compression of the fluid due to its own dead weight.

2.5 Boundary Integral Representation of Volume and Center of Volume

The goal of this approach is that all necessary quantities can be expressed by a boundary integral representation. This allows to formulate all state variables via an integration of the surrounding wetted surface. The fluid volume v and the center $\mathbf{c}$ of the volume can be computed via:

$$v = \frac{1}{3}\int_{\eta}\int_{\xi} \mathbf{x}\cdot{}^{*}\mathbf{n}\; d\xi d\eta \qquad (10)$$

$$\text{and} \quad \mathbf{c} = \frac{1}{4v}\int_{\eta}\int_{\xi} \mathbf{x}\;\mathbf{x}\cdot{}^{*}\mathbf{n}\; d\xi d\eta \qquad (11)$$

A large deformation analysis of the structure including the fluid can be performed using a Newton type scheme for the solution by applying a Taylor series expansion on the governing equations. The following linearization is shown in short for all four cases discussed above. For details we refer to [1], [6], [7].

3 Linearization of the Volume Contribution for Gas and Fluid Models

Within the Newton scheme the deformed state is computed iteratively. Both, the virtual expression and the different additional constraint equations have to be consistently linearized. The linearization of the virtual work expression leads always to three parts, the residual part $\delta^{phg}\Pi_t$, the follower force part $\delta^{phg}\Pi^{\Delta n}$ and the pressure level part $\delta^{phg}\Pi^{\Delta p}$

$$\delta^{phg}\Pi_{lin} = \delta^{phg}\Pi_t + \delta^{phg}\Pi^{\Delta n} + \delta^{phg}\Pi^{\Delta p} \tag{12}$$

$$= \int_\eta \int_\xi ({}^{phg}p_t \, {}^*\mathbf{n}_t + {}^{phg}p_t \Delta^*\mathbf{n} + \Delta^{phg}p^*\mathbf{n}_t) \cdot \delta\mathbf{u} \, d\xi d\eta. \tag{13}$$

3.1 Pneumatic and Hydraulic Model

The follower force part is dependent on the structural displacements $\Delta\mathbf{u}$ respectively the change of the non-normalized normal ${}^*\mathbf{n}$, and thus indirectly on the size of the wetted surface with

$$\Delta^*\mathbf{n} = \Delta\mathbf{u}_{,\xi} \times \mathbf{x}_{t,\eta} + \mathbf{x}_{t,\xi} \times \Delta\mathbf{u}_{,\eta} \,. \tag{14}$$

The pressure change differs only slightly for both models and is only dependent on the volume change

$$\textbf{pneumatic model:} \quad \Delta^p p = -\kappa \frac{p_t}{v_t} \int_\eta \int_\xi {}^*\mathbf{n}_t \cdot \Delta\mathbf{u} \, d\xi d\eta \tag{15}$$

$$\textbf{hydraulic model:} \quad \Delta^h p = -\frac{K}{V} \int_\eta \int_\xi {}^*\mathbf{n}_t \cdot \Delta\mathbf{u} \, d\xi d\eta \quad \text{with } V \equiv v_0 \tag{16}$$

Introducing both into (13) and integrating by parts, we obtain a field and boundary valued problem. The boundary value part vanishes completely for closed structures respectively the parts enclosing the gas/fluid volume. Thus the linearized virtual expression reads:

$$\begin{aligned}
\delta^{p,h}\Pi_{lin} &= \delta^{p,h}\Pi_t \\
&- \kappa \frac{p_t}{v_t} \int_\eta \int_\xi \delta\mathbf{u} \cdot {}^* \mathbf{n}_t \, d\xi d\eta \int_\eta \int_\xi {}^*\mathbf{n}_t \cdot \Delta\mathbf{u} \, d\xi d\eta \qquad \text{pneumatic} \\
&- \frac{K}{V} \int_\eta \int_\xi \delta\mathbf{u} \cdot {}^* \mathbf{n}_t \, d\xi d\eta \int_\eta \int_\xi {}^*\mathbf{n}_t \cdot \Delta\mathbf{u} \, d\xi d\eta \qquad \text{hydraulic} \\
&+ \frac{{}^{p,h}p_t}{2} \int_\eta \int_\xi \begin{pmatrix} \delta\mathbf{u} \\ \delta\mathbf{u}_{,\xi} \\ \delta\mathbf{u}_{,\eta} \end{pmatrix} \cdot \begin{pmatrix} 0 & \underline{\mathbf{W}}^\xi & \underline{\mathbf{W}}^\eta \\ \underline{\mathbf{W}}^{\xi T} & 0 & 0 \\ \underline{\mathbf{W}}^{\eta T} & 0 & 0 \end{pmatrix} \begin{pmatrix} \Delta\mathbf{u} \\ \Delta\mathbf{u}_{,\xi} \\ \Delta\mathbf{u}_{,\eta} \end{pmatrix} d\xi d\eta
\end{aligned} \tag{17}$$

with the skew symmetric tensors

$$\underline{\mathbf{W}}^\xi = {}^* \mathbf{n}_t \otimes \mathbf{e}^\xi - \mathbf{e}^\xi \otimes {}^* \mathbf{n}_t \underline{\mathbf{W}}^\eta = {}^* \mathbf{n}_t \otimes \mathbf{e}^\eta - \mathbf{e}^\eta \otimes {}^* \mathbf{n}_t. \tag{18}$$

Obviously the final linearized expression is a symmetric displacement formulation indicating that the proposed model is conservative as expected.

3.2 Hydrostatic Loading – Incompressible Fluids Under Gravity Loading

The follower force part depends as in 3.1 from the change in the normal and of the gradient of the fluid under gravity loading. The latter comes into the formulation after partial integration resulting in

$$\delta^g \Pi_{lin}^{\Delta n} = \frac{\rho}{2} \int_\eta \int_\xi \delta\mathbf{u} \cdot [\mathbf{g} \cdot \mathbf{e}_\xi \underline{\mathbf{W}}^\xi + \mathbf{g}\mathbf{e}_\eta \underline{\mathbf{W}}^\eta] \, \Delta\mathbf{u} \, d\xi d\eta$$
$$+ \int_\eta \int_\xi \frac{{}^g p_t}{2} \begin{pmatrix} \delta\mathbf{u} \\ \delta\mathbf{u}_{,\xi} \\ \delta\mathbf{u}_{,\eta} \end{pmatrix} \cdot \begin{pmatrix} 0 & \underline{\mathbf{W}}^\xi & \underline{\mathbf{W}}^\eta \\ \underline{\mathbf{W}}^{\xi T} & 0 & 0 \\ \underline{\mathbf{W}}^{\eta T} & 0 & 0 \end{pmatrix} \begin{pmatrix} \Delta\mathbf{u} \\ \Delta\mathbf{u}_{,\xi} \\ \Delta\mathbf{u}_{,\eta} \end{pmatrix} d\xi d\eta. \quad (19)$$

It is obvious that the first part is non-symmetric and disappears if $\mathbf{g}$ is set to zero. The interesting part is the volume conservation and its influence on the pressure in the linearized form. The linearized pressure is a function of the variation of the fluid level $\Delta^0\mathbf{u}$ and the local structural deformation $\Delta\mathbf{u}$

$$\Delta^g p = \rho \mathbf{g} \cdot (\Delta^o \mathbf{u} - \Delta\mathbf{u}). \quad (20)$$

The volume change is zero thus the linearization is zero as well:

$$\Delta^g v = \int_\eta \int_\xi {}^*\mathbf{n}_t \cdot \Delta\mathbf{u} \, d\xi d\eta + \int_{^o\eta} \int_{^o\xi} {}^o\mathbf{n}_t \cdot \Delta^o\mathbf{u} \, d^o\xi d^o\eta = 0. \quad (21)$$

Focusing on the fluid load part – the second part in (21) – we obtain based on the direction of the normal on the fluid level, which is identical to the direction of gravity, the components of the free fluid surface and the corresponding displacement

$$^o n_t = {}^o\mathbf{n}_t \cdot \frac{\mathbf{g}}{|\mathbf{g}|}, \quad (22)$$
$$\Delta^o u = \Delta^o \mathbf{u} \cdot \frac{\mathbf{g}}{|\mathbf{g}|}. \quad (23)$$

Thus the volume change due to the change in the fluid level can be written as

$$\Delta^o v = \int_{^o\eta} \int_{^o\xi} {}^o\mathbf{n}_t \cdot \frac{\mathbf{g}}{|\mathbf{g}|} \Delta^o\mathbf{u} \cdot \frac{\mathbf{g}}{|\mathbf{g}|} \, d^o\xi d^o\eta. \quad (24)$$

Obviously both quantities in the integral are scalars; in addition the fluid level displacement is uniform, thus we obtain

$$\Delta^o v = \Delta^o u \int_{^o\eta} \int_{^o\xi} {}^o\mathbf{n}_t \cdot \frac{\mathbf{g}}{|\mathbf{g}|} \, d^o\xi d^o\eta = \Delta^o u S_t. \quad (25)$$

S_t is the size of the water surface, which can also be computed via a boundary integral over the enclosure of the fluid volume projected onto the direction of gravity

$$S_t = \int_\eta \int_\xi {}^*\mathbf{n}_t \cdot \frac{\mathbf{g}}{|\mathbf{g}|}\, d\xi d\eta. \tag{26}$$

Thus the change in the water level height can be written as

$$\Delta^o u = \frac{\Delta^o v}{S_t} = \frac{1}{S_t} \int_\eta \int_\xi {}^*\mathbf{n}_t \cdot \Delta\mathbf{u}\, d\xi d\eta \tag{27}$$

and the corresponding pressure change becomes

$$\begin{aligned} \Delta p &= \rho \mathbf{g} \cdot \Delta^o \mathbf{u} - \rho \mathbf{g} \cdot \Delta \mathbf{u} \\ &= \rho \frac{|\mathbf{g}|}{S_t} \int_\eta \int_\xi {}^*\mathbf{n}_t \cdot \Delta\mathbf{u}\, d\xi d\eta - \rho \mathbf{g} \cdot \Delta\mathbf{u}. \end{aligned} \tag{28}$$

The linearized variational form of the gravity potential depending on the fluid level is then obtained as

$$\begin{aligned} \delta^g \Pi_{lin}^{\Delta p} &= \int_\eta \int_\xi \Delta p\, {}^*\mathbf{n}_t \cdot \delta\mathbf{u}\, d\xi d\eta \\ &= \rho \frac{|\mathbf{g}|}{S_t} \int_\eta \int_\xi \delta\mathbf{u} \cdot {}^*\mathbf{n}_t\, d\xi d\eta \int_\eta \int_\xi {}^*\mathbf{n}_t \cdot \Delta\mathbf{u}\, d\xi d\eta \\ &\quad -\rho \int_\eta \int_\xi \delta\mathbf{u} \cdot {}^*\mathbf{n}_t\, \mathbf{g} \cdot \Delta\mathbf{u}\, d\xi d\eta. \end{aligned} \tag{29}$$

Obviously the second part of this equation is a non-symmetric term. However, combining both non-symmetric parts of $\delta^g \Pi_{lin}^{\Delta n}$ and $\delta^g \Pi_{lin}^{\Delta p}$, a symmetric expression results for the complete sum

$$\begin{aligned} \delta^g \Pi_{lin} &- \delta^g \Pi_{lin}^{\Delta n} + \delta^g \Pi_{lin}^{\Delta p} \mid \delta^g \Pi_t \\ &= \delta^g \Pi_t + \\ &+ \rho \frac{|\mathbf{g}|}{S_t} \int_\eta \int_\xi \delta\mathbf{u} \cdot {}^*\mathbf{n}_t\, d\xi d\eta \int_\eta \int_\xi {}^*\mathbf{n}_t \cdot \Delta\mathbf{u}\, d\xi d\eta \qquad \text{Term I} \\ &- \frac{\rho}{2} \int_\eta \int_\xi \delta\mathbf{u} \cdot ({}^*\mathbf{n}_t \otimes \mathbf{g} + \mathbf{g} \otimes {}^*\mathbf{n}_t) \Delta\mathbf{u}\, d\xi d\eta \qquad \text{Term II} \\ &+ \int_\eta \int_\xi \frac{{}^g p_t}{2} \begin{pmatrix} \delta\mathbf{u} \\ \delta\mathbf{u}_{,\xi} \\ \delta\mathbf{u}_{,\eta} \end{pmatrix} \cdot \begin{pmatrix} 0 & \underline{\mathbf{W}}^\xi & \underline{\mathbf{W}}^\eta \\ \underline{\mathbf{W}}^{\xi T} & 0 & 0 \\ \underline{\mathbf{W}}^{\eta T} & 0 & 0 \end{pmatrix} \begin{pmatrix} \Delta\mathbf{u} \\ \Delta\mathbf{u}_{,\xi} \\ \Delta\mathbf{u}_{,\eta} \end{pmatrix} d\xi d\eta. \\ &\qquad \text{Term III} \end{aligned} \tag{30}$$

3.3 Compressible Fluids Under Gravity

Considering only the gravity potential of a compressible fluid, we have to integrate over the total volume of the fluid

$$^{hg}\Pi = -\int_v \rho(\mathbf{x})\mathbf{g}\cdot\mathbf{x}dv + \text{const.} \tag{31}$$

In general the density $\rho(\mathbf{x})$ is dependent on the height of the fluid, however, for technical applications with standard heights the density can be assumed to be given by the law of mass conservation. Then for compressible and incompressible fluids the potential is only a function of the form and the volume of the enclosed fluid

$$^{hg}\Pi = -\rho(v)\int_v \mathbf{g}\cdot\mathbf{x}dv + \text{const.} \tag{32}$$

This can be written as a surface integral

$$\begin{aligned} ^{hg}\Pi &= -\rho(v)\int_\eta\int_\xi \mathbf{g}\cdot\mathbf{x}\;\mathbf{x}\cdot{}^*\mathbf{n}\;d\xi d\eta + \text{const.} \\ &= -\rho(v)\mathbf{g}\cdot\mathbf{s}\;, \qquad \mathbf{s}\text{: 1.order volume moment} \end{aligned} \tag{33}$$

The corresponding linearized functional contains two major parts:

$$\delta^{hg}\Pi_{lin} = -\delta\rho(v)\mathbf{g}\cdot\mathbf{s} - \rho(v)\mathbf{g}\cdot\delta\mathbf{s}. \tag{34}$$

From mass conservation $\rho(v)v = \rho_o V$ with ρ_o, V as reference values, we obtain

$$\delta\rho(v) = -\rho_o\frac{V}{v^2}\delta v = -\frac{\rho(v)}{v}\delta v \tag{35}$$

with

$$\delta v = \int_\eta\int_\xi {}^*\mathbf{n}\cdot\delta\mathbf{u}\;d\xi d\eta. \tag{36}$$

The variation of the second part is identical to the variation shown in the previous paragraph. After defining the location of the center of gravity of the fluid

$$\mathbf{c} = \frac{\mathbf{s}}{v}, \tag{37}$$

the variation follows as

$$\delta^{hg}\Pi = \rho(v)\int_\eta\int_\xi \mathbf{g}\cdot(\mathbf{c}-\mathbf{x})\;{}^*\mathbf{n}\cdot\delta\mathbf{u}\;d\xi d\eta. \tag{38}$$

Introducing the compressibility of the fluid in an identical fashion as in 3.1 with Hooke's law

$$^{h}p(v) = -K\frac{v-V}{V}, \tag{39}$$

the final term of a compressible fluid is given as

$$\delta^{hg} W = \ {}^{hg}p(v) \int_\eta \int_\xi {}^*\mathbf{n} \cdot \delta\mathbf{u} \; d\xi d\eta. \tag{40}$$

It is of some help to subdivide the pressure into three parts, ${}^{h}p(v)$ as above and

$$ {}^{x}p = \rho(v)\mathbf{g} \cdot \mathbf{x}, \tag{41}$$

$$ {}^{c}p = \rho(v)\mathbf{g} \cdot \mathbf{c}. \tag{42}$$

Then the form shown in (7) is given and linearization is a straightforward process with

$$\Delta^{hg} p = \Delta^{c} p - \Delta^{x} p - \Delta^{h} p \tag{43}$$

$$\Delta^{c} p = -2\frac{{}^{c}p_t}{v_t} \int_\eta \int_\xi {}^*\mathbf{n}_t \cdot \Delta\mathbf{u} \; d\xi d\eta + \int_\eta \int_\xi \frac{{}^{x}p_t}{v_t} \, {}^*\mathbf{n}_t \cdot \Delta\mathbf{u} \; d\xi d\eta \tag{44}$$

$$\Delta^{x} p = -\frac{{}^{x}p_t}{v_t} \int_\eta \int_\xi {}^*\mathbf{n}_t \cdot \Delta\mathbf{u} \; d\xi d\eta + \rho_t \mathbf{g} \cdot \Delta\mathbf{u} \tag{45}$$

$$\Delta^{h} p = -\frac{K}{V} \int_\eta \int_\xi {}^*\mathbf{n}_t \cdot \Delta\mathbf{u} \; d\xi d\eta \tag{46}$$

The summary of the pressure changes introduced into the linearized virtual work expression results in a symmetric displacement formulation. This implies that the proposed model is conservative

$$\begin{aligned}
\delta^{hg} \Pi_{lin} &= \delta^{hg} \Pi_t \\
&+ \left(\frac{K}{V} - 2\frac{{}^{c}p_t}{v_t}\right) \int_\eta \int_\xi \Delta\mathbf{u} \cdot {}^*\mathbf{n}_t \; d\xi d\eta \int_\eta \int_\xi {}^*\mathbf{n}_t \cdot \delta\mathbf{u} \; d\xi d\eta \qquad \text{part I} \\
&+ \int_\eta \int_\xi \frac{{}^{x}p_t}{v_t} \, {}^*\mathbf{n}_t \cdot \Delta\mathbf{u} \; d\xi d\eta \int_\eta \int_\xi {}^*\mathbf{n}_t \cdot \delta\mathbf{u} \; d\xi d\eta \\
&+ \int_\eta \int_\xi {}^*\mathbf{n}_t \cdot \Delta\mathbf{u} \; d\xi d\eta \int_\eta \int_\xi \frac{{}^{x}p_t}{v_t} \, {}^*\mathbf{n}_t \cdot \delta\mathbf{u} \; d\xi d\eta \qquad \text{part II} \\
&- \frac{\rho_t}{2} \int_\eta \int_\xi \delta\mathbf{u} \cdot ({}^*\mathbf{n}_t \otimes \mathbf{g} + \mathbf{g} \otimes {}^*\mathbf{n}_t) \Delta\mathbf{u} \; d\xi d\eta \qquad \text{part III} \\
&+ \frac{1}{2} \int_\eta \int_\xi {}^{hg}p_t \begin{pmatrix} \delta\mathbf{u} \\ \delta\mathbf{u}_{,\xi} \\ \delta\mathbf{u}_{,\eta} \end{pmatrix} \cdot \begin{pmatrix} 0 & \underline{\mathbf{W}}^{\xi} & \underline{\mathbf{W}}^{\eta} \\ \underline{\mathbf{W}}^{\xi T} & 0 & \\ \underline{\mathbf{W}}^{\eta T} & 0 & 0 \end{pmatrix} \begin{pmatrix} \Delta\mathbf{u} \\ \Delta\mathbf{u}_{,\xi} \\ \Delta\mathbf{u}_{,\eta} \end{pmatrix} d\xi d\eta. \qquad (47) \\
&\qquad \text{part IV}
\end{aligned}$$

The different linearized parts can be interpreted as follows:

I The multiplication of the two surface integrals indicates the volume dependence of the compression level and of the pressure at the center of the fluid volume.

II The change of the local pressure is influenced by changes of the total volume and changes of the location of the center of the volume.
III A hydrostatic pressure generates a nonuniform pressure field, which is represented by a symmetric field equation under realistic boundary conditions, see [11], [12], [2].
IV Follower forces create a symmetric field equation too, considering realistic boundary conditions, see [11], [12], [2], [4], [13].

4 FE-Discretization and Solution Algorithm

The virtual work expression

$$\delta W = \delta^{el} V + \delta^{phg} \Pi - \delta^{ex} \Pi = 0 \tag{48}$$

followed by the linearization process as shown in chapter 3 leads to a residual and a linear term, depending only on the surfaces of the wetted resp. closed volumes. These terms have to be discretized with standard FE shell, membrane or continuum elements. Thus the boundary description is based on the surfaces of the FE elements wetted by the fluid or gas. Further, the discretized and linearized constraint equations as Hooke's law, the mass conservation of the fluid and the computation of the pressure at the center of the structure are included, resulting in general in a hybrid symmetric system of equations for the coupled problem:

$$\begin{bmatrix} {}^{el,phg}\mathbf{K} & -\mathbf{a} & -\mathbf{b} & \mathbf{a} \\ -\mathbf{a}^T & -\frac{K}{V} & 0 & 0 \\ -\mathbf{b}^T & 0 & -2\,{}^{c}pv & v \\ \mathbf{a}^T & 0 & v & 0 \end{bmatrix} \begin{pmatrix} \mathbf{d} \\ \Delta^k p \\ \frac{\Delta \rho}{\rho} \\ \Delta^c p \end{pmatrix} = \begin{pmatrix} {}^{phg}\mathbf{F} \\ 0 \\ 0 \\ 0 \end{pmatrix}. \tag{49}$$

This symmetric system can be reduced to a conventional symmetric displacement representation with the elastic and load stiffness matrix ${}^{el,phg}\mathbf{K} = {}^{el}\mathbf{K} + {}^{phg}\mathbf{K}$, the residual ${}^{phg}\mathbf{F}$ of internal ${}^{el}\mathbf{f}$, external ${}^{ex}\mathbf{f}$ and interaction forces ${}^{phg}\mathbf{f}$, the nodal displacement vector $\mathbf{d}$, a volume pressure gradient ${}^{phg}\alpha_t$ and two rank-one vectors $\mathbf{a}$ and $\mathbf{b}$

$$\begin{aligned} [{}^{el,phg}\mathbf{K} + \mathbf{b} \otimes \mathbf{a} + \mathbf{a} \otimes \mathbf{b}]\mathbf{d} &= {}^{phg}\mathbf{F} \\ &= {}^{ex}\mathbf{f} - {}^{phg}\mathbf{f} - {}^{el}\mathbf{f} \end{aligned} \tag{50}$$

This can be interpreted as a symmetric rank-two update of the matrix ${}^{el,phg}\mathbf{K}$ coupling all wetted degrees of freedom together. Applying the Sherman-Morrison formula an efficient solution can be computed by two additional forward-backward substitutions:

$$\mathbf{d}_1 = {}^{el,phg}\mathbf{K}^{-1}\,{}^{phg}\mathbf{F}, \quad \mathbf{d}_2 = {}^{el,phg}\mathbf{K}^{-1}\,\mathbf{a}, \quad \mathbf{d}_3 = {}^{el,phg}\mathbf{K}^{-1}\,\mathbf{b}. \tag{51}$$

The load stiffness matrix and the other pressure related terms are:

$$
\begin{aligned}
{}^{phg}\mathbf{K} &= -\frac{\rho_t}{2}\sum_e \int_\eta \int_\xi \mathbf{N}^T({}^*\mathbf{n}\mathbf{g}^T + \mathbf{g}\;{}^*\mathbf{n}^T)\mathbf{N}\; d\xi d\eta \\
&+\frac{1}{2}\sum_e \int_\eta \int_\xi {}^{phg}p_t \begin{pmatrix} \mathbf{N} \\ \mathbf{N}_{,\xi} \\ \mathbf{N}_{,\eta} \end{pmatrix}^T \begin{pmatrix} 0 & \underline{\mathbf{W}}^\xi & \underline{\mathbf{W}}^\eta \\ \underline{\mathbf{W}}^{\xi T} & 0 & 0 \\ \underline{\mathbf{W}}^{\eta T} & 0 & 0 \end{pmatrix} \begin{pmatrix} \mathbf{N} \\ \mathbf{N}_{,\xi} \\ \mathbf{N}_{,\eta} \end{pmatrix} d\xi d\eta,
\end{aligned} \tag{52}
$$

$$
\mathbf{a} = \sum_e \int_\eta \int_\xi \mathbf{N}^T\;{}^*\mathbf{n}_t\; d\xi d\eta, \tag{53}
$$

$$
\mathbf{b} = \sum_e \int_\eta \int_\xi {}^{phg}\alpha_t \mathbf{N}^T\;{}^*\mathbf{n}_t\; d\xi d\eta, \tag{54}
$$

$$
{}^{phg}\mathbf{f} = \sum_e \int_\eta \int_\xi {}^{phg}p_t \mathbf{N}^T\;{}^*\mathbf{n}_t\; d\xi d\eta, \tag{55}
$$

$$
{}^{phg}\alpha_t = \frac{K}{2V} - \frac{{}^c p_t - {}^x p_t}{v_t}, \tag{56}
$$

$$
{}^{phg}p_t = {}^c p - {}^x p - {}^h p. \tag{57}
$$

The nodal displacement vector for one iteration step is then given by a linear combination of the three auxiliary solution vectors: $\mathbf{d} = \mathbf{d}(\mathbf{d}_1, \mathbf{d}_2, \mathbf{d}_3)$. For further details and the combination with arc-length schemes, we refer to [9] and [5]. For the cases pneumatics, hydraulics and incompressible fluids with open surfaces only a rank-one update is found. For pneumatics and hydraulics the pressure terms due to gravity disappear with ${}^c p ={}^x p = 0$, as well as the first term in ${}^{phg}\mathbf{K}$, as g is not existent then. Depending on the pressure equation for gas or fluid

$$
{}^h\alpha_t = \kappa \frac{{}^k p_t}{v_t} \qquad \text{or} \qquad {}^h\alpha_t = \frac{K}{V} = \text{const.} \tag{58}
$$

and only a constant pressure term remains ${}^{phg}p_t = -{}^h p_t$.
For incompressible fluids ${}^{phg}\mathbf{K}$ has the same structure as for compressible fluids, but the pressure is dependent on the current coordinate of the fluid only ${}^{hg}p_t ={}^x p = \rho\mathbf{g}(-\mathbf{x}_t)$. However, we have to note that incompressibility can only be considered for fluids with open surfaces. For compressible fluids with open surface the pressure term ${}^h p$ due to overpressure is not present. Then the volume change due to the fluid weight is included in the ${}^c p$-term. However, in this case we have to keep also track of the current height of the fluid level, which is discussed in a forthcoming paper.

5 Numerical Examples

5.1 Pneumatic Multi-Chamber Structure Under Torsional Loading

A pneumatic multi-chamber structure see Fig. 1, two chambers with different initial internal pressure (left $p_{l0} = 0.01$ bar,right $p_{r0} = 1$ bar, with material

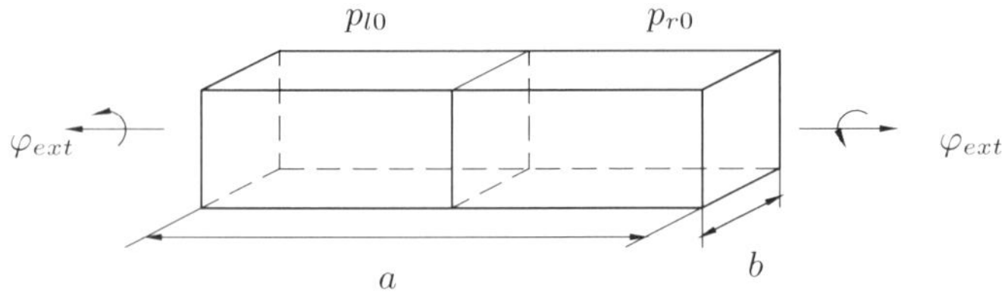

Fig. 1. Pneumatic multi-chamber system with flexible walls; loading by different initial internal pressure and torsion; geometrical data: $a = 10\ cm$, $b = 2\ cm$, $\varphi_{ext} = 45^o$, $p_{l0} = 0.01\ bar$, $p_{r0} = 1\ bar$

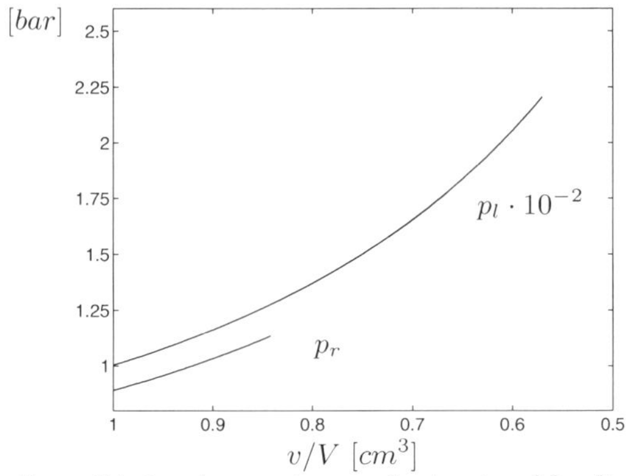

Fig. 2. Pneumatic multi-chamber system under torsional loading; internal pressure vs. relative volume change

law St.-Venant Kirchhoff, Young's modulus $E = 2.4 \cdot 10^4 \frac{N}{cm^2}$, Poisson ratio $\nu = 0.3$ and wall thickness $t = 0.1\ cm$ is first loaded with the different internal pressure. Then a torsional loading is applied by a rotation around the longitudinal axis by $\varphi_{ext} = 45^o$. In the FE model the nodes at one end are moved by prescribed displacements on circles in 20 equal sized steps. The chambers are discretized by 2300 solid-shell elements. The results of the analysis show the stiffening due to the internal pressure. The right chamber with high pressure is behaving like a rigid body, whereas the left chamber with low pressure is strongly deformed. The latter leads to a substantial increase of the internal pressure as a result of the volume reduction, see Figs. 2 and 3.

As the wall between both chambers is deformable, the pressure increase in the left chamber is also communicated to the right chamber resulting in a minor pressure increase there, too. In addition, we have to note that no stability problem arises during the deformation process; the dyadic extension due to the internal pressure somehow stabilizes the structure considerably and leads to a regularization of the equation system. The consistent linearization of the developed algorithm is visible in the quadratic convergence in the last

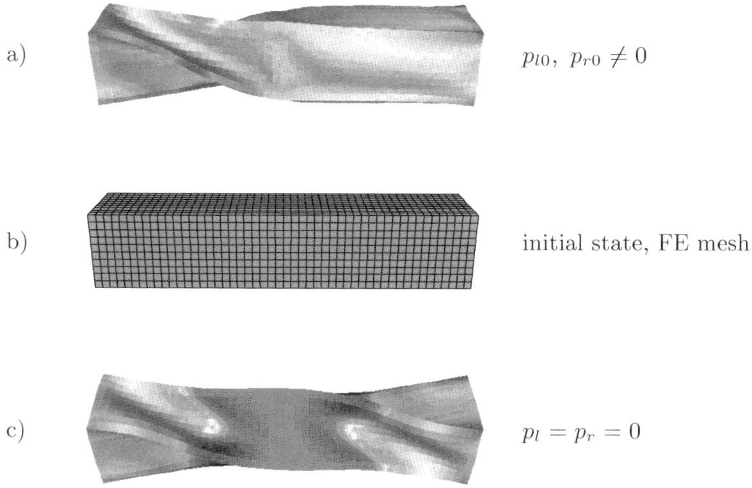

Fig. 3. Pneumatic multi-chamber system under torsional loading; (**a**) deformed high and low pressure chambers, (**b**) undeformed structure, (**c**) deformed structure without internal pressure

Table 1. Pneumatic multi-chamber system under torsional loading; convergence in the last load step at $\varphi_{ext} = 45^o$

iteration step	1	2	3	4
energy	$7.09 \cdot 10^{+1}$	$4.09 \cdot 10^{-1}$	$2.49 \cdot 10^{-3}$	$1.16 \cdot 10^{-4}$
iteration step	5	6	7	8
energy	$3.52 \cdot 10^{-6}$	$9.07 \cdot 10^{-8}$	$2.09 \cdot 10^{-9}$	$4.64 \cdot 10^{-11}$

iterations of the load steps. See e.g. the convergence behavior in the last load step at $\varphi_{ext} = 45^o$ in Table 1.

5.2 Hydrostatics of Partially Filled Multi-Chamber System with Interaction

This multi-chamber structure is chosen to show the effect of the interaction between fluid loaded chambers during a filling process. An open tank structure consisting of two deformable thin-walled chambers, separated by a thin wall is filled, see Fig. 4. Material law: St.-Venant Kirchhoff, Young's modulus $E = 2.0 \cdot 10^4 \frac{N}{cm^2}$, Poisson ratio $\nu = 0.3$, wall thickness $t = 0.1cm$, specific gravity of the fluid $\rho g = 0.1 \frac{N}{cm^3}$). The complete structure is modelled with 1100 solid-shell elements. In the first load step the left chamber is partially filled up to w_{10} ($V_1 = 375cm^3$). Then the right chamber is filled with 30 equal sized volume steps of $\Delta V_2^* = 25\ cm^3$ starting with $w_{20} = 0$. In each load step the

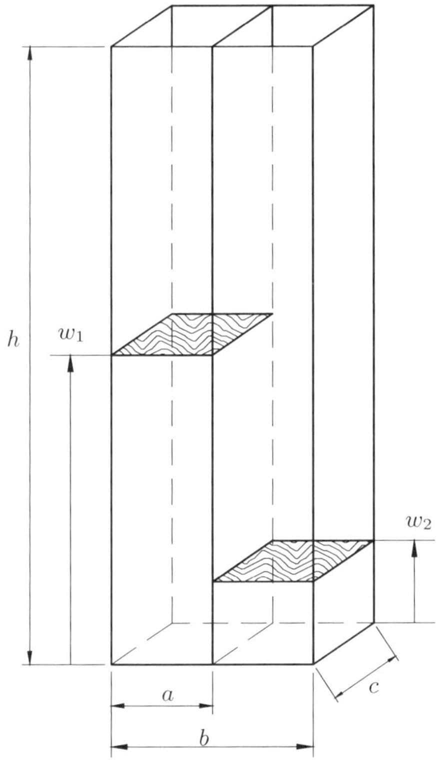

Fig. 4. Hydrostatics of multi-chamber system with interaction during filling; geometrical data: $a = 5cm,\ b = 10cm,\ c = 5cm,\ h = 30cm$

new filling height in the right chamber is determined via ${}^o x_2 = {}^o x_{t2} + \Delta u_2$ with $\Delta u_2 = \frac{V_2^*}{S_{t2}}$ using the free surface S_{t2} computation. The chamber structure is deforming during the filling process and also the fluid level w_1 of the left chamber is rising as a result of the filling of the right chamber, see Figs. 5 and 6.

5.3 Elastic Cylindrical Vessel Fully Filled with Fluid

An elastic cylindrical vessel (weightless, elastic modulus $E = 21 \cdot 10^{10} \frac{N}{m^2}$, Poisson's ratio $\nu = 0.3$) with a very thin wall - close to a membrane - is completely filled with water (density $\rho = 1000 \frac{kg}{m^3}$, bulk modulus $K = 0.5 \cdot 10^9 \frac{N}{m^2}$). In a first load step the vessel is pressurized by 1 bar at the top of the vessel indicating the weight of the plate. In a second step the structure is loaded by a given displacement u_{ext} of the loading plate, see Fig. 7b.

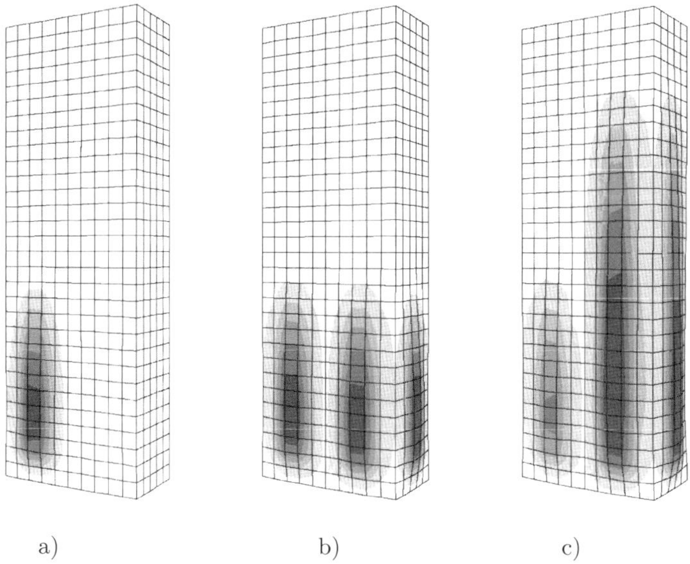

Fig. 5. Hydrostatics of multi-chamber system with interaction during filling; loading states: (**a**) left chamber filled up to w_{10}, right chamber empty, (**b**) right chamber partially filled and (**c**) after last load step (completely filled)

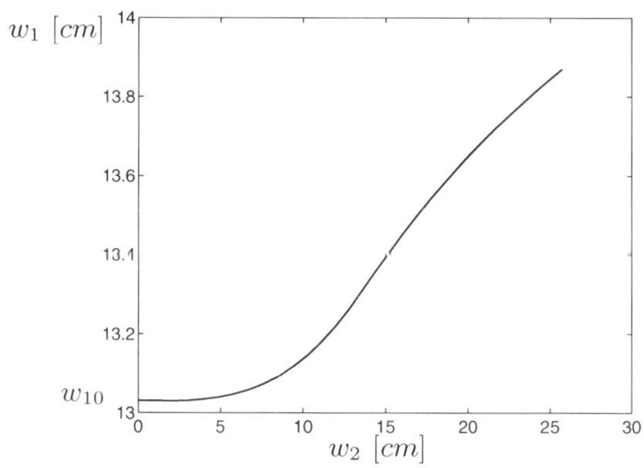

Fig. 6. Hydrostatics of multi-chamber system with interaction during filling; w_{10} fluid level in left chamber before filling of right chamber; fluid level w_1 in left chamber versus fluid level w_2 in right chamber

Due to the deformation density and volume change according to the conservation of mass, see Fig. 8a. The decrease of the volume implicates an increase in the pressure level ${}^{h}p$ in the fluid. For comparison only a gas filling is con-

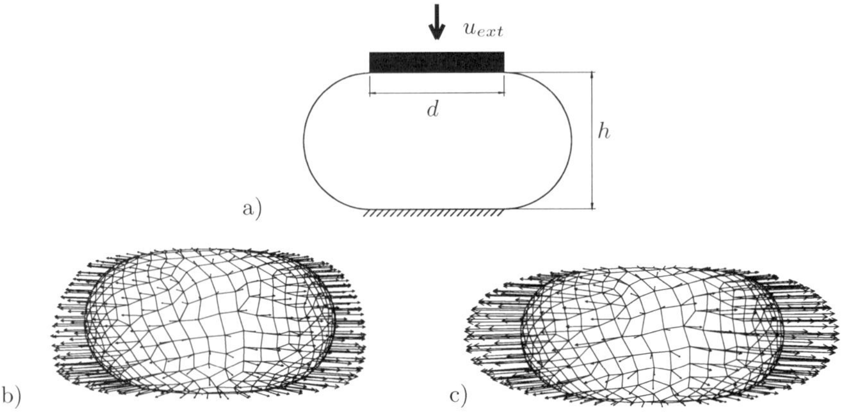

Fig. 7. Elastic cylindrical vessel fully filled (**a**) geometry and loading: height $h = 10m$, diameter $d = 10m$, piston displacement $u_{ext} = 4m$; (**b**) radial displacement vectors - first load step, (**c**) radial displacement vectors - final load step

FE Modelling and Simulation of Gas and Fluid Supported Structures

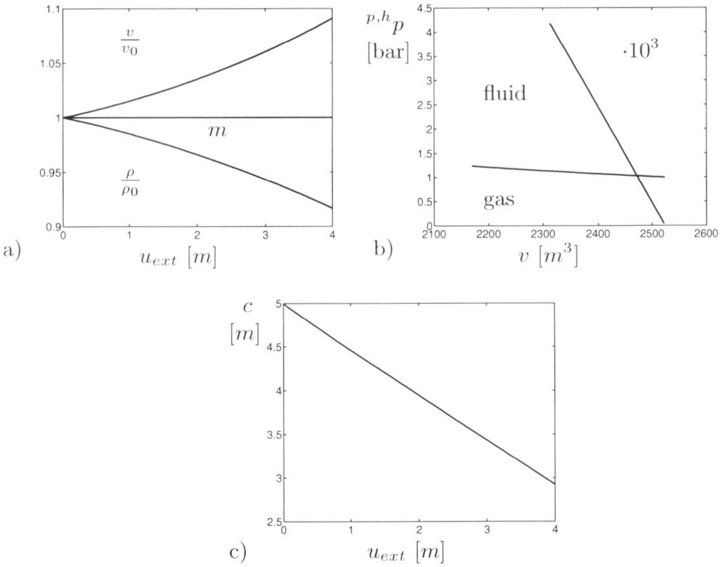

Fig. 8. Elastic cylindrical vessel fully filled (**a**) mass conservation vs. piston displacement u_{ext}, (**b**) fluid pressure ^{h}p / gas pressure ^{p}p vs. fluid volume v, (**c**) location of center c of volume vs. piston displacement u_{ext}

sidered too, see Fig. 8b. A provisional result is the position of the center of the volume, which changes with the displacement of the piston, see Fig. 8c.

5.4 Fluid Filling of Strongly Deformable Thin-Walled Shell

In order to show the ability of the model to capture more numerically difficult situations the filling of a very thin-walled 2D-shell structure showing fairly large deformations almost similar to the deployment of membrane structures is chosen. The shell modelled with 8-node solid-shell elements is assumed to be weightless; material is of Neo-Hooke rubber type with elastic modulus $E = 9.6 \cdot 10^3 \frac{N}{m^2}$, Poisson's ratio $\nu = 0.3$, $t = 1mm$; the water is assumed to be incompressible, specific gravity of the fluid $\rho g = 0.01 \frac{N}{cm^3}$. The filling is performed via an external piston see Fig. 9.

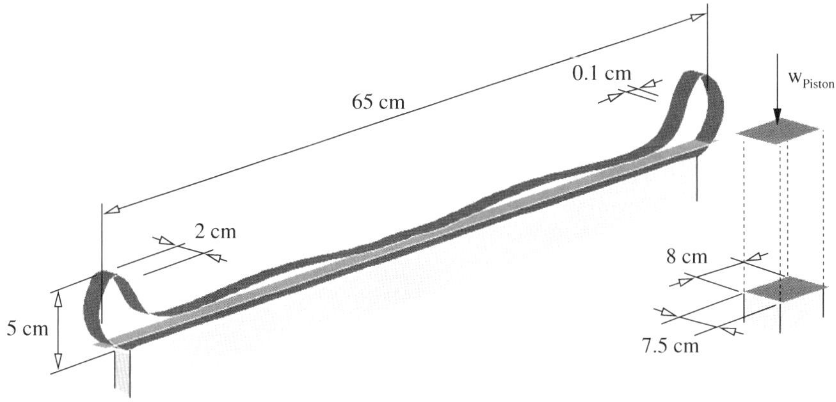

Fig. 9. Fluid filling of a strongly deformable shell, initial geometry and loading

A deformed situation is depicted in Fig. 10 with an almost filled structure. The shell is discretized with three different meshes of equal sized elements.

During the filling the water level is rising and touches the elements often only partially. As then the water pressure is checked at the Gauss-points of the surface (2*2 integration) convergence becomes difficult. In particular, when the load level comes into contact with the almost horizontal element faces, the number of iterations is rising, as minor changes in the water level lead to a major change in the wetted surface. As depicted in Fig. 11 the water level shows a rather linear dependence of the piston motion resp. the added water volume.

From the diagram in Fig. 12 we see that the algorithm allows a separation of the water surface without any problem which is more visible for the coarser mesh. The stiffer behavior of coarser mesh leads to a rather early separation whereas the finer meshes show a separation at an almost filled state, see e.g. Fig. 10. For a complete filling it would be necessary to take the compressibility of the water into account which is the subject of a forthcoming paper [9].

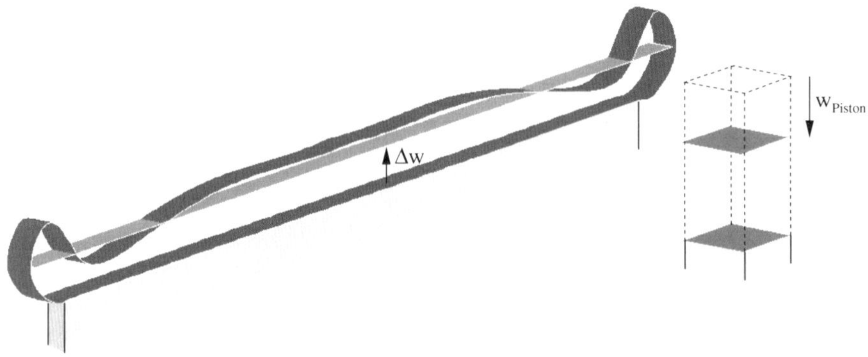

Fig. 10. Fluid filling of a strongly deformable shell; deformed state with almost fully filled structure

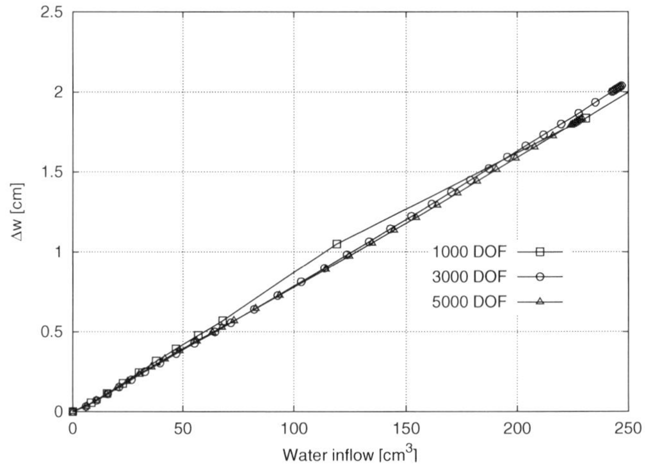

Fig. 11. Fluid filling of a strongly deformable shell; water level vs. added water volume

6 Conclusions

The proposed approach to describe the gas and fluid effects and their interaction with deforming structures by state equations has several advantages. First, the mesh-free modelling of the fluid resp. the gas allows to perform large deformation analysis without remeshing. Second, contact models between fluid and structure are not necessary. Third, stability investigations can be carried out taking the specific decomposition of the stiffness matrix

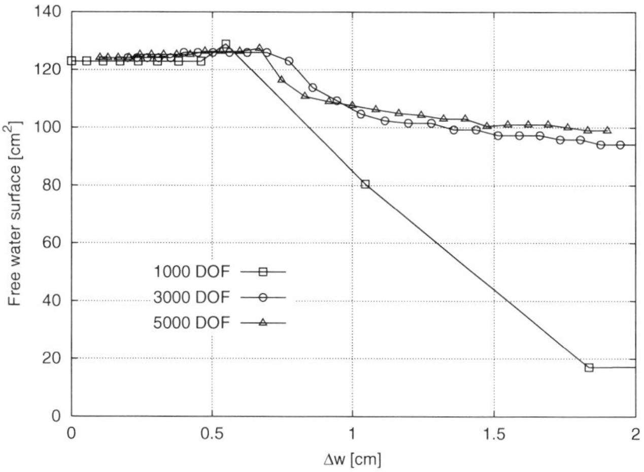

Fig. 12. Fluid filling of a strongly deformable shell; free water surface vs. water height

of the coupled problem into account, see [7]. Fourth, the solution of the coupled equation can be efficiently performed based on the subsequent use of the Sherman-Morrison formula involving only the triangular decomposition of the structural matrix. Summarizing all, the computational effort is significantly lower and better adjusted than in conventional methods based on full discretization. The numerical examples show the efficiency of the applications to thin-walled structures though due to the highly nonlinear behavior convergence is often rather difficult to achieve. Some of our forthcoming work is devoted to the static unfolding and filling of very flexible folded membrane like structures.

References

1. J. Bonet, R.D. Wood, J. Mahaney, and P. Heywood. Finite element analysis of air supported membrane structures. *Comput. Methods Appl. Mech. Engrg.*, 190:579–595, 2000.
2. H. Bufler. Konsistente und nichtkonsistente druckbelastungen durch flüssigkeiten. *ZAMM*, 72(7):T172 – T175, 1992.
3. H. Bufler. Configuration dependent loading and nonlinear elastomechanics. *ZAMM*, 73:4 – 5, 1993.
4. G. Romano. Potential operators and conservative systems. *Meccanica*, 7:141–146, 1972.
5. T. Rumpel. *Effiziente Diskretisierung von statischen Fluid-Strukturproblemen bei grossen Deformationen.* Dissertation (in German), Universität Karlsruhe, 2003.

6. T. Rumpel and K. Schweizerhof. A hybrid approach for volume dependent fluid-structure-problems in nonlinear static finite element analysis. In J. Eberhardsteiner H.A. Mang, F.G. Rammerstorfer, editor, *Fifth World Congress on Computational Mechanics (WCCM V)*, 2002.
7. T. Rumpel and K. Schweizerhof. Volume-dependent pressure loading and its influence on the stability of structures. *Int. J. Numer. Meth. Engng*, 56:211–238, 2003.
8. T. Rumpel and K. Schweizerhof. Hydrostatic fluid loading in nonlinear finite element analysis. *Int. J. Numer. Meth. Engng*, 59:849–870, 2004.
9. T. Rumpel and K. Schweizerhof. Large deformation fe analysis of thin walled shell and membrane structures fully filled with compressible fluids. *(in preparation)*, 2004.
10. H. Schneider. *Flüssigkeitsbelastete Membrane unter großen Deformationen.* Universität Stuttgart, 1990. Dissertation.
11. K. Schweizerhof. *Nichtlineare Berechnung von Tragwerken unter verformungsabhängiger Belastung.* Universität Stuttgart, 1982. Dissertation.
12. K. Schweizerhof and E. Ramm. Displacement dependent pressure loads in nonlinear finite element analyses. *Computers & Structures*, 18(6):1099–1114, 1984.
13. M. J. Sewell. On configuration-dependent loading. *Arch. Rational Mech. Anal.*, 23:327–351, 1966.

Widespan Membrane Roof Structures: Design Assisted by Experimental Analysis

M. Majowiecki[1]

IUAV
Facoltà di Architectura
University of Venice, ITALY
majo@mail.asianet.it

Summary. *Wide span structures are today widely applied for sport, social, industrial, ecological and other activities. The experience collected in last decades identified structural typologies as space structures, cable structures, membrane structures and new - under tension - efficient materials which combination deals with lightweight structural systems, as the state of art on long span structural design. In order to increase the reliability assessment of wide span structural systems a knowledge based synthetical conceptual design approach is recommended. Theoretical and experimental in scale analysis, combined with a monitoring control of the subsequent performance of the structural system, can calibrate mathematical modelling and evaluate long term sufficiency of design.*

Key words: Wide span structures , snow and wind loading, experimental analysis, reliability

1 Introduction

Considering the statistical results of [1], the unusual typologies, new materials and the "scale effect" of long span structures, several special design aspects arise as:

- the snow distribution and accumulations on large covering areas in function of statistically correlated wind direction and intensity;
- the wind pressure distribution on large areas considering theoretical and experimental correlated power spectral densities or time histories;
- rigid and aeroelastic response of large structures under the action of cross-correlated random wind action considering static, quasi-static and resonant contributions;
- the time dependent effect of coactive indirect actions as pre-stressing, short and long term creeping and temperature effects;
- the local and global structural instability;
- the non linear geometric and material behaviour;
- reliability and safety factors of new hi-tech composite materials;

E. Oñate and B. Kröplin (eds.), Textile Composites and Inflatable Structures, 173–194.

- the necessity to avoid and short-circuit progressive collapse of the structural system due to local secondary structural element and detail accidental failure;
- the compatibility of internal and external restrains and detail design, with the modeling hypothesis and real structural system response;
- the parametric sensibility of the structural system depending on the type and degree of static indeterminacy and hybrid collaboration between hardening and softening behaviour of substructures.
- In the case of movable structures, the knowledge base concerns mainly the moving cranes and the related conceptual design process have to consider existing observations, tests and specifications regarding the behaviour of similar structural systems. In order to fill the gap, the IASS working group no. 16 prepared a state of the art report on retractable roof structures [2] including recommendations for structural design based on observations of malfunction and failures.

From the observations of the in service performance, damages and collapses of all or part of structural systems, we have received many informations and teachings regarding the design and verification under the action of ultimate and serviceability limit states. Limit state violation for engineered structures have lead to spectacular collapses as the Tay (1879) and Tacoma bridges (1940). Sometimes an apparently "unimaginable" phenomenon occurs to cause structural failure. The Tacoma Narrows Bridge previously cited was apparently one such a case. It was also a design which departed considerably from earlier suspension bridge design.

Long span coverings were subjected to partial and global failures as that of the Hartford Colisseum (1978), the Pontiac Stadium (1982) and the Milan Sport Hall (1985) due to snow storms, the Montreal Olympic Stadium due to wind excitations of the membrane roof (1988), the Minnesota Metrodome (1983) air supported structure that deflated under water ponding, etc. Those cases are lessons to be learned from the structural failure mechanism in order to identify the design and construction uncertainties in reliability assessment. Many novel projects of long span structures attempt to extend the "state of the art". New forms of construction and design techniques generate phenomenological uncertainties about any aspect of the possible behavior of the structure under construction service and extreme conditions.

Fortunately, structures rarely fail in a serious manner, but when they do it is often due to causes not directly related to the predicted nominal loading or strength probability distributions. Other factors as human error, negligence, poor workmanship or neglected loadings are most often involved (Ref 1). Uncertainties related to the design process are also identified in structural modelling which represents the ratio between the actual and the foreseen model's response.

According to Pugsley (1973), the main factors which may affect "proneness to structural accidents" are:

- new or unusual materials;
- new or unusual methods of construction;
- new or unusual types of structure;
- experience and organization of design and construction teams;
- research and development background;
- financial climate;
- industrial climate;
- political climate.

Cause	%
Inadequate appreciation of loading conditions or structural behaviour	43
Mistakes in drawings or calculations	7
Inadequate information in contract documents or instructions	4
Contravention of requirements in contract documents or instructions	9
Inadequate execution of erection procedure	13
Unforeseeable misuse, abuse and/or sabotage, catastrophe, deteriora tion (partly "unimaginable"?)	7
Random variations in loading, structure, materials, workmanship, etc.	10
Others	7

Table 1. Prime causes of failure. Adapted from Walker (1981)

All these factors fit very well in the field of long span structures involving oftenly something "unusual" and clearly have an influence affecting human interaction.

In Table 1, the prime cause of failure gives 43% probability (Walker, 1981) to inadequate appreciation of loading conditions or structural behaviour. Apart from ignorance and negligence, it is possible to observe that the underestimation of influence and insufficient knowledge are the most probable factors in observed failure cases (Matousek & Schneider, 1976).

Performance and serviceability limit states violation are also directly related to structural reliability. Expertise in structural detail design, which is normally considered as a micro task in conventional design, have an important role in special long span structures: reducing the model and physical uncertainties and avoiding chain failures of the structural system.

According to the author, knowledge and experience are the main human intervention factors to filter gross and statistical errors in the normal processes of design, documentation, construction and use of structures.

The reliability of the design process in the field of special structures must be checked in the following three principal phases: the conceptual design, analysis, and working design phases.

Fig. 1. Montreal Olympic Stadium - A cable stayed roof solution

1.1 Some Wide Span Enclosures

Long span structures needs special investigations concerning the actual live load distribution and intensity on large covering surfaces. Building codes normally are addressed only to small-medium scale projects. The uncertainties relate to the random distribution of live loads on long span structures imply very careful loading analysis using special experimental analysis.

Due to the lack of space, only some design & analysis illustrations of wide span enclosures, where the author was directly involved, will be included in the present paper with the intention to transmit some experiences that today may be part of the knowledge base.

From the direct author's experience in designing large coverings, the most important experimental investigation regarding live load distribution concerns the snow drift and accumulation factors and the dynamic action of wind loading.

2 Design Assisted by Experimental Analysis

2.1 Snow Loading Experimental Analysis on Scale Models

Olympic Stadium in Montreal. During the design of the new roof for the Montreal Olympic Stadium (Fig. 1) a special analysis of snow loading was made considering three roof geometries varying the sag of the roof from 10 m, 11.5 m and 13 m., in order to find a minimization of snow accumulation.

The experimental investigation was carried out by RWDI [3] to provide design snow according to FAE (Finite Area Element) method, representing up to day a state of the art on the matter.

The FAE method uses a combination of wind tunnel tests on a scale model and computer simulation to provide the most accurate assessment possible to estimate 30 year snow loads.

Snow loads depend on many cumulative factors such as, snowfall intensity, redistribution of snow by the wind (speed and direction), geometry of the building and all surroundings affecting wind flow patterns, absorption of rain in the snowpack, and depletion of snow due to melting and subsequent runoff. The current NBCC (National Building Code of Canada) provides minimum design loads for roofs which are based primarily on field observations made on a variety of roofs and on a statistical analysis of ground snow load data. There are, however, numerous situations where the geometry of the roof being studied and the particulars of the site are not well covered by the general provisions of the code. In these situations, a special study, using analytical, computational and model test methods, can be very beneficial since it allows the specific building geometry, site particulars and local climatic factors to all be taken into account. The National Building Code allows these types of studies through its "equivalency" clause and various references to special studies in its commentary.

The model of the three new roof shapes were each constructed at 1:400 scale for the wind tunnel tests. The three model roof designs were each instrumented with 90° directional surface wind velocity vector sensors covering the surface. On the console roof, an additional 90 sensors were installed. Measurements of the local wind speed and direction, at an equivalent full-scale height of 1 m above the roof surface, were taken for 16 wind directions. The wind speed measurements were then converted to ratios of wind speed at the roof surface to the reference wind speed measured at a height equivalent at full scale to 600 m.

The 30 year ground snow prediction is obtained by interpolation of the data using the Fisher-Typett type I extreme value distribution method (Fig.2), including both snow and rain ($S_s + S_r$), to be 2.8 kPa, which is in agreement with the code value.

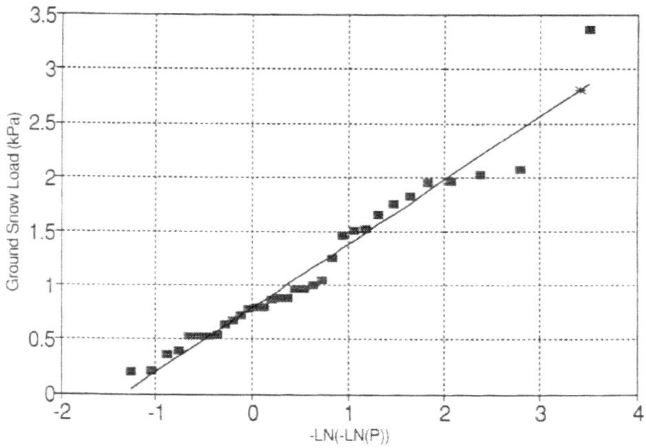

Fig. 2. Fisher-Typett Type 1 extreme values plot ground snow load prediction

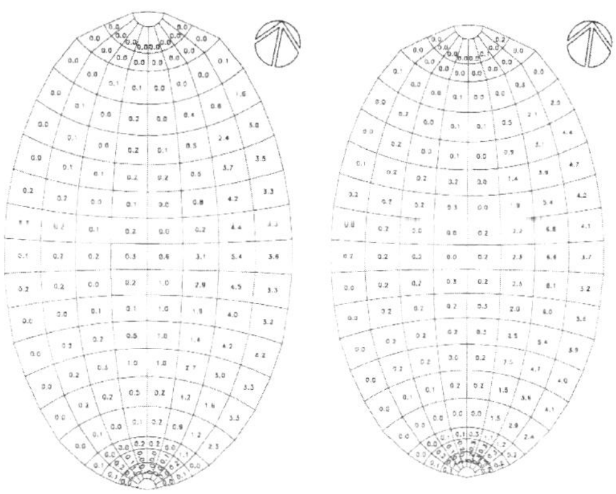

Fig. 3. Comparative analysis of snow loading distribution in function of roof shape (10-13m)

Results of structural load cases and local peak loading, not to be considered as acting over the roof simultaneously are shown in Figs. 3-4. The shape of the roof with a sag of more than 12m. gives separation of the air flow and turbulence in the wake increasing considerably the possibility of snow accumulations. The order of

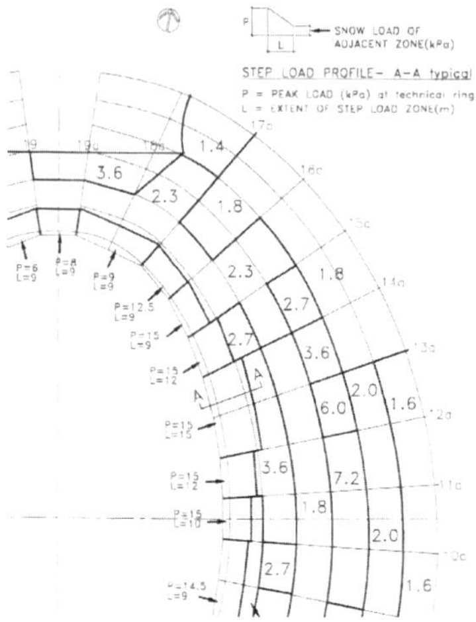

Fig. 4. Sliding and wind snow accumulations step loads

magnitude of the leopardized accumulations in the roof are of 4-15 kN!; local overdimensioning was necessary in order to avoid progressive collapse of the structural system.

2.2 Wind Loading-Experimental Analysis on Scale Models: Rigid Structures-Quasi Static Behaviour

The Cp factors: the Olympiakos Stadium in Athens

Tests have been performed in two distinct phases, the first phase has been devoted to the characterization of the appropriate wind profile in the BLWT, the second one has been dedicated to the identification of the pressure coefficients on the roofing of the new stadium. Because of the great number of pressure taps on the roofing (252), the second phase consisted of three distinct measurement sets.

The stadium is located near to the sea, as a consequence a "sea wind profile" with the parameters listed below and taken from literature and laboratory expertise, seems to be a good approximation of the wind profile in the area (Fig. 5):

profile exponent	$\alpha = 0.15/0.18$ (level ground, with few obstacles, sea),
roughness length	$z_0 = 5/15$ cm (cultivated fields),
integral length scale	$L_U = 50/100$ m.

In the following paragraph the characteristics of the wind profile actually obtained in the BLWT are examined, and the consistency of the choice in the chosen geometric scale (1:250).

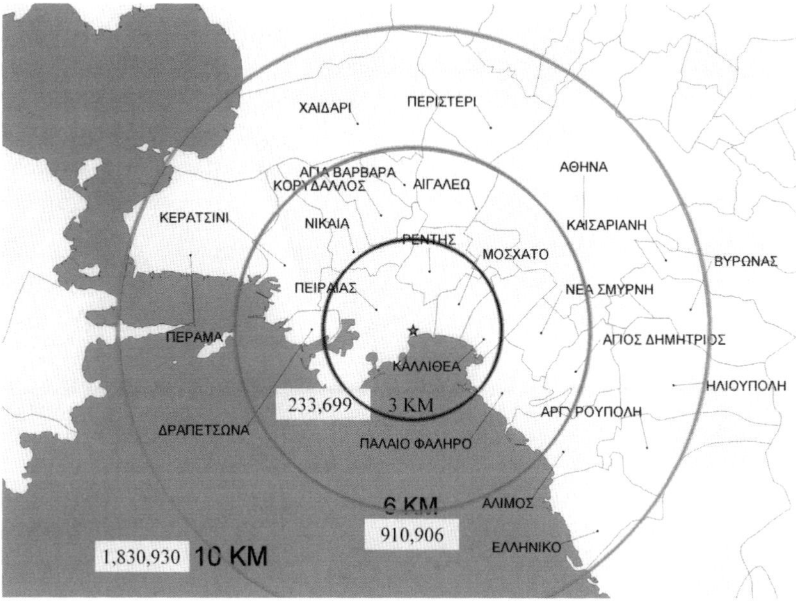

Fig. 5. Geographic location of the stadium

The model has been made in a geometric scale of 1:250 and includes: the roofing, the stands, all the structures of the stadium, and other private and public buildings not far then 250 m (in full scale) Figs. 10-11 from the centre of the stadium. The geometric scale has been chosen in order to fulfil the similitude laws (Figs. 6-9). In turn the extension of the model around the stadium was dictated by the chosen scale and by the diameter (2m) of the rotating platform over which the model has been placed in the wind tunnel.

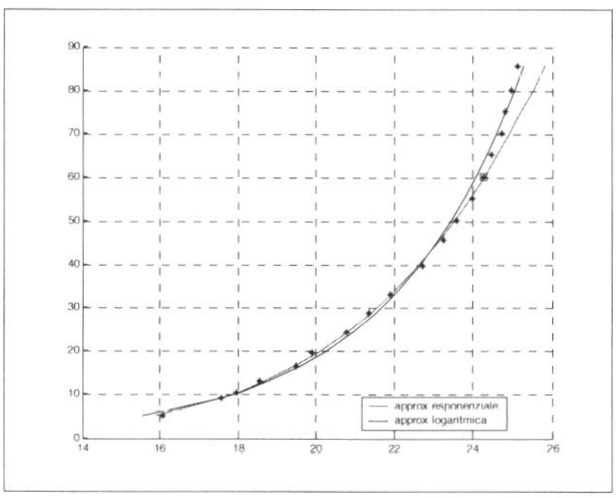

Fig. 6. Profile of mean wind velocity

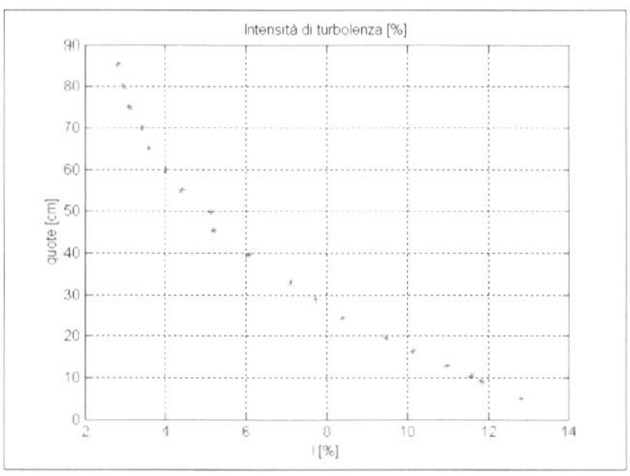

Fig. 7. Profile of the turbulence intensity

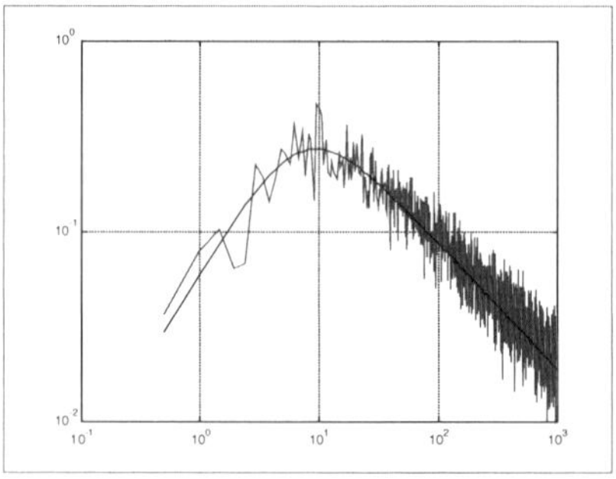

Fig. 8. Spectral density of the longitudinal component of the wind velocity ("fitting" with Von Karmán spectral density)

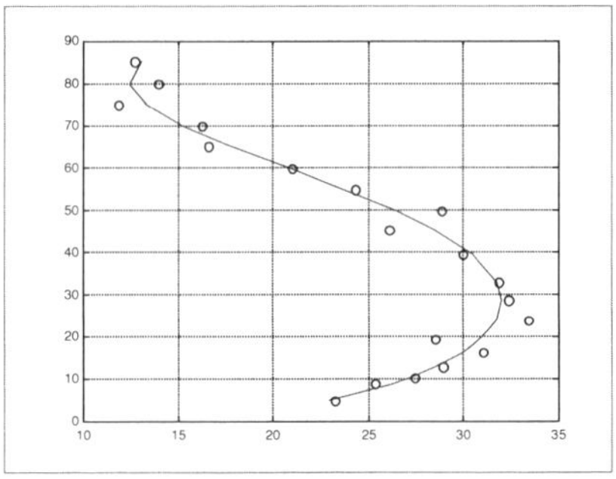

Fig. 9. Integral length scale at different levels ("fitting" with Von Karmán spectral density)

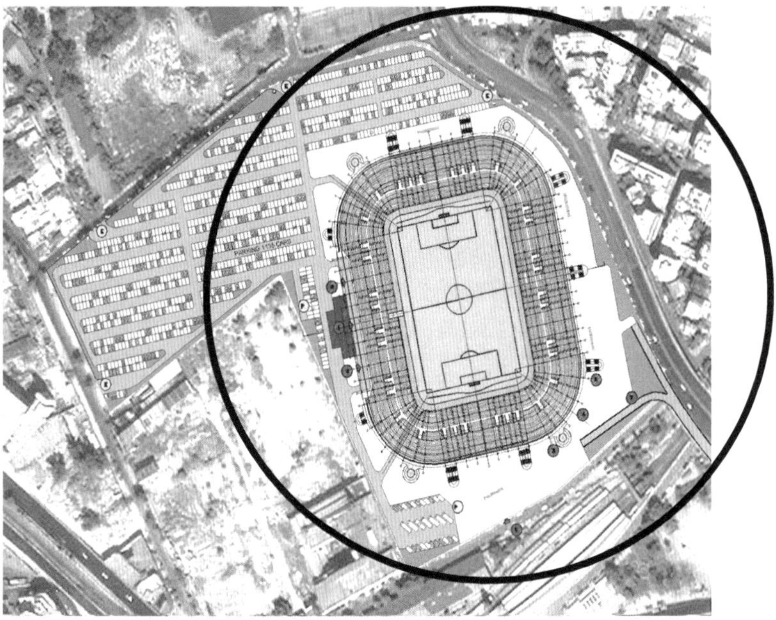

Fig. 10. Circle which identifies the location of the buildings included in the model

The roofing has been equipped with 252 pressure taps, of which 126 at the extrados and 126 at the intrados, in order to get the net pressures on the roofing. In the model the roofing of the stadium (Fig. 12) has a box structure in order to allow for the settlement of the pressure taps inside. A minimum thickness of about 7 mm has been required for the roofing structure to allow for the insertion of the pneumatic

Fig. 11. 3D Renderings

connections. The location of the pressure taps has been chosen to cover the whole roofing surface according to the Fig. 13, which shows also the influence area of each pressure tap. These areas have been obtained performing a triangulation among the pressure taps and linking together the barycentres of the identified triangles.

In the above figure the positions of the pressure taps are shown together with their influence areas; each position identifies the position of both the tap at the intrados and the tap at the extrados, which lay on the same vertical and are spaced out by the thickness of the box structure of the roofing.

The pressure measurements have been performed using piezoelectric transducers linked to the pressure taps through Teflon pipes (Fig. 14).

Fig. 12. Wind tunnel scale model

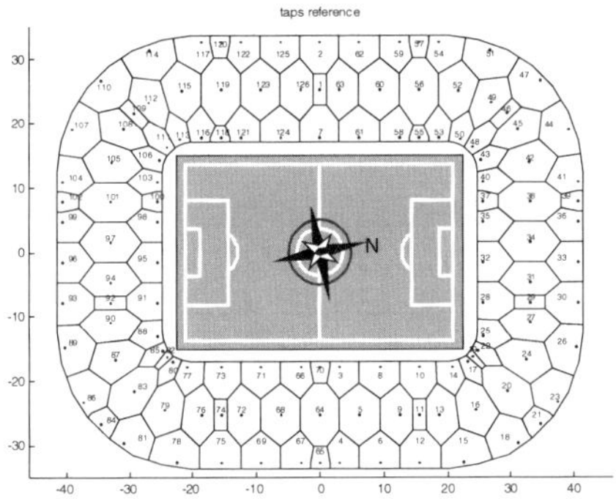

Fig. 13. Position of the pressure taps (each position corresponds to two pressure taps, one at the extrados and the other at the intrados of the roofing)

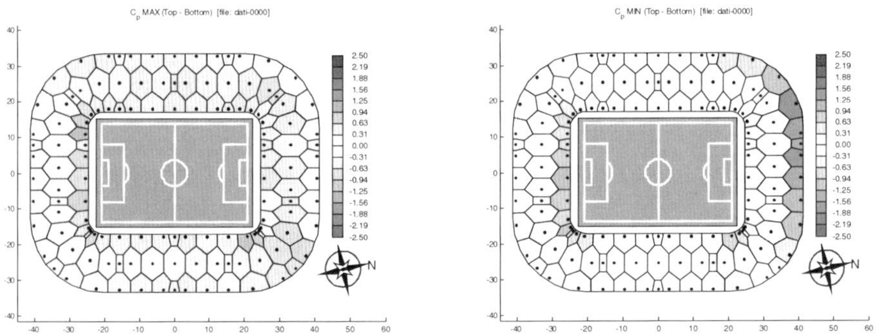

Fig. 14. Maximum and minimum values of net pressure coefficients (wind direction: 0°)

Measurement and use of load time histories: The Thessaloniki Olympic sport complex

The integration of the wind tunnel data into the design process presents significant problems for wide span sub-horizzontal enclosures; in contrast to buildings (high rise buildings) where knowledge of the base moment provides a sound basis for preliminary design, there is not single simple measure for the roof. The study of the Stadium of the Alpes and the Rome stadiums [4,5,6] drew attention to the inability of the measuring system employed to provide data in a form that could readily be based as input to the sophisticated dynamic numerical model developed by the designer and lead to discussion between the designer and the wind tunnel researchers to examine alternate techniques that might be used in future projects [7].

The discussions centered on the use of high speed pressure scanning systems capable of producing essentially simultaneous pressure measurements at some 500 points at rates of perhaps 200 Hz per point. With such a system it would be possible to cover in excess of 200 panels and produce a complete description of the load. Such a system would produce roughly 1 to 2×10^6 observations for a single wind direction and it is clear that some compression of the data would be required. One possible approach would be to produce a set of load histories, $Q_j(t)$, such that:

$$Q_j(t) = \int_A p(x, y, t)\phi_j(x, y)dA \tag{1}$$

where
$p(x, y, t)$ nett load per unit area at position (x, y);
$\phi_j(x, y)$ weighting function

For a series of pressure taps of the approximation to $\phi_j(t)$ would be

$$Q_j(t) - \sum_{i=1}^{N} \bar{p}_i(\bar{x}_i, \bar{y}_i, t)A_i\phi_j(\bar{x}_i, \bar{y}_i) \tag{2}$$

A_i area of ith panel;
p_i pneumatic average of pressure at the taps in the ith panel;
x_i, y_i geometric centre of the taps on the ith panel;
N number of panels.

The requirements of a system designed to produce the load histories, $\phi_j(t)$, is discussed in the following section.

In collaboration with the Boundary layer wind tunnel laboratory of the University of Western Ontario, a new very practical method to obtain the structural response under the random wind action and small displacements (linear response) has been applied under the name of the "orthogonal decomposition method".

If the weighting functions, $\phi_j(t)$, are chosen as mode shapes then $\phi_j(t)$ is a modal load and its use in conjunction with a dynamic model is clear; either as a set of time histories or a set of modal force spectra and cross-spectra. In the initial stages of a design the roof shape is probably known with reasonable accuracy but mode shapes not so. In such cases it might be appropriate to choose a suitable set of ϕ_j from which modal loads corresponding to shapes ψ_i can be estimated when the design is more advanced. In such a case we can approximate ψ_j as:

$$\psi_j \simeq \psi_j^i \sum_i^M a_{ij}\phi_j \tag{3}$$

the values of a_{ij} can be evaluated by minimizing the discrepancy between ψ_j and ψ_j, ie:

$$\frac{\partial}{\partial a_{ij}} \int \left(\psi_j - \sum_i a_{ij}\phi_i\right)^2 dA = 0 \qquad i = 1, M \tag{4}$$

If the functions ϕ_i are chosen as a set of orthogonal shapes $\int \phi_i\phi_j dA = 0;\ i \neq j$ then the coefficients are given as

$$a_{ij} = \frac{\int \phi_i\phi_j dA}{\int \phi_i^2\phi_j dA} \tag{5}$$

For a finite panel sizes the corresponding relationship is:

$$a_{ij} = \frac{\sum_{k}^{N} \phi_i(\bar{x}_k, \bar{y}_k)\phi_j(\bar{x}_k, \bar{y}_k)A_k}{\sum_{k} \phi_i^2(\bar{x}_k, \bar{y}_k)A_k} \tag{6}$$

where

$$\sum_{k}^{N} \phi_i(\bar{x}_k, \bar{y}_k)\phi_j(\bar{x}_k, \bar{y}_k)A_k = 0 \qquad i \neq j$$

The experiment would involve the recording of the local histories $\psi_j(t)$ from which the model time histories could be constructed and the analysis conduced in either the time or frequency domain (Figs. 15–18). For the type of structure under consideration resonant effects are small and the response is largely a quasi-static to a spatially varied load. The deflections induced are closely related to the imposed loads and their distribution differs significantly from the Gaussian form [7]. In such a case the time domain solution, which preserves the extreme value distribution, is to be preferred over a frequency domain approach.

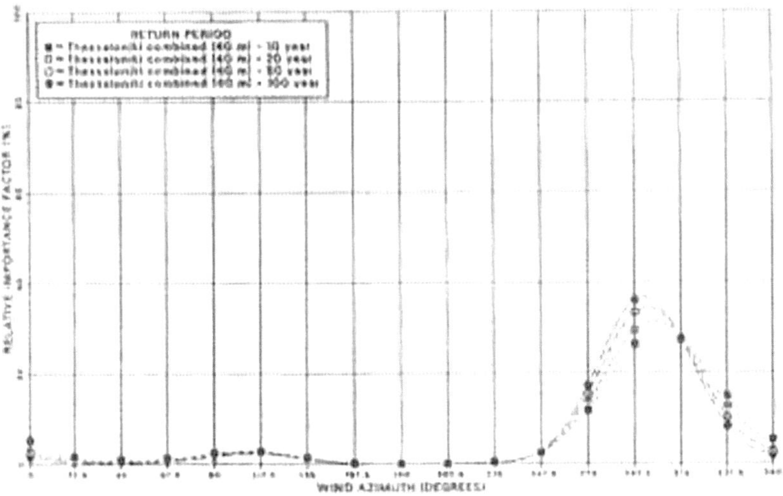

Fig. 15. Relative contribution of Azimuthal Direction to the exceedance probability of various return period wind speeds for Thermi, Thessaloniki, Greece

For the seismic analysis a frequency domain approach was adopted. The Kanai-Tajimi PSD was used under the design response spectra prescribed by Eurocode 1; under strong-motion, an acceleration time history was artificially generated according to site and durability characteristics [8].

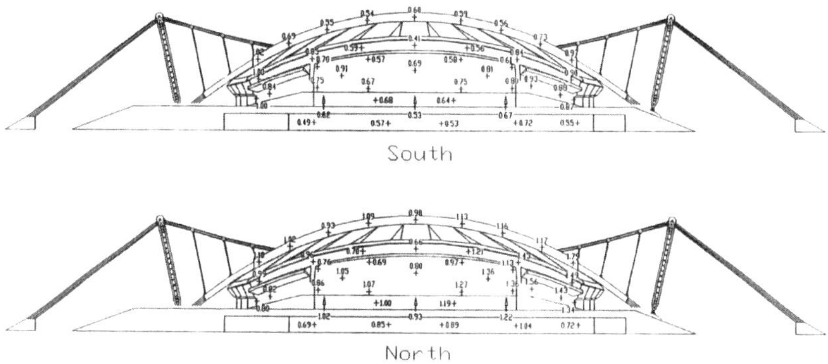

Fig. 16. Taps location

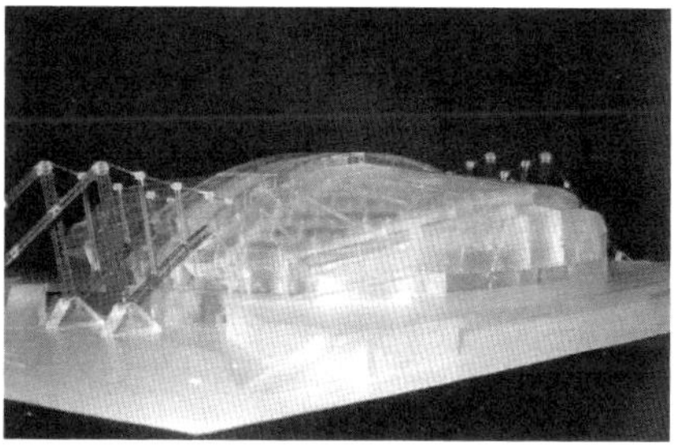

Fig. 17. Views of pressure model

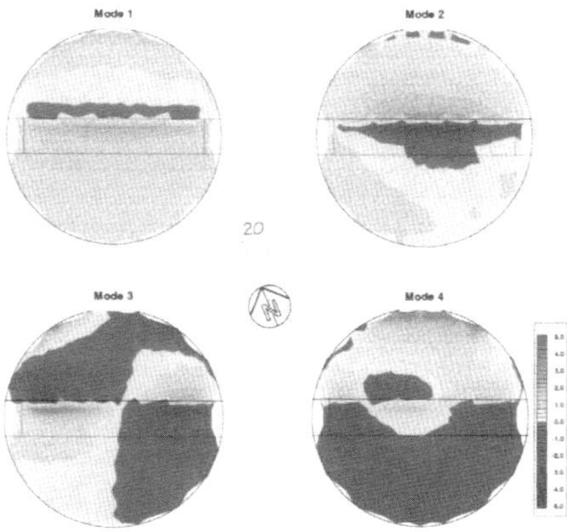

Fig. 18. Orthogonal decomposition: pressure mode shapes

2.3 Wind Loading-Experimental Analysis on Scale Models: Flexible Structures-Aerodynamic Behaviour: The olympic stadium in Rome

The wind induced response of the cable supported stadium roof was analysed by a non linear model and a field of multicorrelated artificial generated wind loading time histories [6].Wind tunnel tests have been carried out at the BLWT Lab. of UWO on a model of 1:200 Fig. 19 scale determining:

- time histories of the local pressures for every 10° of incoming flow direction; the maximun, minimun and average values of the wind pressure have then been evaluated, as well as the root mean square of its fluctuating part;
- presssure coefficients (maxima,minima and average) for every 10° of incoming direction;
- auto and cross-spectra of the fluctuating pressure (averaged on every single panel).

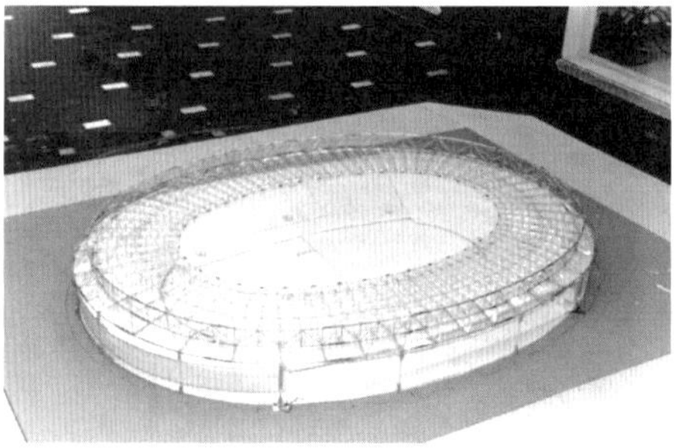

Fig. 19. Aeroelastic model for Rome Olympic Stadium

Fig. 20. Aeroelastic model for the Braga Stadium

The aerodynamic behaviour shows a clear shedding phenomenon. The external border of the structure, constituted of the trussed compression ring with triangular section and tubular elements and by the roofing of the upper part of the stands, disturbs the incoming horizontal flow in such a way so that vortex shedding is built up. This causes the roofing structure to be subjected to a set of vortices with a characteristic frequency. This is confirmed by the resulting Power Spectra Density Function of the fluctuating pressures, which shows a peak at about 0.15Hz even if the values rapidly decrease with increasing distance Fig. 21.

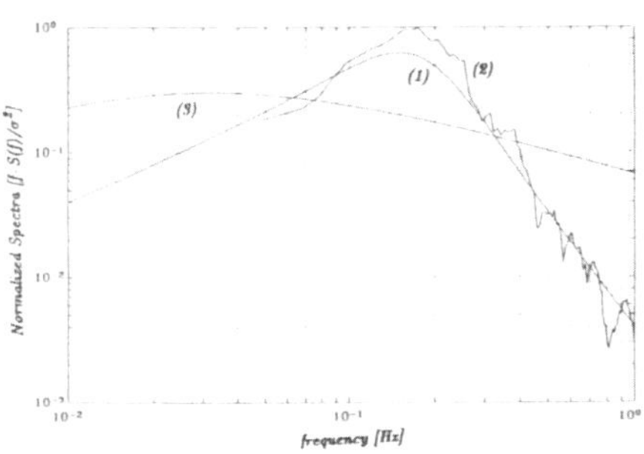

Fig. 21. Target (1), simulated (2) and Kaimal's (3) normalized spectra of wind velocity

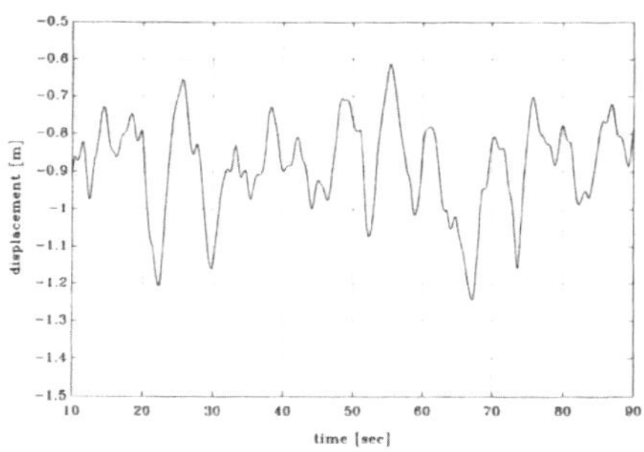

Fig. 22. Time History of the displacement (leeward side at tension ring, run #2)

A fluid-interaction non linear analysis in time domain, made for the checking of La Plata stadium design [9] shows a better agreement between theoretical model and experimental values.

3 Reliability Analysis

The sensibility analysis regarding the new suspended cable roof of Braga (Portugal).

3.1 Reliability Analysis of the Roof Structural System. Cable Strain Parametric Sensibility

Considering that in the basic solution the roof will be covered by a long span structural system with only uplift gravitational stabilization (Fig. 20) it is essential to proceed to the analysis of the response of the structural system to loading patterns and wind induced oscillations.

The analytical process will be organized in order to be controlled by experimental investigations in reduced and full scale.

The reduced scale experimental analysis on rigid and aeroelastic models are concerned with the determination of the dynamic loading on the roof surface and of the stability of the structural system.

The full scale experimental investigations are addressed to check, by a monitoring program, the validity of the global analysis process.

The uncertainties on the elastic modulus of the cable, geometrical and elastic long term creeping, tolerances of fabrication and erection, differences with design prestress, non uniform distribution of temperature, non linear behaviour, created a sensitive response on the suspended roof hanging from a set of suspended cables. The sensibility analysis showed that the response is sensitive to the standard deviation of the cable strain ($\Delta\varepsilon$) variations. The failure probability is given by the probability that an outcome of the random variables ($\Delta\varepsilon$) belongs to the failure domain D. This probability is expressed by the following integral [10]:

$$P_f = \int_{D_f} f_{\Delta\varepsilon}(\Delta\varepsilon) \cdot d\Delta\varepsilon \tag{7}$$

and the most probable failure mechanism will involve primarily the border cables.

The sensibility analysis was, therefore, extremely important to detect the weak points of the structural system and permits proper local dimensioning to prevent chain failure, as illustrated with the failure simulation of same sensitive cable elements.

The roof is composed by a structural concrete plate sustained by **n** prestress cables. In the analysis the roof, the bending moments at **m** points will be considered. For a particular load combination, the **n** cables have computed strains given by the vector $\boldsymbol{\varepsilon}$. Considering that these effects are represented by the vector of random variables $\boldsymbol{\Delta\varepsilon}$ with mean values $\boldsymbol{\mu}$ and standard variations $\boldsymbol{\sigma}$, the problem is to estimate the probability, P_f, that the generated random bending moments $\mathbf{M}$ will be larger than the plate ultimate resistance moments, $\mathbf{M}_u$, at any of the **m** points of the structural plates system.

3.2 Roof Structural System Data

The following probabilistic description was considered for the random variables $\boldsymbol{\Delta\varepsilon}$.

$\boldsymbol{\mu}$ = Vector of mean values of $\boldsymbol{\Delta\varepsilon} = 0$ (i.e., all possible actions on the cables are considered by the load combination itself).

$\boldsymbol{\sigma}$ = Vector of standard deviations of $\boldsymbol{\Delta\varepsilon} = 0$. The $\boldsymbol{\sigma}$ values were varied from 0.5×10^{-3} to 0.1×10^{-3} so that the sensibility of the system can be studied. These values were selected to cover the range of failure probabilities of practical significance.

$\mathbf{f}_{\boldsymbol{\Delta}\varepsilon}(\boldsymbol{\Delta}\varepsilon)$ = Probability density function = Normal distribution with parameters $\boldsymbol{\mu}$ and $\boldsymbol{\sigma}$.

3.3 Failure Condition

For load case "i" the bending moments, M_x, M_y and M_{xy} in the 130 points of the plate can be computed as follow:

$$\begin{aligned} M_x &= M_{Gx_i} + \sum_{j=1}^{34} A_{x_{i,j}} \cdot \Delta\varepsilon_j \\ M_y &= M_{Gy_i} + \sum_{j=1}^{34} A_{y_{i,j}} \cdot \Delta\varepsilon_j \\ M_{xy} &= M_{Gxy_i} + \sum_{j=1}^{34} A_{xy_{i,j}} \cdot \Delta\varepsilon_j \end{aligned} \tag{8}$$

Considering the bending moments in each direction, the failure functions at each point of the plate ($1 \leq r \leq 130$), $G_r\ (\epsilon)$, are the following hyperplanes,

$$\begin{aligned} M_{Upx} - (M_{Gx_i} + \sum_{j=1}^{34} A_{x_{i,j}} \cdot \Delta\varepsilon_j) < 0 \\ M_{Upy} - (M_{Gy_i} + \sum_{j=1}^{34} A_{y_{i,j}} \cdot \Delta\varepsilon_j) < 0 \end{aligned} \tag{9}$$

$$\begin{aligned} M_{Unx} - Abs(M_{Gx_i} + \sum_{j=1}^{34} A_{x_{i,j}} \cdot \Delta\varepsilon_j) < 0 \\ M_{Uny} - Abs(M_{Gy_i} + \sum_{j=1}^{34} A_{y_{i,j}} \cdot \Delta\varepsilon_j) < 0 \end{aligned} \tag{10}$$

$$M_{Uxy} - Abs(M_{Gxy_i} + \sum_{j=1}^{34} A_{xy_{i,j}} \cdot \Delta\varepsilon_j) < 0 \tag{11}$$

where $G_r \leq 0$ is failure and M_{Uxy} is computed from the Johanssen Theory as the smallest of the following expressions

$$M_{Uxy} = (M_{Upx} + M_{Upy})/2 \qquad M_{Uxy} = (M_{Unx} + M_{Uny})/2 \tag{12}$$

In these formulas, $M_{Upx}, M_{Upy}, M_{Unx}, M_{Uny}$ and M_{Uxy} are considered always positive.

The failure condition is obtained when failure is reached at any point of the plate, i.e., the structural failure can be defined as

$$(G_1 \leq 0) \cup (G_2 \leq 0) \cup \_ \cup (G_{130} \leq 0) \tag{13}$$

3.4 Solution Method

Since a closed form solution is not possible for the integral in (7) the failure domain defined by equations above, Montecarlo Simulation must be used. By Montecarlo Simulation, the failure probability is obtained by computing $\mathbf{G}_r(\boldsymbol{\Delta\varepsilon})$ for several values of $\boldsymbol{\Delta\varepsilon}$ generated with normal distribution. An approximation to the failure probability is obtained by counting the number of times that $\boldsymbol{\Delta\varepsilon}$ belong to the $\mathbf{D}_f$ with respect to the total number of simulations. For small failure probabilities, however, direct application of Montecarlo Simulation is not possible because of the large number of needed iterations to get enough accuracy. To avoid this problem, the Orientated Simulation Method was used in this report. A complete description of the method can be found in the paper [10].

Load Case	Beta	Phi (-Beta)
1	5.8739	2.14E-09
2	5.7957	3.42E-09
3	5.9555	1.31E-09
4	5.5733	1.26E-08
5	4.1218	1.87E-05
6	4.8436	6.41E-07
7	*1.6658*	*4.79E-02*
8	5.7281	5.11E-09
9	5.5396	1.53E-08
10	*2.6269*	*4.31E-03*
11	*2.3812*	*8.63E-03*
12	4.3046	8.37E-06
13	4.3045	8.37E-06
14	5.8201	2.96E-09
15	5.7479	4.55E-09
16	5.8415	2.61E-09

Table 2. Table of Betas

3.5 Results and Conclusions

All the load cases were analysed and the following preliminary conclusions are described as follows.

In order to identify the most dangerous load case the minimum reliability index β for each load cases were calculated for a standard deviation $\sigma = 0.5 \times 10^{-3}$ for

$\Delta\varepsilon$ of all cables. The following table (Table 2) summarizes the index β (computed with $\sigma = 0.5 \times 10^{-3}$).

The load cases 7, 10 and 11 have the lowers β, i.e., the higher failure probability, and therefore they are the critical load condition. Particularly critical is the load case 7.

3.6 Failure Probability and Sensibility Analysis

The Fig. 23 shows the failure probability for load combination 7 as a function of the standard deviation, σ, of the cable strain variations, $\boldsymbol{\Delta\varepsilon}$.

a. The problem is extremely sensitive to the standard deviation, σ, of the cable strain variations, ϵ. For example for load case 7, if σ is increased from 2×10^{-4} to 3×10^{-4}, Pf is increased from 2×10^{-5} to 480×10^{-5}.
b. Cable standard deviation, σ, should be maintained below 2×10^{-4} for the designed ultimate bending moment.
c. Larger cable standard deviation, σ, could be allowed increased the slab reinforcement along x-direction in the critical roof zone.

The Fig. 24 shows the most probable values of $\epsilon(\times 10^{-3})$ in each cable at failure for load combination 7.

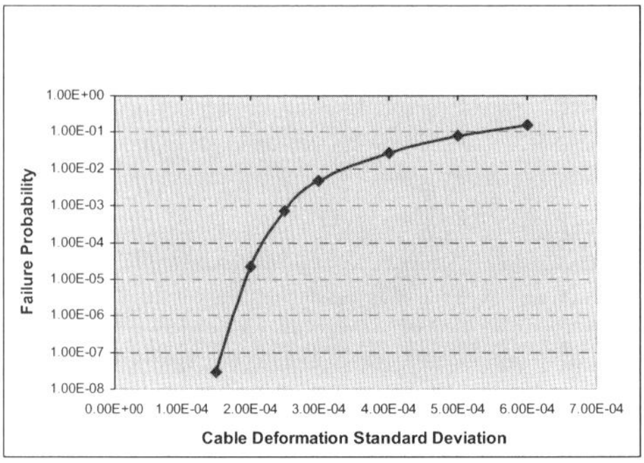

Fig. 23. Failure probability in function of cable deformation standard deviation

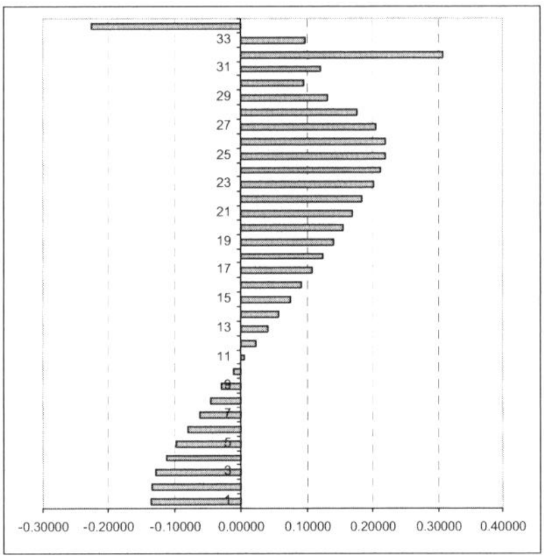

Fig. 24. Most probable $\Delta\varepsilon$ meach cable at failure for load comb. 7

The following comments can be done.

a. The most probable values of ϵ are practically independent of the standard deviation σ. In other words, the configuration at failure is constant. This configuration is reached with more probability as the standard deviation of ϵ increases.
b. The most probable configuration at failure is mainly due to variations in the strains of cables 32 and 34. Since elongations of cables can be computed as $L = L\epsilon$, the elongation at failure of cables 32 and 34 are approximately $L_{32} = 210m \times (-0.2 \times 10^{-3}) = 4.2$ cm and $L_{34} = 210m \times (0.3 \times 10^{-3}) = 6.3$ cm.

4 Monitoring

The roof structures of the Torino Stadium have been built in 1990. According to the quality control and maintenance program, the in service subsequent performance of the structure has been controlled by site inspection, experimental measurement and spot monitoring of representative structural parameters. The anchorage forces in the cable stays have been controlled during 1992 and considerable differences in average and peak values of pre-stressing have been observed between experimental and expected theoretical values. The authors proceeded, with a computer simulation of the actual observed anchorage force values, to determine a new pre-stressing sequence in order to fit the original design cable force. Actually the structure is under normal monitoring observation according to a special maintenance program.

4.1 Displacement and Force Control

The first tensional controlling operations were performed on the cable anchorage forces in 08.05.1992. The geometrical control of the central ring was performed during and after the completion of the operations, dated 13.05.1992. On the 19.05.1992 the designers were requested to conceive an intervention plan to eliminate the detected errors related to the intensity and distribution of the state of initial stress.

Therefore, on the 3.08.1992, it was presented a proposal related to a theoretical simulation of the experimentally observed stressing state and to the definition of a correct new pre-stressing. The comparative analysis between the results, referred to the actual service state anchorage forces (whose permanent loads were updated according to the final consumptive analysis) and the experimental data, revealed average and peak values' differences bigger than the allowable ones. The average prestressing level is reduced of 30%, with peak differences of 40-50% (Figs. 25-26).

4.2 Interpretation of Results

In order to have a better scenario of the pre-stressing level and distribution, a new control of all anchorage forces under uniform temperature condition (night-time) had been produced. The results confirmed the spot control measurements. Considering the main characteristics of the structural typology adopted or the stadium the variation of the pre-stressing state can be related principally to:

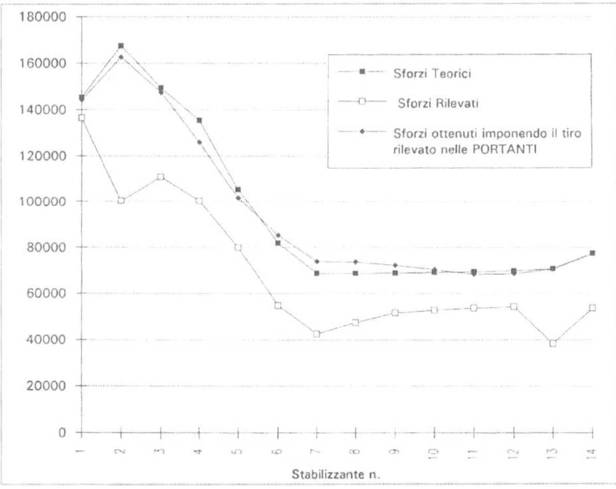

Fig. 25. Stabilizing cable. Anchorage forces. Computed actual values. Measured values. Restored values

- non uniform temperature variation;
- elastic an anelastic foundations' settlements and soil interactions;
- random errors in cutting or marking the cables;
- geometrical, elastic or anelastic short and long term creeping of cables;
- errors in pre-stressing procedure.

The theoretical simulations, which include the compsuntive updating of the permanent loads, provide to the eliminations of the uncertainties related to the loading variation's influence. The soil conditions and measurements of settlements during 4 years permit to remove any consideration of influence of soil interaction in variation of pre-stressing level. Measurements taken during night-time show no influence for temperature variations.

Considering that peak values are random distributed it is also possible to disregard the influence of errors in cutting and marking of cables, normally respecting the symmetry of construction. The logical answer, therefore, confirmed by a parallel experience on the structural system of the Olympic Stadium in Rome, is that the variation noted in pre-stressing level of the system can be related to a combinations of creeping and to uncertainties concerning the pre-stressing procedure.

"Geometrical" equivalent creeping of spiral strands is supposed to be removed by 50% breaking load shop prestressing cycles. Many cable structures showed creeping in short and long term. Creeping of cables is not well documented in technical literature. Some fabricators give $3\,^0/_{00}$ as "possible values" of anelastic creeping which imply the impossibility to guarantee any value of E-modules ($3\,^0/_{00}$ of E imply a variation of 500 N/mm^2 !).

Measurements of the geometry of the central ring show a separation between expected theoretical and measured values, but the experimental Z-configuration of the ring had only small variations in time during 3-year observations. The consequence of the above mentioned observations is that the changements observed in the pre-stressing state must be related principally to the procedure of pre-stressing under high gradient of daily temperature variations. The tensostructure was prestressed by an iterative "step by step" sequence. The number of steps have to be determined in function of the characteristics of the structural system as the parametric sensibility, the correlation degree between incremental values and the verification of the level of the transitory sequential overstressing. The reliability of the process is increased by the number of iterations, allowing to calibrate and automatically control the pre-stressing according with a predictor-corrector techniques.

As a matter of fact, probably in order to reduce the time needed for the operation, it was decided to make the pretensioning in just one step. This choice allowed the minimization of the reliability of the process even though the geometric, mechanic, thermical variations along with those related to the mathematical modeling of the pretensioning simulation could contribute to increase, effectively, the varying between the project's values and the ones related to the expected forces, actually induced. The designers recommended the functional restore of the in service state by bringing the pre-stressing values back to the average expected ones.

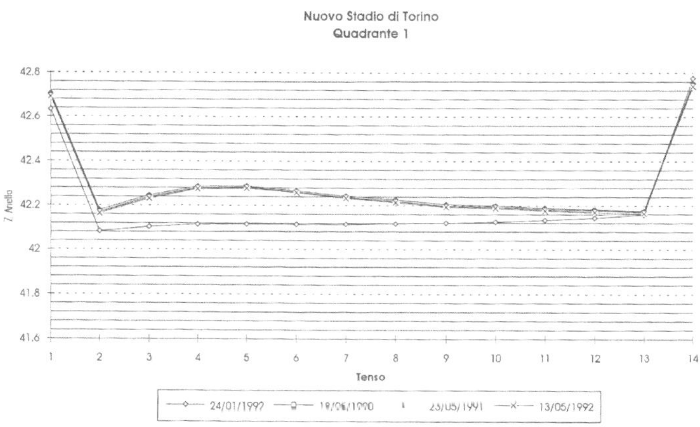

Fig. 26. Z coordinate of the inner ring according to experimental measurement

4.3 Measurements and Monitoring

A special monitoring program was adopted during transitory time in order to control the displacements of some points of the roof.

Before and after the pre-stressing procedure a dynamic load was applied to the structure and, with the use of a special accelerometer giving efficient possibility to integrate accelerations in range of low frequencies ($< 10 - 1$ Hz) , power spectral densities of response could be plotted. With the frequencies and associated mode shapes it was possible to observe the increase of geometrical stiffness of the structural system by the new pre-stressing level. Accelerations and displacements, before and after prestressing, are shown in Table 3.

	Acceleration m/sec^2	
	Before	After
A3	0.93 - 1.19	0.86 - 0.96
A4	0.50 - 1.00	0.32 - 0.44
	Displacements mm	
	Before	After
A3	7.8 - 10.0	6.8 - 7.5
A4	6.8 - 9.2	5.2 - 7.1

Table 3. Accelerations and displacements under dynamic load before and after pre-stressing

5 Conclusions

It has been noted the influence of knowledge base on conceptual design in removing gross human intervention errors from initial design statements.

Design assisted by experimental investigation is a useful integration of the design process of wide span structures.

Sensibility analysis is an extremely powerful tool to determine the influence of parametric design uncertainties for unusual long span structural systems.

References

1. Melchers RE (1987) Structural reliability. Elley Horwood ltd.
2. Structural Design of Retractable Roof Structures. IASS working group no. 16, WIT Press, (2000).
3. RWDI: Roof snow loading study-roof re-design Olympic Stadium Montreal. Quebec. Report 93-187F-15, (1993).
4. Majowiecki M (1990) Observations on theoretical and experimental investigations on lightweight wide span coverings. International Association for Wind Engineering, ANIV.
5. Vickery BJ and Majowiecki M (1992) Wind induced response of a cable supported stadium roof. Journal of Wind Enginerring and Industrial Aerodynamics 1447-1458.
6. Vickery BJ (1993) Wind loads on the Olympic Stadium - orthogonal decomposition and dynamic (resonant) effects. BLWT-SS28A.
7. Majowiecki M (1998) Snow and wind experimental analysis in the design of long span sub-horizontal structures. J. Wind Eng. Ind. Aerodynamics.
8. Majowiecki M, Zoulas F and Ermopoulos J (1999) The new sport centre in Thermi Thessaloniki: conceptual design of the structural steel system. IASS Congress Madrid, September.
9. Lazzari M, Majowiecki M, Saetta A and Vitaliani R (1998) Analisi dinamica non lineare di sistemi strutturali leggeri sub-horizzontali soggetti all'azione del vento. 5o. Convegno Nazionale di Ingegneria del vento, ANIV, Perugia.
10. Puppo AG and Bertero RD (1992) Evaluation of Probabilities using Orientated Simulation. Journal of Structural Engineering, ASCE 118(6), June.

Fabric Membranes Cutting Pattern

Bernard Maurin[1] and René Motro[2]

[1] Laboratoire de Mécanique et Génie Civil maurin@lmgc.univ-montp2.fr
[2] Laboratoire de Mécanique et Génie Civil motro@lmgc.univ-montp2.fr

1 Introduction

Tensile fabric membrane design implies successive stages, each of one related to particular problems requiring well adapted approaches and appropriated results.
The first step of the analysis deals with the form-finding process that corresponds with the coupling in lightweight structures between the form (geometry) and the forces (initial tension). The objective is to determine the shape of the membrane associated to its prestress distribution. A good control on the tension in the fabric must be obtained in order to have suitable stresses, for instance that ensure the absence of compressive zones.
The following stage focuses on the realization of the tensile membrane calculated during the form-finding. More precisely, the objective is to determine the starting configuration (set of plane *strips*) which, once assembled on the site according to specified anchoring conditions, will lead closely to the required surface, that is to say to the theoretical one (target strip) calculated during the shape-finding procedure with its associated characteristics of form and prestress. The erection process indeed generated deformations on each strip that will define in the end a mechanically equilibrated geometry coupled with a prestress distribution. The purpose is to minimize the differences between the target state and the therefore obtained state. It corresponds with the *cutting pattern stage*.
In case of low deviances, the prejudice will be essentially aesthetic such as disgraceful folds (Fig. 1 left) but, in case of higher differences, the integrity of the whole structure could be affected since large membrane zones may be less or not tensioned, this leading to major risk of failure (wind fluttering, horizontal areas with stagnant rain water..., Fig. 1 right). The cutting pattern necessitates the specification of the surface cutting lines also called *strip edges*. This procedure has to take into consideration several parameters likewise technology, geometry, mechanics and aesthetics.
Each strip being so identified, the designer must next calculate the associated

E. Oñate and B. Kröplin (eds.), Textile Composites and Inflatable Structures, 195–212.

plane fabric cutting patterns. Most of the used methods split the process into two different stages. In the first one, the 3D strip is flattened onto a projection plane; in the second, the pretension of the membrane is considered so as to reduce its dimensions.

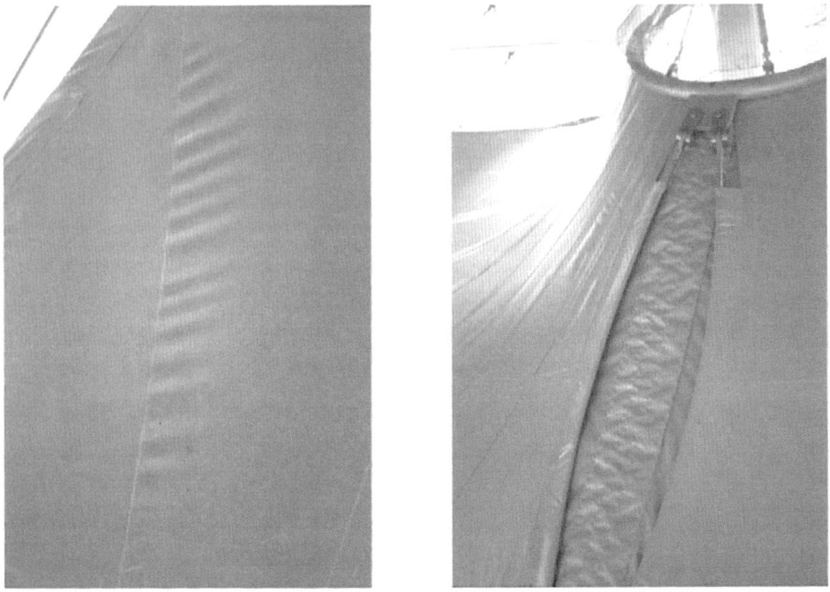

Fig. 1. Folds at strip edges; compressive zones

2 Strip Edges Determination

This process results in determining the balance between various, and sometimes opposite, requirements.

2.1 Technological Issues

The design must firstly take into account the maximum strip widths in connection with the products available from fabric manufacturers. Generally, membranes are supplied in the form of 2m width rolls [5]. After cutting, the strips are assembled by thermo-welding (fusing of the material between high frequency electrodes and pasted by applying a pressure); the resulting membrane is next transported to the erection site.

2.2 Geometrical Issues

Some designers consider as necessary to have strip edges along geodesic curves ([2] and [14]). It allows indeed, in the particular case of surfaces that are developable onto a plane (on the mathematical meaning), to generate straight lines in accordance with an economical objective of minimal material wastes. This approach may however be relativized since, in the case of double curvature geometries, the surfaces are not developable: we know that such operation leads to unavoidable distortions. It is then judicious to use low dimension strips on a surface zone with high total curvatures. A numerical method devoted to the calculation of membrane curvatures is presented in appendix. Nevertheless, this consideration has to be balanced with a resulting increase of the cutting operations and welding lengths and therefore of the total cost.

2.3 Mechanical Issues

The production of the fabric does not end in a perfect isotropy between the warp and weft directions (higher strength and stiffness for the warp) even if improvements in production processes aim to reduce this difference. The low shear strength of the fabric has also to be taken into account. Thus, an ideal configuration will be related to the positioning of the strip edges, that correspond after cutting approximately with the warp direction, along the direction of the main forces, that is to say the maximum principal stresses. In that case, the fabric weft is thus turned on to the minimum stresses directions with resulting shear forces close to zero. All of these theoretical considerations have however to be balanced with practical aspects: inexistence of exact solution, knowledge of stresses due to the initial stresses in the fabric and to climatic effects. If the action of wind is paramount (pressure normal to the surface), then the directions of maximum stresses correspond with the directions of the membrane maximum curvatures. For snow (vertical action) the answer is much more complex but some basic cases such as the radial positioning of the strip edges at the top of anchoring masts (Fig. 2). In addition to these requirements dealing with the surface, others considerations are related to the membrane edges. Since the initial pretension is applied by progressively tensioning edge cables, it is necessary that strip edges be orthogonal to these cables.

However, so as to point out the problems associated to particular situations, we quote the case of the design of Mina Valley tents in Mecca build for pilgrims [10]. The project, realised in two stages in 1997/98, is composed of 40000 tents with a rectangular frame (from 4x4m to 8x12m) with a vertical mast at middle. The membranes build during the first stage are based upon the basic radial positioning of strip edges, but the difficulties in tensioning the fabric with the mast have lead to prohibited folds on the surface. The designers of the second team have then decided to set the strip edges parallel to the anchoring sides.

Fig. 2. Radial strip edges

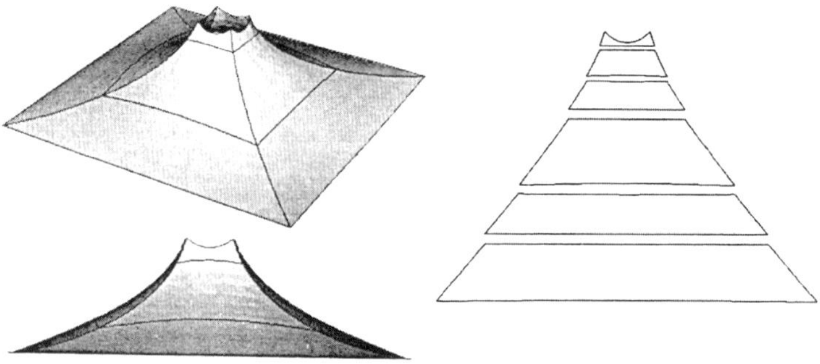

Fig. 3. Strips used for the Mina project stage 2

It was allowed by the absence of snow and has resulted in the vanishing of the folds (Fig. 3).
We emphasize herewith on the fact that small structures design may generate more difficulties than wider membranes design since the dimension of the fabric rolls appears as important with reference to the dimensions of the structure.

2.4 Aesthetical Issues

The approach could however be modified when architects play a role. Their creativity may for instance leads to the making of geometric drawings by using fabric samples of different colours. Moreover, since the visual perception of the surface is dependant on the strip edges positioning, mainly at night, this architectural feature could lead to specific patterning strategies.

3 Cutting Shapes Determination

3.1 Background

Before presenting several used methods, we aim to point out some significant principles.

Since the objective is to have a good adequacy between the target state and the real state, it is thus necessary for each strip to evaluate the result in the light of the morphological parameters of forms and forces:

- If the geometry of the strip put into place is close to those theoretically determined during the form-finding stage, we will say that it exists a *geometrical equivalence* between the two strips. However, one point has to be respected: an edge belonging to two strips must have the same length on the plane cutting shapes so as to allow their future assembly.

- Similarly, if the prestress field generated in the strip is close to the required one, the *sthenical equivalence* is ensured. We may observe that it implies the perfect knowledge of the selfstress state determined during the form-finding process.

Nevertheless, these two principles only represent a virtual reality since it is illusory to expect a complete equivalence but very particular cases. A patterning method without taking into account all the geometric and sthenic data will however not offer the possibility to have an optimal solution to the problem. The same comment is also relevant if these considerations are not seen as indissociable and so envisaged as two separate steps (flattening and then reduction). We remark that, as far as we can know, most of the used methods are based upon such splitting.

Let's now have a look on the existing flattening processes.

The first technique is the *simple triangulation method* (Fig. 4). The 3D strip (a) determined by form-finding is mapped with a series of triangles between the longest edges, leading to the geometry (b). The obtained triangles are next successively flattened onto a plane by keeping identical the lengths of their sides (c). Since this method is quick and easy to apply, it is at the core of numerous CAD tools.

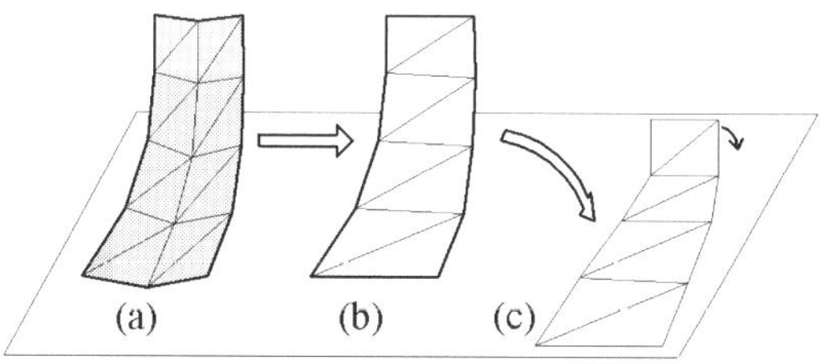

Fig. 4. Simple triangulation method

We may however note that it does not take into consideration a lot of points: those located inside the strip and some on top and bottom edges. Its use therefore requires caution so as to avoid important errors.
Several authors have thus proposed improved processes. L. Gründig has put forward a method which takes into account all the points belonging to the strip edges [7]; this objective being also one of H. Tsubota's aims [13]. In both cases, the geometry of edges is calculated by using minimization error processes. We may regret that data of internal points are still avoided. The method proposed by T. Shimada ([12] and [1]) offers some improvements on that purpose. It consists in determining a plane domain composed of triangular surface elements which, once transformed into the 3D strip, leads to a minimal strain energy. The material characteristics are used in the mechanical formulation. However, parameters related to the prestress of the membrane are not considered.
In each situation, the development of the strip must be followed by an operation so as to take into account the initial stresses. Three main strategies may be pointed up:
- The strip is not modified. Stresses in the fabric are generated by the displacements of anchoring zones (mast erection, shortening of edge cable lengths...).
- If the strip is triangulated, every element is reduced along two directions in relation with the selfstresses determined by shape-finding. In the case of uniform stresses within the strip (minimal area surface for instance) the solution in not difficult. On the contrary, specific methods are to be used to determine an accurate result.
- The most commonly used technique results in considering a reduction scale factor for the developed shape. The designer applies it generally along the longer direction of the strip (the warp direction, with a factor from 2 to 3%) and the transverse reduction along the weft is obtained during the strips assembling (bilayer of the welded zone close to 2cm width).
The experience and a good knowledge of the material, mainly obtained by mechanical testing (stiffness, creep... [6]), play nevertheless a major role in these methods and the designer must proceed carefully.
If every cutting pattern process unavoidably leads to errors, we may remark that the greatest part of them are directly related to the splitting of the technique into two separate operations of flattening (for the geometry) and reduction (for the prestress). Hence, it appears that a better solution obviously relies in an *integrated approach* which takes into account at the same time the considerations of form, forces and material. The target state has to be determined in a comprehensive process without splitting of these parameters. Several research teams have thus proposed innovative patterning methods based upon such integrated approach. We may quote the works of J. Kim [8] and the method developed at the Mechanics and Civil Engineering Laboratory at Montpellier University and called *stress composition* method [9].

3.2 Stress Composition Method

We consider a 3D target strip Ω^{L} calculated by form-finding process and with every elementary prestress tensor known $\{\sigma^{\mathrm{ff}}_{loc}\}_L = \{\sigma_x \ \sigma_y \ \sigma_{xy}\}_{\mathrm{L}}$. The method relies in the determination of the plane cutting shape Ω^0 such as its exact transformation into Ω^{L} (that is to say by considering the geometrical equivalence as respected) generates these stresses (i.e. the sthenical equivalence as an objective). We will in the end go back on the relevance of this starting assumption.

On that purpose, a starting domain Ω^* is defined (determined by the orthogonal projection of the target strip on the development plane) and is deformed into Ω^{L} hence generating the prestress $\{\sigma_{loc}\}_{\mathrm{L}}$. If these values are different from $\{\sigma^{\mathrm{ff}}_{loc}\}_L$ then Ω^* is modified into Ω^0, so as the resulting stresses $\{\sigma^{\mathrm{mod}}_{loc}\}_*$ balance the observed deviation (Fig. 5). The calculation of the tensor $\{\sigma_{loc}\}_{\mathrm{L}}$ is achieved in accordance with the hypothesis of large displacements. The mechanical characteristics of the fabric are taken into account during every transformation.

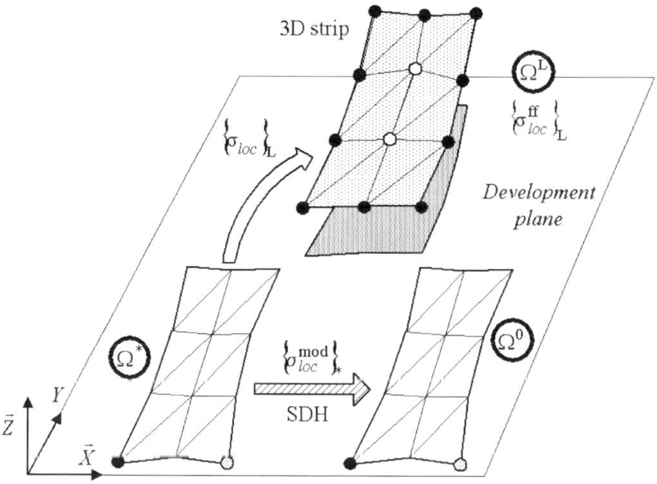

Fig. 5. Used configurations

The modification of Ω^* into Ω^0 is achieved according to the small displacements and small strains hypothesis (SDH), by acting on the boundary conditions of the frontier nodes of Ω^* in order to have $\{\sigma^{\mathrm{mod}}_{loc}\}_*$ close to the deviation $\{\sigma_{loc}\}_{\mathrm{L}} \quad \{\sigma^{\mathrm{ff}}_{loc}\}_{\mathrm{L}}$.

The associated displacement $\{d^{\mathrm{mod}}_{\mathrm{f}}\}^0_*$ (Fig. 6) may be determined with reference to the matrix relationship:

$$[A_\sigma]_* \{d^{\mathrm{mod}}_{\mathrm{f}}\}^0_* = \{\sigma^{\mathrm{mod}}_{\mathrm{loc}}\}_* \tag{1}$$

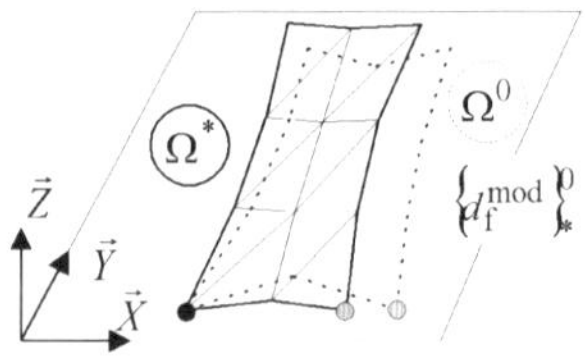

Fig. 6. Transformation of Ω^* into Ω^0

The equation is solved by using a least square method that provides a first approximate solution of Ω^0 called $\Omega^{0(1)}$. We consider next a second "starting" domain $\Omega^{*(2)} = \Omega^{0(1)}$; it allows therefore to calculate a second approximation $\Omega^{0(2)}$ of Ω^0. This iterative process constitutes the background of the stress composition method (Fig. 7).

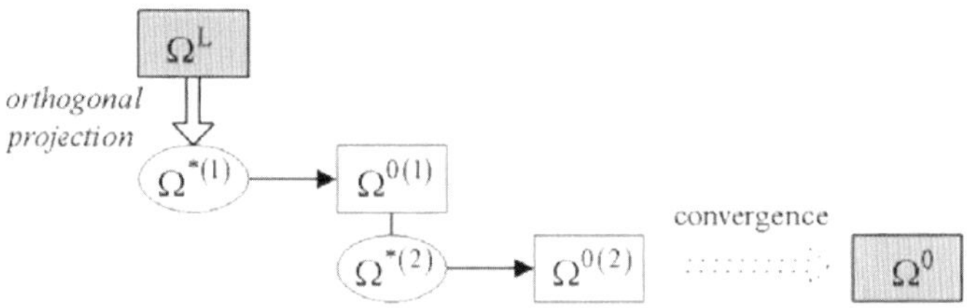

Fig. 7. Stress composition method

It converges in p iterations into the domain $\Omega^{0(\mathrm{p})}$ characterized by a stress deviation $\Delta\sigma_0^{\mathrm{L(p)}}$ according to:

$$\Delta\sigma_0^{\mathrm{L(p)}} \, \|\{\sigma_{loc}^{\mathrm{ff}}\}_{\mathrm{L}}\| = \|\{\sigma_{loc}^{\mathrm{ff}}\}_{\mathrm{L}} - \{\sigma_{loc}\}_{\mathrm{L}} + \{\sigma_{loc}^{\mathrm{mod(p)}}\}_*\| \tag{2}$$

The sthenic convergence criterion is set by the designer accordingly to the required maximum value of $\Delta\sigma_0^{\mathrm{L(p)}}$. The vectorial norm $\| \ \|$ corresponds to the euclidian norm.

3.3 Application

The searched 3D strip belongs to a minimal area surface characterised by an isotropic and uniform prestress $\{\sigma_{loc}^{\mathrm{ff}}\}_L = \{\sigma_0 \ \sigma_0 \ 0\}_{\mathrm{L}}$ with $\sigma_0 = 250$ daN/m. The dimensions of the strip in plane projection are 10m x 2m with an elevation of 1m in the top vertex (Fig. 8). The mechanical features of the material are those of a manufactured fabric: warp direction Young modulus equal to 24900 daN/m and 23000 daN/m for the weft direction with Poisson coefficients 0,097 and 0,090.
The stress composition method converges in seven iterations with a stress distortion $\Delta\sigma_0^{\mathrm{L(7)}} = 3,70\%$ with reference to the target state.

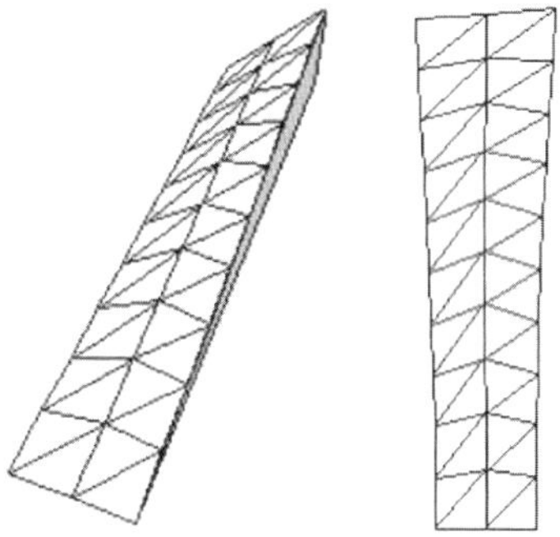

Fig. 8. Strip determined with the stress composition method

The following graph represents the values of the principal stresses σ_1 and σ_2 for every 40 triangular elements used for the modelling; the ideal solution is drawn by the horizontal line $\sigma_1 = \sigma_1 = \sigma_0 = 250$ daN/m (target state). We observe a regular stress distribution within the surface and the solution appears as quite satisfactory.

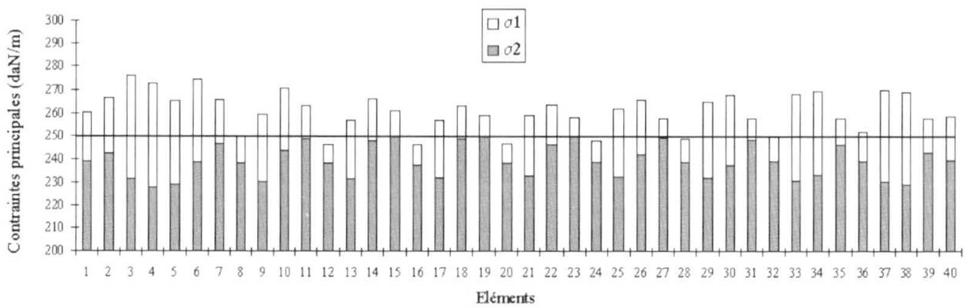

Fig. 9. Principal stresses obtained with the integrated method

If the same strip is now calculated with the simple triangulation method followed by a reduction, the determined shape may be characterised by a stress deviance $\Delta\sigma_0^{L(7)} = 26,62\%$ (evaluated after remeshing the plane domain with 40 triangular elements by adding internal nodes without altering the geometry of edges).
The values of the resulting principal stresses are presented in the following graph.
We point out that, when all the geometry of the domain is not taken into consideration likewise during the mapping with simple triangulation, increasing deviances occur. Such result is however not so surprising. A more detailed analysis of the graph allows noticing a higher distortion between the principal stresses for the elements located in the top zone of the surface (right part of the graph). A possible explanation relies in the fact that these elements are

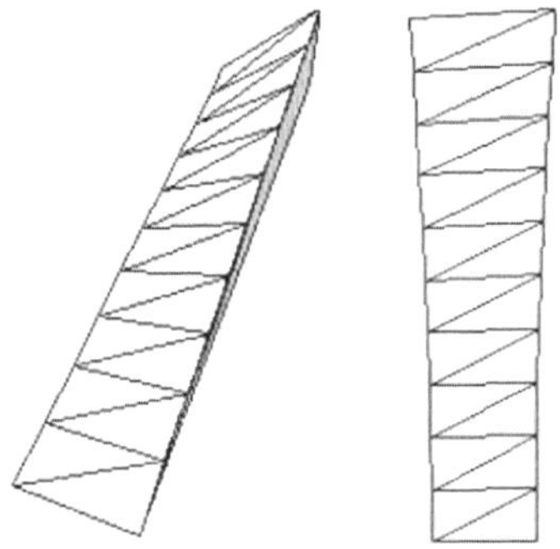

Fig. 10. Strip determined with simple triangulation method

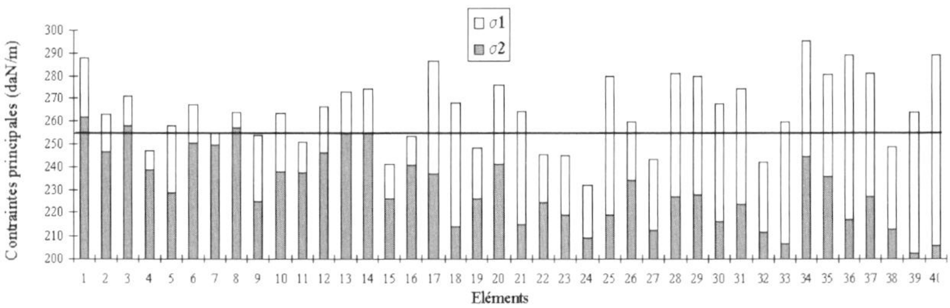

Fig. 11. Principal stresses obtained with simple triangulation method

subjected to the highest strains, which may generate shear forces and thus differences between the principal stresses.

4 Modelling of the Strip Prestressing

All the cutting pattern methods previously presented are based upon an implicit hypothesis: the plane strip will be exactly transformed into the target 3D strip, that is to say the surface determined by form-finding will be, in the end, exactly obtained. Some comments may be pointed up. Firstly, we notice that such postulate is a strong one with reference to possible consequences. However it must also be kept in mind the difficulty to avoid such assumption in order to be able to determine the cutting shapes. Nevertheless, a rigorous process has to check both the geometrical and sthenical deviances between the target membrane and those obtained after prestressing the assembled membrane in the site. Cutting pattern and prestressing stages may be thus regarded as a "back and return" operation.
The modelling of the prestressing operation could be envisaged according to two levels of complexity:
- A global approach in which the designer studies the deployment of the whole

membrane. Beside the difficulty in writing the mechanical formulations (large displacements context), main difficulties are associated with the modelling of the assembled membrane (mainly the folds [4]) and various interaction phenomena (friction between fabric and tensioning edge cables, sequence of erection...).
- A local strategy aiming to analyze separately each tensioned strip. Even if its adequacy with the reality is somehow hypothetic, this approach is more easily achievable and may leads to relevant and pertinent informations. One possible method is presented in the following paragraphs.
It consists in considering the transformation of Ω^0 into a tensile configuration by prescribing nodal displacements on the boundary (nodes on strip edges) until a perfect coincidence with the target strip is obtained. This prestressing leads in p steps to a domain $\Omega^{\mathcal{P}}$ with an estimated matching with Ω^L by considering both the form (geometrical equivalence of inner points) and the stress distribution (sthenical equivalence). If a good correspondence is verified, we may thus conclude in the good adequacy of the cut shape.
The formalism relies on a total lagrangian formulation considering the reference configuration Ω^0 and successive steps. The transformation between two steps (for instance from Ω^1 to Ω^2, assumed as very close) is achieved with an increment of the nodal displacements $\{d\}_1^2$ (Fig. 12).

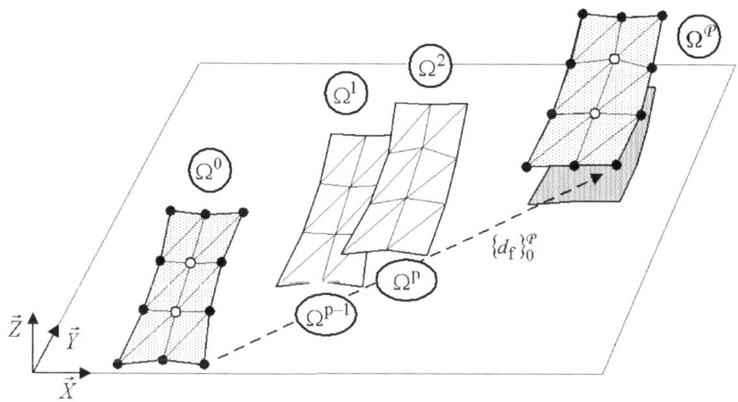

Fig. 12. Prestressing of the strip Ω^0

By considering Ω^1 as already determined (equilibrium obtained), Ω^2 must necessarily verify the virtual works theorem [3]:

$$\delta W_0^{2-1} = \langle \delta d \rangle_0 \, [k_T]_0^2 \, \{d\}_1^2 = 0 \tag{3}$$

The tangent stiffness matrix appears in this writing; it could be splitted according to its linear, non linear (initial displacements) and geometric components:

$$[k_T]_0^2 = [k_L]_0^2 + [k_{NL}]_0^2 + [k_\sigma]_0^2 \tag{4}$$

The non linear problem may be iteratively solved with the Newton-Raphson method. When the convergence is obtained (low residual on $\Omega^{\mathcal{P}}$), it is then possible to determine a level of geometric distortion $\Delta g_{\mathrm{L}}^{\mathcal{P}}$ related to the deviance in internal nodes positions and a level of sthenic distortion $\Delta\sigma_{\mathrm{L}}^{\mathcal{P}}$.
We consider as an illustrative example the prestressing of the strip previously calculated with the integrated method. The figure 13 represents the target strip Ω^{L} and the associated plane domain Ω^{0}. In ten incrementation steps, it leads to a configuration $\Omega^{\mathcal{P}} = \Omega^{10}$ close to Ω^{L}; we verify moreover that $\Delta g_{\mathrm{L}}^{10} = 0,21\%$. This result is in accordance with the geometrical equivalence used during the patterning process. We note in addition that the distortion between the target and the obtained stresses is $\Delta\sigma_{\mathrm{L}}^{10} = 3,81\%$ (with a distribution of principal stresses on $\Omega^{\mathcal{P}}$ close to those represented in Fig. 9).

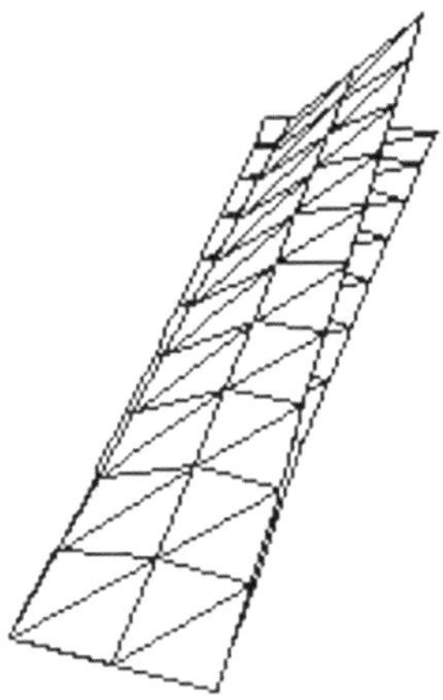

Fig. 13. "Back and return"

A second illustration deals with the analysis of a strip extracted from a minimal area shape ("chinese hat" type) with dimensions 20m and 4m for circle diameters and 4m for the height. The plane cutting shape determined with the integrated method is characterized by an error of of 10,48% for the pretension and 0,92% for the geometry (Fig. 14 left). A calculation with the simple triangulation method leads to higher deviances with approximately 45% for stresses (Fig. 14 right). Such result is however only the consequence of the imperfections of this process that avoids a lot of points inside the strip and an important boundary point located on the bottom circle.
The presented two tests allow to point out the importance of the errors due to the flattening of a doubly curved surface. They have to be considered in parallel with the possible errors generated during the form-finding stage. It emphasizes on the fact that distortions generated during the patterning stage play the main role.

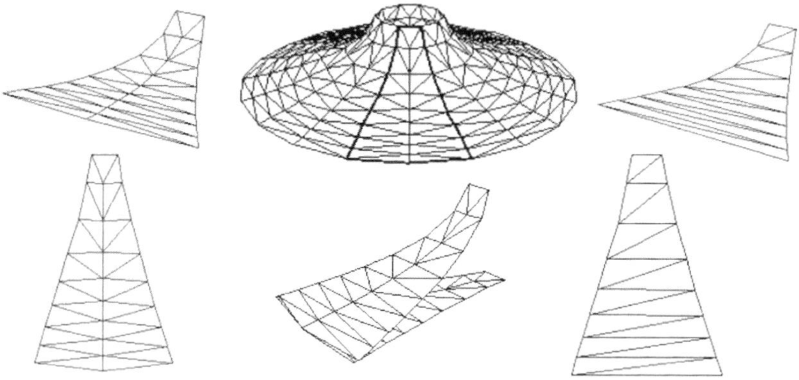

Fig. 14. Strip determined with integrated method (left) and simple triangulation (right)

5 Conclusion

The cutting pattern of strips for tensile membranes is more often achieved by considering the stage of flattening and the stage of reduction as separated. Instead of this splitting, it can be used an integrated approach such as the stress composition method that takes into account the characteristics of the fabric. It allows to take more appropriately into consideration the coupling between the parameters of form, forces and material and, thus, to respect the different requirements of technology, geometry, mechanics and architecture.

6 Appendix: Membrane Local Curvatures Computation

Tensile fabric membrane design implies several stages which require the use of geometrical characteristics.
Firstly, the designer has to verify that the shape calculated by form-finding analysis satisfies numerous requirements.

- The maximum curvature radius must be less than prescribed values. It traduces the relationship between the membrane stiffness at one point and the curvatures; therefore the behaviour of the structure when climatic loads are active. This requirement could affect the membrane itself and reinforcing edge cables as well. Most generally, the limits are defined by national codes. For instance, in France, the membrane maximum curvature radius is 35m and the edge cable maximum curvature radius is 25m.
- The minimal slope must be superior to specified values. The objective is to avoid too "horizontal" areas where rainwater could possibly stay and thus damage the fabric because of local lower stiffness. A suitable criteria may be put forward by determining at the studied zone the vector perpendicular to the surface.

Secondly, a geometrical analysis is necessary so as to provide the designer with important parameters during the cutting pattern procedure:

- The value of the gaussian curvature is associated to the difficulty to develop a curved surface onto a plane (flattening). The designer should take it into account so as to control the width of the strips over the membrane. Hence, a "map" representing the gaussian curvature value at every point of the surface could be a useful tool.
- The specification of the strip edges which divide the membrane is approached under several criteria. The use of geodesic curves could be of importance and therefore their calculation which imply numerous geometrical informations such as surface normal vectors.

The objective of this appendix is thus to submit a numerical approach devoted to the determination of membrane local curvatures. It deals with the case of mapped surfaces that define the height of the points as a function of the two other plane coordinates, that is to say $z = f(x, y)$ with f representing a bijection between z and the pair (x, y). In that case, theoretical relationships allow to calculate at every point the values of the mean and gaussian curvatures, and next, to determine the maximum and minimum curvature radii. We propose to solve all partial differential equations by using the polynomial shape functions of finite elements.

6.1 The Calculation Strategy

In the case of mapped surface, a vertical line with fixed x and y coordinates intersects the domain at only one point for which we can write the relationship $z = f(x, y)$.
For every surface it exists two orthogonal planes that define the principal curvature directions θ_1 and θ_2 associated to the main curvatures ρ_1 and ρ_2 (minimum and maximum values).
With function z it is then possible to calculate at every point the partial derivative terms [11]:

$$p = z_{,x} \ ; \ q = z_{,y} \ ; \ r = z_{,xx} \ ; \ s = z_{,xy} \text{ and } t = z_{,yy} \tag{5}$$

Thus the surface gaussian curvature G and the mean curvature H are defined by:

$$\begin{gathered} G = (rt - s^2)\,(1 + p^2 + q^2)^{-2} \text{ and} \\ H = (t(1 + p^2) - 2pqs + r(1 + q^2))\,(1 + p^2 + q^2)^{-3/2} \end{gathered} \tag{6}$$

The curvatures radii R_1 and R_2 are next determined with:

$$\begin{gathered} G = \rho_1 \rho_2 \ ; \ H = 0,5(\rho_1 + \rho_2) \text{ and} \\ R_1 = \frac{1}{\rho_1} = (H + (H^2 - G)^{1/2})^{-1} \ ; \ R_2 = \frac{1}{\rho_2} = (H - (H^2 - G)^{1/2})^{-1} \end{gathered} \tag{7}$$

The two principal directions θ_1 and θ_2 may be calculated according to the system (with $\mathcal{D} = tg\theta = \frac{dy}{dx}$):

$$\begin{cases} \mathcal{D}_1 + \mathcal{D}_2 = (-t(1+p^2) + r(1+q^2))(pqt - s(1+q^2))^{-1} \\ \mathcal{D}_1\mathcal{D}_2 = (s(1+p^2) - pqr)(pqt - s(1+q^2))^{-1} \end{cases} \tag{8}$$

And the components of vector $\overrightarrow{n}_s$ normal to the surface are:

$$\overrightarrow{n}_s(-hp\,;\,-hq\,;\,h) \text{ with } h = (1+p^2+q^2)^{-1/2} \tag{9}$$

In fact the main difficulty is related to the determination at every point of the function $z = f(x, y)$ and its partial derivatives. Thus, we propose to use finite element polynomial shape functions N in order to interpolate the geometry of the surface within the area "close to" a chosen point. The elevation z is written $z = N_i z_i$; the index i indicates a summation over the n nodes element ($i = 1$ to n). We note that a not too large surface surrounding a selected point can always be described as a mapped domain by using a local referential coordinate transfer.

Since partial derivative equations involve an order two for derivatives, the finite element must have at least a quadratic polynomial interpolation. We propose to use 6 nodes triangle (T6) and 9 nodes rectangle (R9) for a quadratic interpolation and 10 nodes triangle (T10) for a cubic interpolation. Theoretically, a higher interpolation degree should lead to more accurate results.

The associated standard elements are represented in Fig. 15 in their cartesian natural axis ξ and η that are used to define shape functions $N_i(\xi, \eta)$.

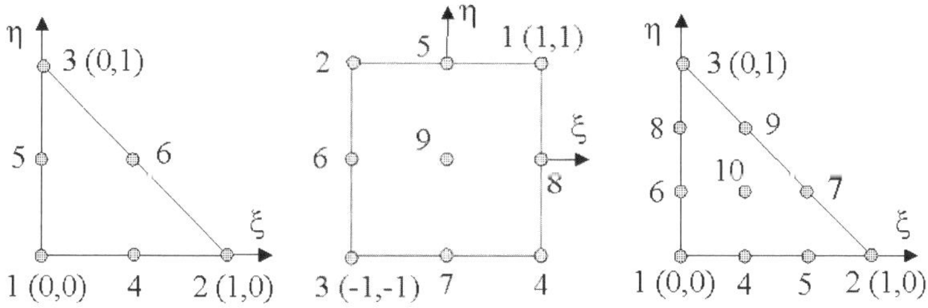

Fig. 15. Elements (T6, R9 and T10)

These elements are not dependent on those used for the mesh during the form-finding analysis. The only matching up parameter is the nodal coordinates of the discrete surface. The objective is to calculate the interpolated values:

$$p = N_{i,x}\, z_i\,;\; q = N_{i,y}\, z_i\,;\; r = N_{i,xx}\, z_i\,;\; s = N_{i,xy}\, z_i \text{ and } t = N_{i,yy}\, z_i \tag{10}$$

We must write the jacobian matrix corresponding to the transformation between the standard element and the element over the surface and calculate its determinant $D = N_{i,\xi}\, N_{i,\eta}\, (x_i y_j - x_j y_i)$ and its first derivatives:

$$\begin{cases} D_{,\xi} = (N_{i,\xi\xi}\, N_{j,\eta} + N_{i,\xi}\, N_{j,\xi\eta})\,(x_i y_j - x_j y_i) \\ D_{,\eta} = (N_{i,\eta\xi}\, N_{j,\eta} + N_{i,\xi}\, N_{j,\eta\eta})\,(x_i y_j - x_j y_i) \end{cases}$$

$$\begin{cases} (D^{-1})_{,\xi} = D^{-1}_{,\xi} = -D^{-2}\, D_{,\xi} \\ (D^{-1})_{,\eta} = D^{-1}_{,\eta} = -D^{-2}\, D_{,\eta} \end{cases} \qquad (11)$$

If we calculate then the components of the inverse jacobian matrix:

$$\begin{gathered} a = \xi_{,x} = D^{-1}\, N_{i,\eta}\, y_i \;;\; b = \eta_{,x} = -D^{-1}\, N_{i,\xi}\, y_i \text{ and} \\ c = \xi_{,y} = -D^{-1}\, N_{i,\eta}\, x_i \;;\; d = \eta_{,y} = D^{-1}\, N_{i,\xi}\, x_i \end{gathered} \qquad (12)$$

The partial first derivatives are defined according to:

$$N_{i,x} = a\, N_{i,\xi} + b\, N_{i,\eta} \text{ and } N_{i,y} = c\, N_{i,\xi} + d\, N_{i,\eta} \qquad (13)$$

And the partial second derivatives are:

$$\begin{cases} N_{i,xx} = a^2 N_{i,\xi\xi} + b^2 N_{i,\eta\eta} + 2ab N_{i,\xi\eta} + (a a_{,\xi} + b a_{,\eta}) N_{i,\xi} \\ \qquad + (a b_{,\xi} + b b_{,\eta}) N_{i,\eta} \\ N_{i,xy} = ac N_{i,\xi\xi} + bd N_{i,\eta\eta} + (cb + ad) N_{i,\xi\eta} + (c a_{,\xi} + d a_{,\eta}) N_{i,\xi} \\ \qquad + (c b_{,\xi} + d b_{,\eta}) N_{i,\eta} \\ N_{i,yy} = c^2 N_{i,\xi\xi} + d^2 N_{i,\eta\eta} + 2cd N_{i,\xi\eta} + (c c_{,\xi} + d c_{,\eta}) N_{i,\xi} \\ \qquad + (c d_{,\xi} + d d_{,\eta}) N_{i,\eta} \end{cases} \qquad (14)$$

With the following coefficients:

$$\begin{cases} a_{,\xi} = -D^{-1}_{,\xi} N_{i,\eta} y_i + D^{-1} N_{i,\xi\eta} y_i \\ a_{,\eta} = -D^{-1}_{,\eta} N_{i,\eta} y_i + D^{-1} N_{i,\eta\eta} y_i \\ b_{,\xi} = -D^{-1}_{,\xi} N_{i,\xi} y_i - D^{-1} N_{i,\xi\xi} y_i \\ b_{,\eta} = -D^{-1}_{,\eta} N_{i,\xi} y_i - D^{-1} N_{i,\xi\eta} y_i \end{cases}$$

$$\begin{cases} c_{,\xi} = -D^{-1}_{,\xi} N_{i,\eta} x_i - D^{-1} N_{i,\xi\eta} x_i \\ c_{,\eta} = -D^{-1}_{,\eta} N_{i,\eta} x_i - D^{-1} N_{i,\eta\eta} x_i \\ d_{,\xi} = -D^{-1}_{,\xi} N_{i,\xi} x_i + D^{-1} N_{i,\xi\xi} x_i \\ d_{,\eta} = -D^{-1}_{,\eta} N_{i,\xi} x_i + D^{-1} N_{i,\xi\eta} x_i \end{cases} \qquad (15)$$

Since derivatives $N_{,\xi} ... N_{,\xi\eta}$ are dependent on ξ and η values, the calculation could be achieved at any chosen point within the element, for instance a point located in the middle of two nodes.

6.2 Applications

Test

The first application allows the verification of formulations. It deals with an hyperbolic paraboloid (HP) defined by $z = k\, xy$. The gaussian and mean curvatures are hence:

$$G(x,y) = -(k^{-2} + x^2 + y^2)^{-1}\,;\; H(x,y) = -k^3\,xy\,(1 + k^2(x^2 + y^2))^{-3/2} \quad (16)$$

We observe that $H \neq 0$ for non zero x and y: contrary to a common idea, the HP is not a minimal area surface!

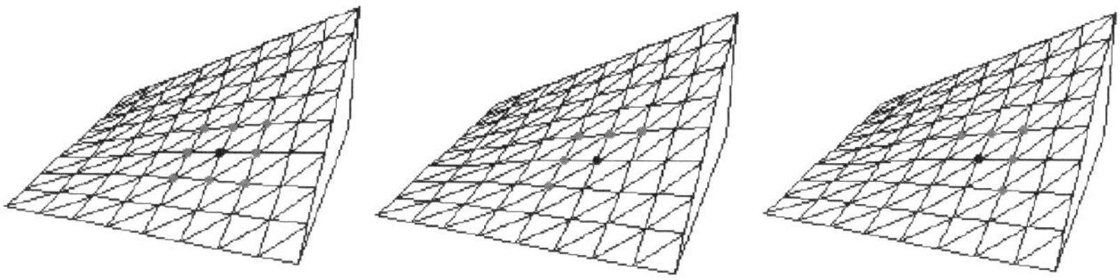

Fig. 16. HP curvatures computation: with R9 (unique possibility) and T6 (several possibilities)

The calculation at one selected point could be envisaged according to different approaches: a quadratic formulation with a R9 rectangle surrounding the point or with a T6 triangle. In that second case, several positioning of the triangle may be used as represented in Fig. 16 (b and c). A cubic interpolation could be as well employed with T10 triangles (not represented).
For R9 element it comes $\xi = \eta = 0$ and $\xi = \eta = 0,5$ for T6. Computations have led to accurate values which are moreover independently on the used element (R9, T6 or T10) and also independently on its positioning in the case of triangles. However, such a result is just a consequence of the HP being a ruled surface. It will not be the same if “non analytic” surfaces (that is to say surfaces which can not be analytically defined) are considered as in the following example.

Membrane example

This application deals with a fabric membrane generated with a shape finding process (force density method, dynamic relaxation...). The aim could be for instance to evaluate some characteristics at one selected point: the maximum curvature radius, the gaussian curvature and the normal vector.
Figure 17 illustrates the several options related to the choice of the element and to its positioning (diverse possibilities excepted for rectangle R9 - a). It is obvious that the associated results will be different and the issue deals with the choice to make.
The problem of positioning may be considered as quite analogous to the stress determination issue at one node after a finite element analysis. Several strategies are indeed achievable: the average value of gauss point values (integration points surrounding the node); the average value of the nodal values obtained on each element comprising the node etc...

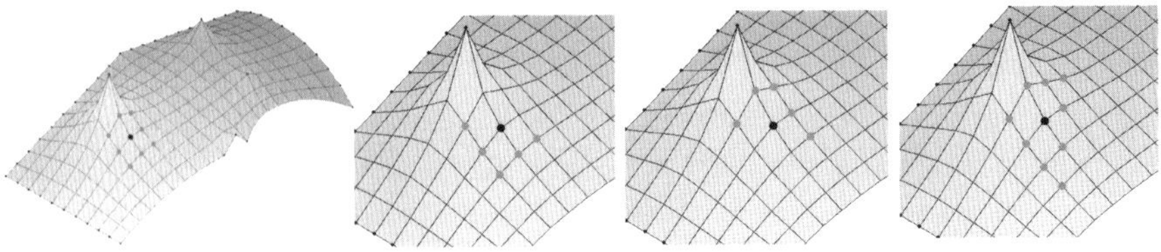

Fig. 17. Different calculation strategies (R9, T6, T6 and T10)

However, we emphasize on the necessity to define approaches based upon the objective of minimal errors. The evaluated geometrical parameter certainly plays a part in the determination of the suitable method.

References

1. Allera R (1992) Mise en forme des structures textiles tendues. Thèse de doctorat, Institut National Polytechnique de Grenoble
2. Barnes MR, Wakefield DS (1988) Form-finding, analysis and patterning of surface-stressed structures. 1st O. Kerensky Memorial Conf., London
3. Bathe KJ (1982) Finite element procedures in engineering analysis. Prentice Hall
4. Dinkler D, Wiedemann B (2001) A finite element concept for wrinkling membranes. Proc. of the IASS symposium, Nagoya,TP04 1–8
5. Ferrari Tissage et Enduction (1989) Rapport documentaire sur les produits
6. Fujiwara J, Ohsaki M, Uetani K (2001) Cutting pattern design of membranes structures considering viscoelasticity of the material. Proc. of the IASS symposium, Nagoya, TP047 1–8
7. Gründig L, Bäuerle J (1990) Automated cutting pattern determination and control for prestressed membranes. Textile Composites in Buil. Cons., part. 2, Ed. Pluralis, 109–120
8. Kim JY, Lee JB (2002) A new technique for optimum cutting pattern generation of membranes structures. Engineering Structures, vol. 24, 745–756
9. Maurin B, Motro R (1999) Cutting pattern with the stress composition method. Int. Journal of Space Structures, vol. 14 N2, 121–129
10. Moncrieff E, Gründig L, Ströbel D (1999) The cutting pattern of the pilgrim's tents for phase 2 of the Mina valley project. Proc. of the IASS 40th anniversary congress, Madrid, C1 129–136
11. Osserman R (1997) Geometry V. Springer Verlag
12. Shimada T, Tada Y (1989) Development of a curved surface using a finite element method. Int. Conf. on Comp. aided optimum Design of Structures, Southampton, 23–30
13. Tsubota H, Yoshida A (1989) Theoretical analysis for determining optimum cutting patterns for membrane structures. IASS Int. Symp. on tensile structures, Madrid, 512–536
14. Xia X, Meek J (2000) Computer cutting pattern generation of membranes structures. Int.Journal of Space Structures, vol 15 N2, 95–110

Inflated Membrane Structures on the Ground, in the Air and in Space - A Classification

Bernd Kröplin[1]

Institute of Statics and Dynamics of Aerospace Structures
University of Stuttgart
bkroeplin@isd.uni-stuttgart.de

Summary. *In the following an introduction to basic principles of inflated structures, their potential and their limitation is given. Geometry and Shapes are investigated as well as structural differences within high and low pressure structures and their typical pressure control problems. A brief morphological sketch of inflated structures is presented, which opens the room to further development in the future.*

Key words: Membrane structures, inflated structures

Introduction

Membranes are known as very thin walled flexible surfaces, shaped and stabilised by pretensioning. They are a promising solution for covering and spanning wide spaces. In contrast to thin shells, which also carry their loads mainly by in-surface forces, they are not able to withstand compression loads, which result in folds. The pretension can be built up either by an external tension field or by internal pressurisation. Membranes have the advantage of foldability. They can be packed and trans- ported easy and can be erected by inflation to serve as a permanent or a temporary building. The variability in construction can be classified as follows:

1. Permanent membrane structures with external pretension
2. Permanent and temporary membrane structures with inflated walls
3. Structures with full gas volume
4. Buoyancy structures
5. Interesting combinations

1 Permanent Membrane Structures with External Pretension

The outside tension field is normally erected by cables, struts and bearings. Examples therefore are the Olympic roof in Munich, see Fig. 1 or the cooling tower of Schmehausen, see Fig. 2. This type of construction results in huge struts and bearings, since all pretension forces have to be grounded.

E. Oñate and B. Kröplin (eds.), Textile Composites and Inflatable Structures, 213–220.

Fig. 1. Olympic stadium roof Munich [1], permanent

Fig. 2. Cooling tower Schmehausen [2], permanent

2 Permanent and Temporary Membrane Structures with Inflated Walls

Another class are the inflated structures. There are two different types, the structures with inflated walls and the structures with inflated volumes. The walls can be inflated by high or low pressure. For high pressure closed wall-chambers are used with the pressure probably exeeding 1 bar requiring rein- forced membrane materials. An example is the Festo-building at Stuttgart, see Fig. 3. Pressure and manufacturing required a rubber coated Polyester fabric. In the case of low pressure, about 50 mbar, an open wall can be used, whereby the pressure is permanently imposed by a blower, which is capable to erect the structure. See e.g. the temporary Builtair-roof in Barcelona Fig. 4. Recently the inflated wall principle was also introduced by Russian reserchers to reentry vehicles for the return from space to earth.

Fig. 3. Festo-building in Stuttgart [3], permanent

Fig. 4. Build-Air-building in Barcelona [4], temporary

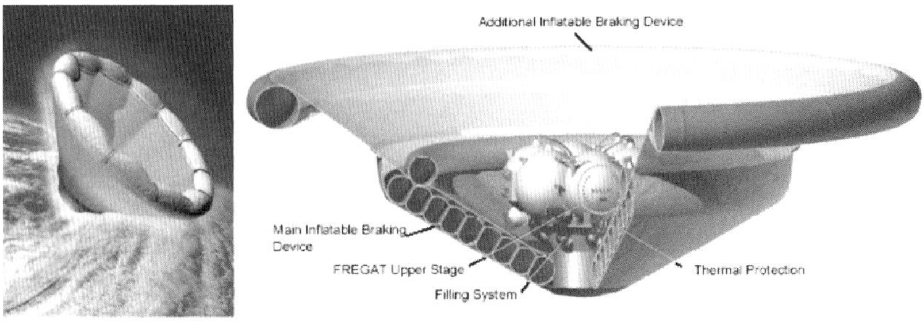

Fig. 5. Inflated reentry capsule [5]

3 Structures with Full Gas Volume

The inflated tennis court-halls of the last century are a well known example for this class of structures, see Fig. 6. Since the air pressure has to be maintained permanently for the stability of the structure, the doors are designed as double door pressure lock. Another example to overcome the totally inflated space inside is to inflate only the roof. Fig. 7 gives an example. For adaptation to different load cases, e.g. snow, a spherical pressure chamber is included.

Fig. 6. Tennis court cover [6], permanent

4 Buoyancy-Structures

If the inside space of the structure is filled with a gas lighter than air, an upward force is created, which is called buoyancy. The buoyancy depends on the density difference between the internal and the external medium. The principle has been used since the 18th Century. 1 m^3 buoyancy gas, e.g. helium, can roughly carry 1 kg

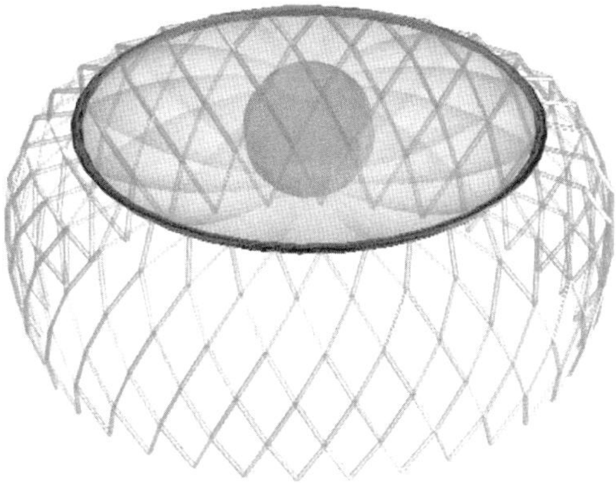

Fig. 7. Inflated roof with pressure chamber [7]

Fig. 8. High altitude balloon [8]

at mean sea level. The buoyancy is a natural energy potential. Its activation needs no energy. Nevertheless the potential for "flying structures" seems to be not fully explored yet.

The classical example is the high altitude balloon. It can be seen from Fig. 8, that in this case the shape depends on the expansion of the buoyancy gas, which expands when the outside air density diminishes during the ascent.

If the shape of the body should be kept constant at all altitudes, e.g. for aerodynamic drag reasons, a ballonet-technique can be used, see Fig. 9. Inside the body air inflated balloons are placed, which release their volume to the outside in the same manner as the air pressure decreases during ascent and the buoyancy gas expands. During decent air has to be pumped into the ballonets in order to keep the gas pressure in the body at a certain level. The differential pressure level is typically around 500 Pascal.

The ballonet principle can be used as pressure control in ground structures also as external ballonet outside the structure. A hall of his type has been built 1992 at the International Garden Exhibition in Stuttgart, see Fig. 10. The full roof of the structure, which can be opened including the supporting ring-structure, is carried by buoyancy.

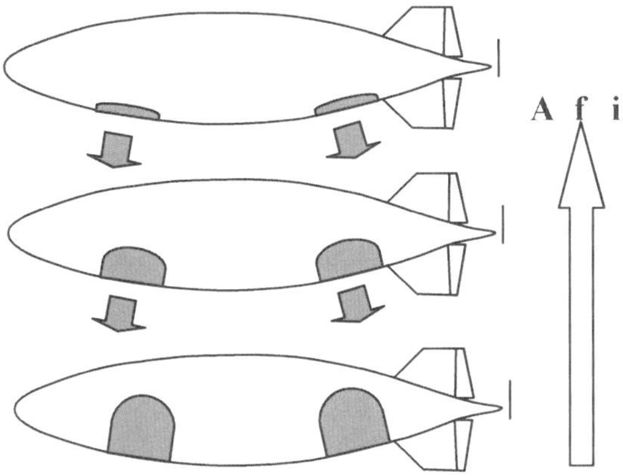

Fig. 9. Ballonet system in an airship [9]

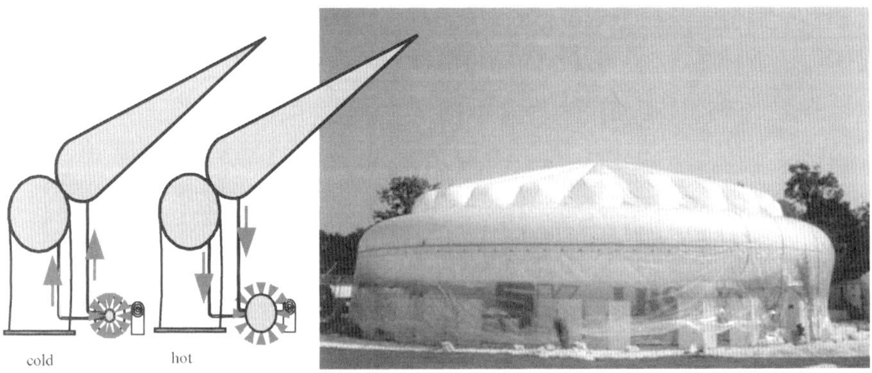

Fig. 10. HELION - Buoyancy structure Stuttgart 1993 [10]

5 Interesting Combinations

In order to optimize the structure further different principles might be coupled. For this three examples are given:

The AirChain An interesting and efficient concept is obtained, if a number of high altitude balloons is coupled to a chain and covered by a aerodynamically smooth surface. The vehicle flies controlled by engines in each segment, see Fig. 11.

Fig. 11. AirChain Stuttgart, Test Flight 1999 [11]

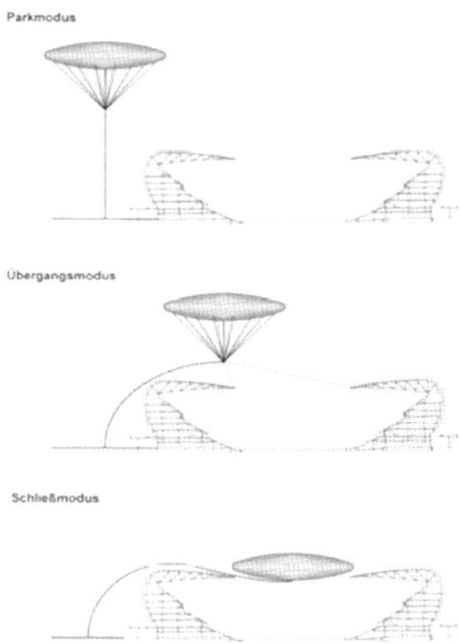

Fig. 12. Flying Roof, Munich 2000 [12]

The Flying Roof The mixture of the buoyancy principle and the Japanese Kite tradition lead to the design of a station roof. The stadium can be closed or opened within half an hour. The roof is parked in the air outside of the station and is stabilized like a kite, see Fig. 12.

The Paraglider Finally the good old parachute combines the externally prestessed membrane and the inflated membrane and receives from an intelligent combination of both its stability and its flight control see Fig. 13.

Fig. 13. Double Wing Paraglider at automatically guided flight [13] Stuttgart 2003

References

1. Leonhardt F, Schlaich J (1973) Vorgespannte Seilnetzkonstruktionen, Das Olympiadach in München, SFB 64 Weitgespannte Flächentragwerke, Universität Stuttgart, Mitteilungen 19/1973, Werner-Verlag, Düsseldorf.
2. Schlaich J, Mayr G. (1974) Naturzugkühlturm mit vorgespanntem Membranmantel. Der Bauingenieur 49:41–45.
3. http://www.festo.com/airtecture/
4. Proceedings, First European Workshop on Inflatable Space Structures, 21–22 May, (2002).
5. http://buildair.com
6. http://www.traglufthallen.biz/
7. Final Report, (2001). Brite EuRam Project Nr. BE96-3015 "New methodology for Design and Manufacturing of Inflated Structures (INFLAST)", 1997-2000.
8. www.usgcrp.gov/usgcrp/seminars/ 000714FO.html
9. internal report ISD
10. http://www.isd.uni-stuttgart.de
11. http://www.isd.uni-stuttgart.de
12. Feasibility study of a flying roof, internal report.
13. Double Wing Paraglider, Automatic Guidance, Stuttgart (2003).

Post-Tensioned Modular Inflated Structures

Romuald Tarczewski[1]

[1]Wroclaw University of Technology
romuald.tarczewski@pwr.wroc.pl

Pneumatic structures are usually considered as inalterable, with predetermined features, e.g. shape etc. According to this approach, they cannot be easily rebuilt or modified. Application of techniques current in other sectors of construction industry leads to the structures that are much more flexible and adaptable.

1 Introduction

The prestressing technique was applied to concrete and steel constructions in first decades of 20th century. It appeared after a period of development of these constructions, and its usage is obvious and common at present. Yet in reference to pneumatic constructions, the conception of prestressing appeared at the beginning of their development. The first known documentary evidence of the conception to use structural pneumatic elements comes from engineer Joachim A. Sumovski. He obtained, in 1893, an American patent on air-inflated structures [1]. One of the drawings included in patent specification presents the structure that obtains its shape and loading capacity as a result of prestressing, Fig. 1.

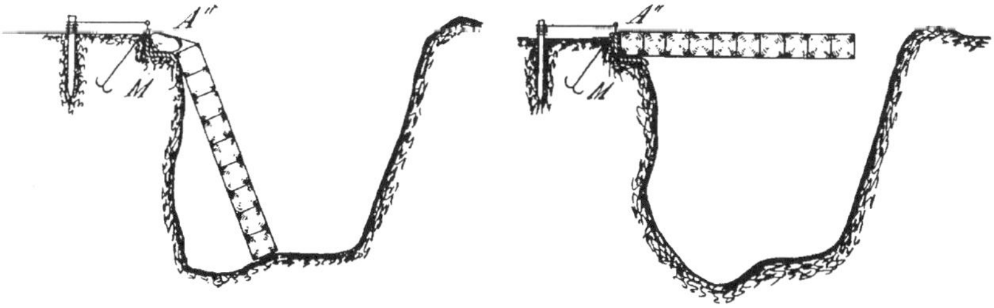

Fig. 1. An example of air-inflated structure invented by J.A. Sumovski

E. Oñate and B. Kröplin (eds.), Textile Composites and Inflatable Structures, 221–239.

However in air-supported structures, particularly of large span, strengthening cables are nowadays often applied [2] – prestressing is not yet obvious in air-inflated structures.

Another, well-known structural concept, which provides with many advantages in terms of creating structures – is the use of repeatable, modular units, that can be assembled in a larger, complex structure. The spectacular example of application of repeatable air-inflated units for constructing a large structure – was the pavilion of Fuji Group, at Expo '70 exposition in Osaka [3].

The advantages of modular structures are especially exposed, when units of relatively small dimensions are used. This allows constructing structures of various shapes, while limited quantity of different units are applied. The units can be used repeatedly many times. Shape of the structures created in this way is limited only by topological restrictions of division of considered surface. This applies in particular to the shell structures.

Post-tensioning is a way of splicing small elements but also a way of shaping the structure. Coupling of these operations raises the efficiency of solution. Because of specific properties of pneumatic structures, suitable technical solutions are required, especially concerning supplying the elements with air, stabilization of their shape as well as method of connection.

2 Modular Air-Inflated Elements Applied to Shell Structures

2.1 Principles of Composition

Depicted structures present a class of spatially curved surface girders. Their rigidity and loading ability are strongly related with the shape. Spatial curvature itself, is not a permanent, generic feature of the structure (as it is in concrete shells for example), but is achieved and maintained by means of post-tensioning, causing very large initial deformation. Thus, the final shape and properties of the structure are function of initial configuration and the course of post-tensioning process.

Internal structure of modular air-inflated shells distinguishes them from other air-inflated structures and from other shell structures. They consist of the following basic elements:

- air cushions (main modular elements)
- tension cables
- cross-braces (optional)

Air cushions and cables appear in shells of all types, while cross-braces only in the shells with increased structural height [4]. After assembling, structure forms complete roofing and does not require any additional membranes to cover space. General view and components of an exemplary structure is shown on Fig. 2.

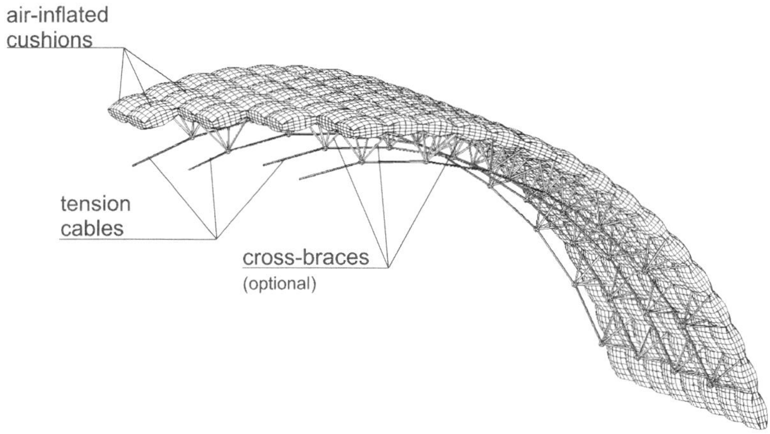

Fig. 2. Composition of modular air-inflated shell

2.2 Structural Components

Air cushions

Pneumatic modular elements have a form of cushions shaped in order to fit geometrical constrains of the prospective surface. Basic shapes are: rectangle, square, rhomb, hexagon and triangle. Following conditions must be fulfilled in order to enable effective application of modular system:

- shape of the cushions must correspond with final shape of the structure; it is obvious that the usage of as few varying shapes as possible is profitable
- elements must be small enough in relation to the final structure, to form a smooth, easily deformable surface (at least ten times smaller); additional criteria can be also applied, such as easiness of in-site manipulation e.g. one workers should be able to carry the cushion
- the connections between the elements have to assure their suitable integration as well as continuity of transmission of internal forces
- the cushions should be suitably equipped with guides allowing the usage of post-tensioning cables

Other components

Bar members (i.e. cross-braces) are optional – they appear only in some types of shells. These additional bars are used in order to increase the structural height. They can be made of any lightweight material capable of carrying compression, like steel, aluminum, wood or composite. Naturally, closed sections (tubular) are better than the others, for this purpose. Connection of bars and cushions must prevent damages (e.g. by means of strengthened pockets).

Post-tensioning cables are always placed at the internal side of curved shell. In case of anticlastic shells, cables are placed in two layers – at the opposite sides of the shell. Direction of cables in each layer corresponds with the curvature of the shell, Fig. 6. The cables can be placed directly below the cushions or below cross-braces. In both cases the cable should be able to slide freely through the nodes. Both, fiber ropes (made of natural or man-made fibers) and wire ropes can be used as post-tensioning cables. It is significant that the cushion was protected from damages caused by ropes, e.g. by means of protective jackets.

2.3 Transmission of Forces

After completion, modular elements must transmit internal forces induced in the structure. It is possible due to the compression of cushions' sides (touching each other) and tension of their external cover and post-tensioning cables, Fig. 3. Thus, the way of assembling the cushions must ensure a full contact of side surfaces and continuity of external cover.

These basic principles allow setting a geometrical configuration of the shells of various size and shape, designed for various purposes.

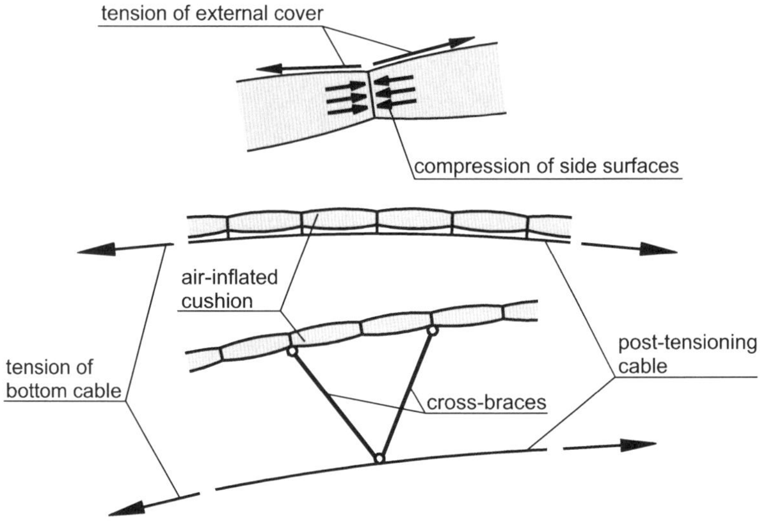

Fig. 3. Transmission of forces in modular air-inflated shells

3 Application of Post-Tensioning and Self-Erection

The structure is stabilized by means of post-tensioning. This process induces internal forces that are shown on Fig. 3. Distribution of these forces is invariable during exploitation, though their values can change. Alteration of forces' direction effects in destruction of the structure – in consequence of opening of the gaps between cushions or "compression"of their external cover, Fig. 4.

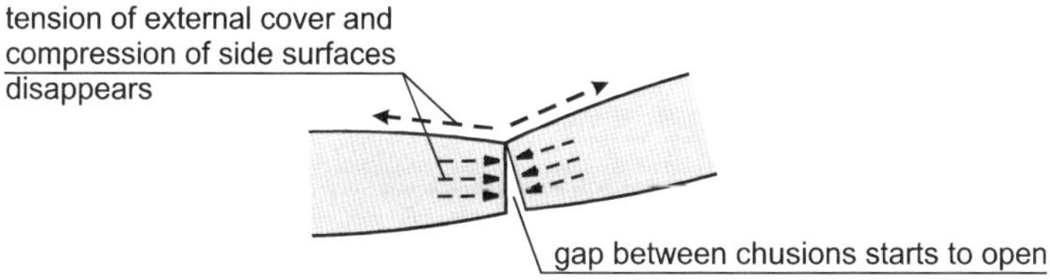

Fig. 4. Failure mode of the shell

There are two ways of realizing the post-tensioning procedure:

- structure is post-tensioned and erected simultaneously (self-erection)
- post-tensioning is applied to previously erected structure

3.1 Self-Erection Procedure

In that case, the flat structure is assembled at ground level as a near mechanism. It is stabilized and finally shaped by means of self-erection. This process unifies operations of post-tensioning, erection (i.e. construction) and spatial curving of the structure. The essence of the process is the introducing into the structure forces that cause its large deformation [5]. In practice, the process starts simply with a shortening of the bottom cables. The cables are attached to the fixed supporting points while going through all the other joints to the opposite, mobile, supporting points.

As a result – supporting points are brought closer to each other. Thus the deformation is introduced to the structure and it starts to erect. The process is continued till required position is obtained. Then the cables are fixed in the mobile supporting points. Fig. 5 presents successive stages of self-erection process.

Air-inflated shell can be post-tensioned either in one direction or in two directions. Unidirectional post-tensioning is applied in order to get structures with zero Gausian curvature (cylindrical), while bidirectional - when structures with negative Gausian curvature (anticlastic) are required, e.g. hypar surfaces, Fig. 6.

Bidirectional post-tensioning is performed successively. At the beginning, cables of the first direction are tensioned to 50–60 % of assumed value. Then

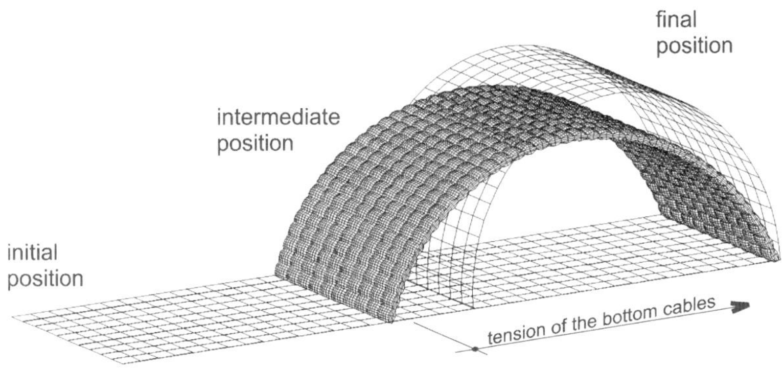

Fig. 5. Scheme of self-erection procedure

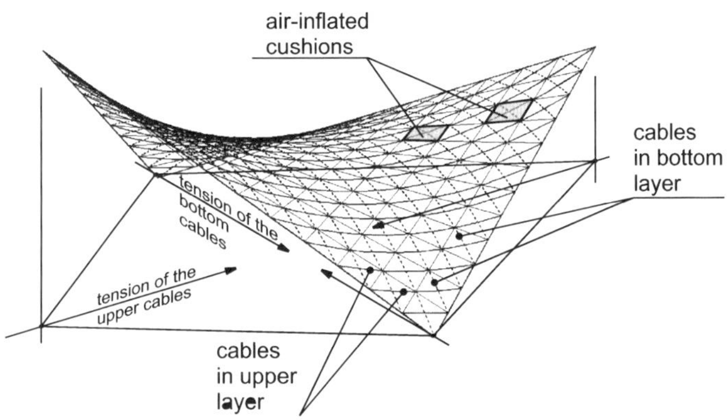

Fig. 6. Bidirectional post-tensioning of air-inflated shell

cables of the second direction are tensioned to the same value. Finally, the cables are alternatively rectified, to reach requested values.

3.2 Post-Tensioning Applied to Completed Structure

Another way of assembling is to put cushions in position one by one – as in an igloo and then tension the cables to stiffen the structure. In that case, shell must be shaped as a self-stable before post-tensioning is performed. Certainly, scaffoldings can be used, however, this overthrows the whole system.

Structures with positive Gausian curvature (synclastic), surfaces of revolution in particular, can be achieved in that way. Fig. 7 presents such a structure, shaped as a hemisphere, made of hexagonal elements.

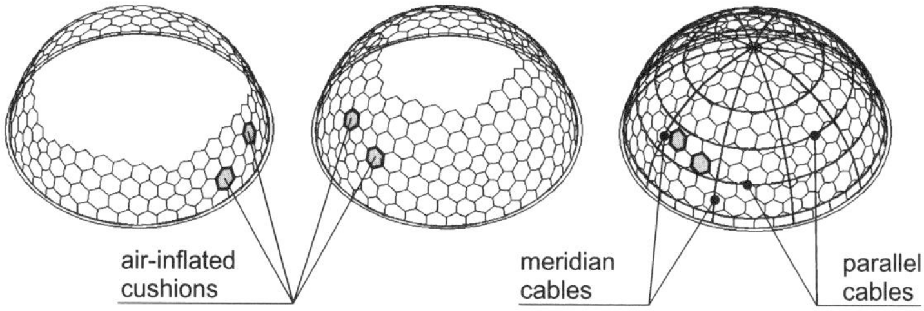

Fig. 7. Construction of synclastic air-inflated modular shell

4 Adjustment of Rigidity and Hardening Systems

4.1 Structures with Variable Rigidity

Curvature of air-inflated shells formed in self-erection process can be controlled by means of changing their rigidity along the span [6]. It is an effective way of shaping this kind of structures, which allows fulfilment of the requirements. Initial stiffness of inflated shell is defined mostly by its structural height. The change of height causes the change of stiffness. This can be achieved in two ways:

- either by use of additional rods, i.e. cross-braces, moving tension cables down from the cushions
- or by change of cushions' thickness

Figure 8 presents those two methods.

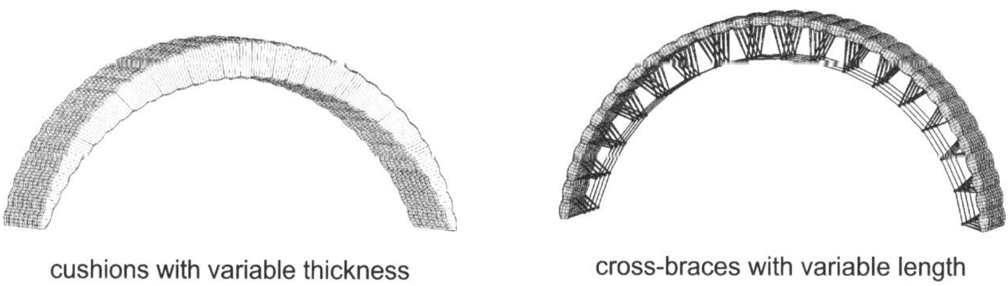

Fig. 8. Methods of rigidity alteration in modular shells

Two compared shells, of the same initial length and subjected to the same upthrust, i.e. nearing of supports, but with various, variable rigidities, demonstrate distinctly different final geometry. If the rigidity of the shell is increased in the central part – the curvature is smaller in the center than in peripheres (structure is more flat). On the other hand – the curvature is smaller on sides, when rigidity is decreased in the central part of the shell (structure is more scarped). Fig. 9 presents a comparison of shapes of cylindrical shells with variable rigidity. More sophisticated shapes can be achieved when rigidity is changing according to a parametric function.

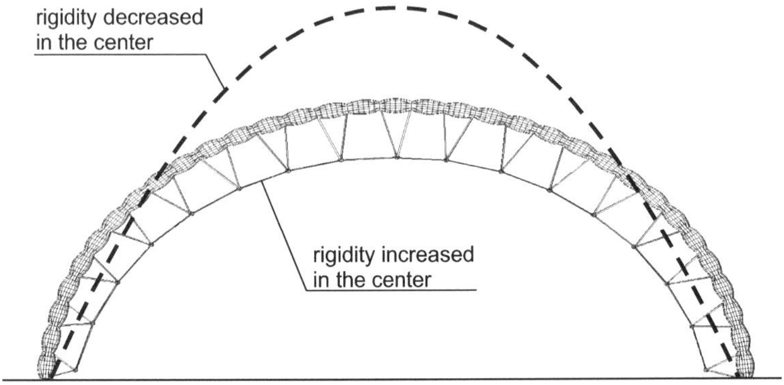

Fig. 9. Comparison of shells with variable rigidity

4.2 Adaptable – Hardening Structure

Variation of rigidity allows constructing a structure that actively adapts to external loadings. If the internal lever arm increases together with increasing external loadings – deformations increment slowly. Load–deformation relation in this case is an exponentially growing function. If the internal lever in the structure can be self adjusted in order to find position of equilibrium, then the structure reveals a "hardening" characteristic. In numerous situations this specific type of structures is more advantageous than any other, with a linear response. An example of hardening system is described in [7]. Its application in pneumatic shells [8] is shown on Fig. 10.

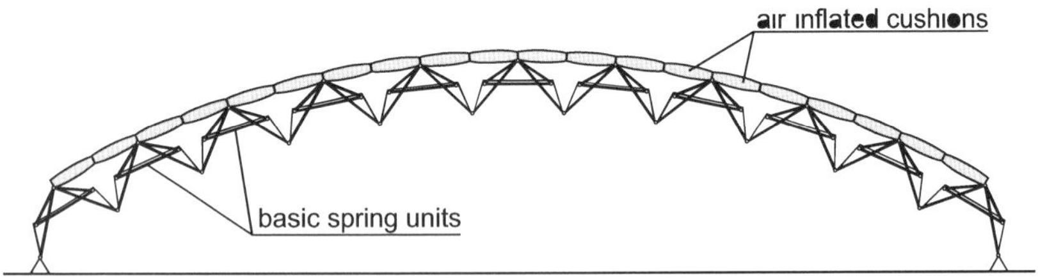

Fig. 10. Example of air-inflated hardening structure

5 Technological Concepts

5.1 Air-Inflated Cushions

The cushions are the basic components of the shell. The air supplying system described below maintains internal pressure. Cushion is usually a flat element – its thickness is smaller than dimensions in plane. Two main surfaces (upper and bottom) and several side surfaces can be distinguished in the cushion. It is generally made of soft textile or foil, suitable for pneumatic structures.

The cushions are equipped with elements that allow attaching a tension rope. These are the flexible hoops enabling the rope to slide easily. The rope is dragged through the hoops during assembling. The hoops are placed at the cushion's corners, on one or on both main surfaces (e.g. for hypar shells).

If the cross-braces are to be placed in the structure, strengthened pockets are made in the cushions. Strengthening prevents damage of soft fabric caused by bar edges.

Cushions with flaccid main surfaces must be equipped with internal elements ensuring flat shape after inflation. Fabric diaphragms (with openings) or set of threads can be applied in this case. These elements are not necessary if both main surfaces are made of rigid panels. Fig. 11 presents a general scheme of the cushion entirely made of fabric.

For the structures with variable rigidity, cushions with variable thickness can be prepared. Main surfaces of such a cushion are not parallel and side surfaces have various width, as shown on Fig. 8.

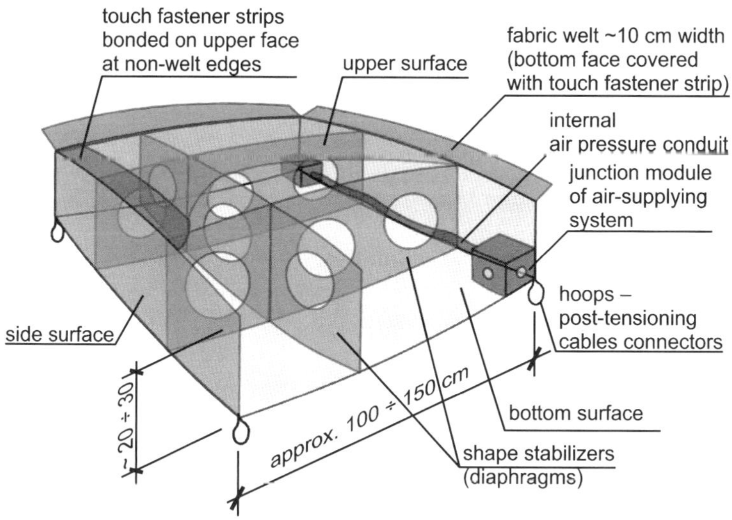

Fig. 11. Internal structure of air-inflated cushion

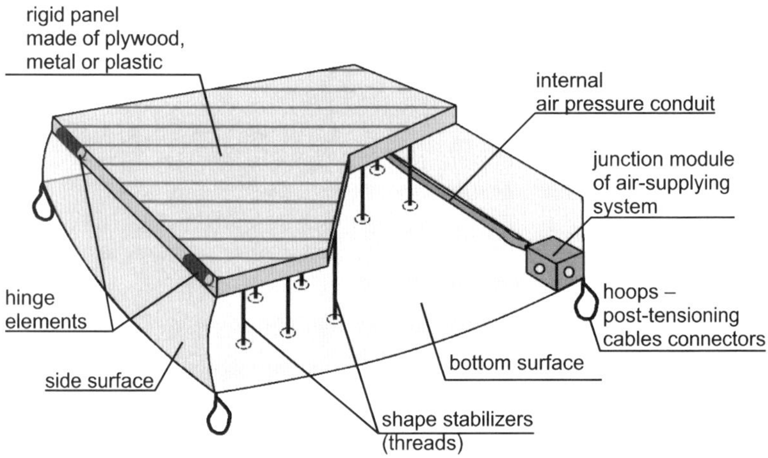

Fig. 12. Internal structure of semi-rigid air-inflated cushion

For some applications, semi-rigid cushions are necessary. In this case, one or both main surfaces can be made of rigid panels (metal, plywood or plastic), Fig. 12. Side surfaces are always made of textile, to assure a proper contact between cushions after assembling.

5.2 Connections of Cushions

There are two points of connection:

- connection of main surfaces
- connection of side surfaces

Side surfaces are connected by means of direct clamping, as it is shown on Fig. 3. In order to ensure a proper connection, contact of surfaces cannot be restricted in any way (e.g. by protruding rigid panels, cross-braces or connectors of air supplying system).

Connection of main surfaces should assure continuity of transmission of tensile forces in the shell, in all directions. Additionally, if the shell is used as a cover protecting against weather conditions, these connections should assure the tightness of the shell.

If the main surfaces of the cushion are soft, a convenient type of connection is "touch fastener" (e.g. Velcro). Welts with a bottom face covered with fastener strips are placed along some of cushion's edges. The edges at the reverse sides of the cushion are also covered with touch fastener strips. The continuity and tightness of the shell can be easily obtained during assembling, Fig. 13.

Another feasible connection is a "zipper"– a wire strand or flexible bar dragged through the small, tight-fitting eyes placed along edges, Fig. 14. Both

Fig. 13. Connections of soft cushions – touch fasteners

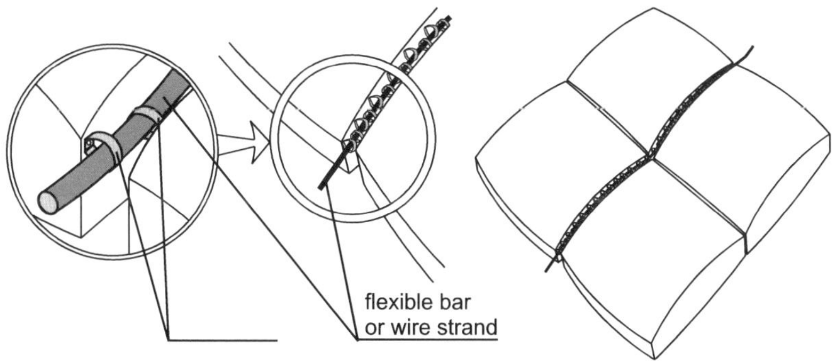

Fig. 14. Connections of soft cushions – "zipper"

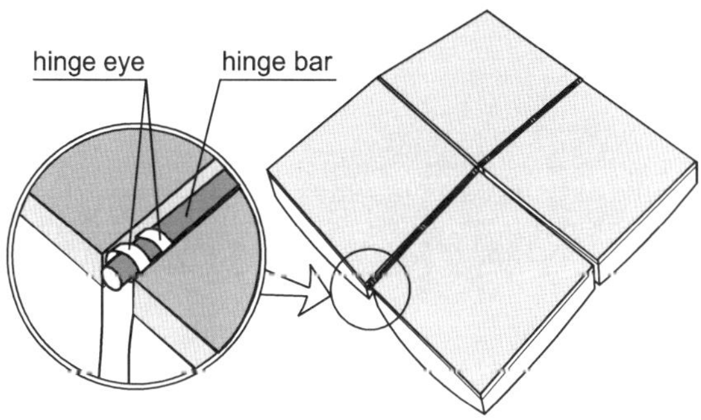

Fig. 15. Hinge connections of semi-rigid cushions

concepts can be used together, when high value internal forces are induced in the shell. In that case, the "zipper" transmits the forces, while touch fasteners are tightening the cover.

Connection of semi-rigid cushions can be formed as a two-piece hinge, completed during assembling, Fig. 15.

5.3 Air Supplying System

Air-inflated shell can be exploited only, when equipped in an efficient air supplying system. This system should assure unfailing inflation and deflation of cushions and maintenance of internal working pressure. It should be foolproof.

An exemplary solution is a system integrated with cushions. It consists of the web of internal pressure conduits, placed inside cushions. These conduits are linked by means of self-tightening, male connectors. Junction modules are situated at side–bottom corners of the cushion. This facilitate coupling eliminates any external part of the conduits. Air pressure inside the conduits is significantly higher than the working pressure in the cushions. The mentioned above junction, contains control unit (valves) regulating internal pressure. When the pressure decreases, the valve opens and air from the conduit is forced inside. When the pressure increases, e.g. in the case of elevated temperature, the valve ejects needless air outside. The same function is used for deflation.

It is possible to use special explosive cartridges for very quick (instant) inflation. In that case, a suitable socket is used in the junction of air-supplying system. The cartridge, placed in this socket, can be detonated manually or by remote control. Explosion products fill the cushion and the control valve maintains the pressure. Fig. 16 presents a functional diagram of junction module in the air supplying system.

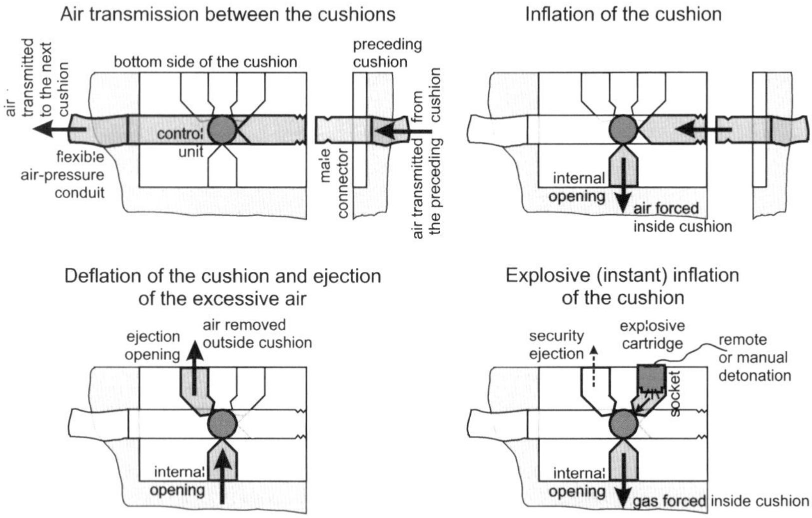

Fig. 16. Functional diagram of junction module in the air supplying system

6 Examples of Application

Modular inflated shells are suitable for wide range of applications. Some examples given below are aimed to emphasize general concepts rather then describe particular implementations. These shells can be successfully combined with many other well-known types of structures.

6.1 Cylindrical Shells

The simplest possible shape of modular air-inflated shell is a cylinder. This can be used directly for some applications, or as a part of a bigger, complex structure. However, even in this case, structure can be constructed in various ways. This can be composed only of cushions and cables, Fig. 17, or with optional use of cross-braces, Fig. 18 and Fig. 19.

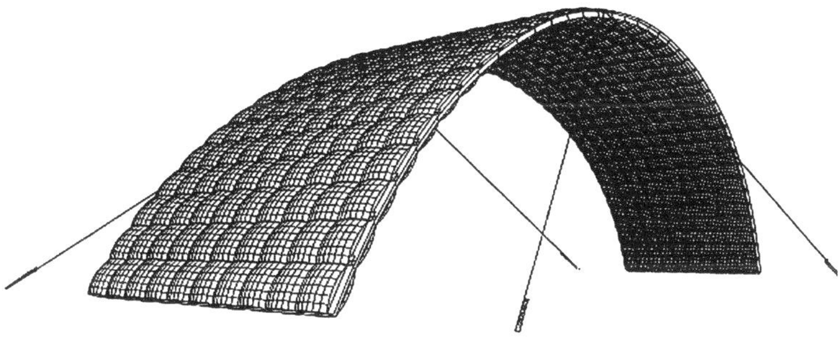

Fig. 17. A simple cylindrical shell composed only of cushions and cables

Cross-braces can be placed inside the structure or in reverse position, over cushions, in order to enlarge internal space, Fig. 18.

For a specific architectural effect, semi-rigid cushions are very promising. The external rigid panel can be made of translucent or painted plastic and some printed patterns can be added, Fig. 19.

6.2 Construction Site Shelter

In many cases, the construction site should be prevented from the weather conditions. It concerns, for example, concrete works as well as earth works. An exemplary structure is presented on Fig. 20. It consists of the three simple cylindrical shells that form a shelter over rectangular area. Exactly the same structure is suitable for sport facilities, e.g. tennis court.

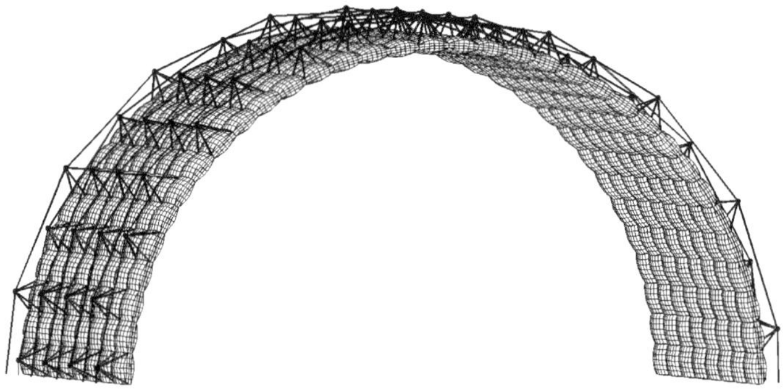

Fig. 18. Cylindrical shell with cross-braces placed in reverse position, over cushions

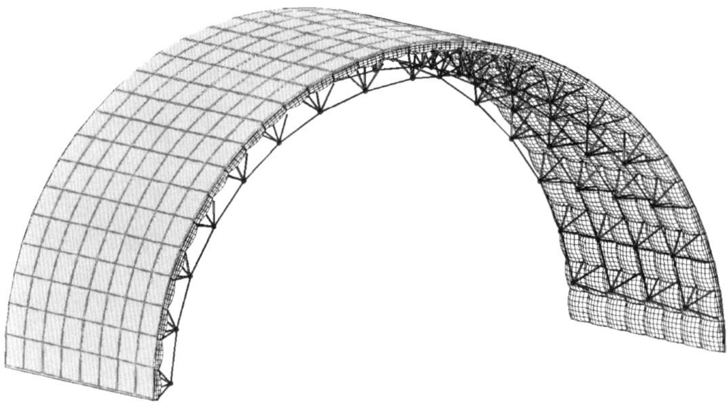

Fig. 19. Cylindrical shell composed of semi-rigid, translucent cushions

6.3 Multiple Hypar Roof

Interesting shapes can be obtained as a result of bidirectional post-tensioning. A single hypar surface, as shown on Fig. 6, can be multiplied to form a complex structure, Fig. 21. This is a good solution for itinerant theatre troops, temporary markets, exhibitions etc.

6.4 Deployable Inflated Bridge

Easily assembled and quickly erected structures, as described above, are predestinated for applications in military bridging systems. In this case the bridge

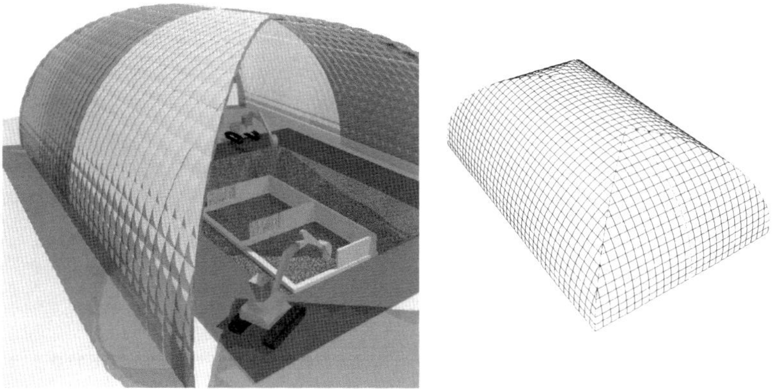

Fig. 20. General view and scheme of construction site shelter

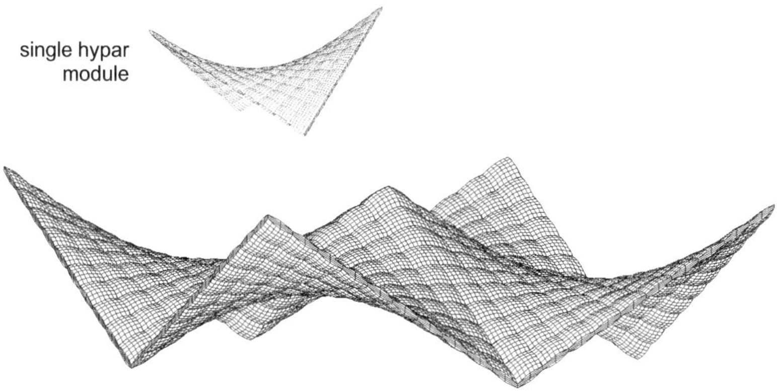

Fig. 21. Multihypar structure composed of four modules

should be launched in short time, with reduced number of in-site operations. The stowage space required for transportation should be reasonably small.

The idea to apply air-inflated elements for structural systems of portable bridges appeared at early stage of development of pneumatic structures, in sixties. Beam-type elements were tested in these early studies [9]. However, adoption of such elements caused some problems with wrinkling collapse under concentrated load. Another disadvantage of these structures is strictly predetermined span of the bridge, which cannot be changed after manufacturing. Problems recognized in beam-type bridges can be avoided by means of applying modular, post-tensioned structural systems [10].

Self-erecting inflated structures applied to deployable bridge

Deployable bridge can be constructed as a narrow sector of modular air-inflated shell. Use of semi-rigid cushions is necessary in this case, as they form a carriageable surface of the bridge. The same principles of shaping as described for modular inflated roofing are valid. Some exemplary configurations presented here are named as:

- type 1, which consists of cushions with constant thickness and cross-braces
- type 2, which consists of cushions with variable thickness
- type 3, which is basically a type 2 with modified method of deployment

Some main features are common for all these types. Rigidity of the structure is always increased in the center. Rigid panels are hinge connected on all edges. Bottom joints allow that cable is freely sliding through. A transverse steel pipe is fixed to the terminal cushions. It couples the cushions in order to properly support the structure on the ground and allows fastening of bottom cables. Air-supplying unit is placed in the transportation vehicle. Inflation by means of explosive cartridges is particularly useful in these applications.

Even after assembling, the modular inflated bridge is not an inalterable structure. The span of the bridge can be easily adjusted to the obstacle's size, simply by means of insertion or removal of required amount of cushions. This operation can be done without disassembling whole the structure, however it requires a modified construction of cable connectors.

Loading capacity of the bridge can be adjusted by means of changing its structural height and value of post-tensioning force.

Deployable inflated bridge – type 1

Deployable inflated bridge of type 1 consists of:

- rectangular semi-rigid cushions of constant thickness
- cross-braces arranged as a half of octahedron
- bottom chord (rope)

Figure 22 presents principles of composition of the structural system. Cross-braces are longer in the center and shorter in peripheries in order to adjust rigidity of the bridge. Cross-braces can be placed below every row of cushions or every second row – according to the track width and considered loadings.

General principles of assembling of the type 1 bridge are similar to those applied for self-erected roofing structures. Assembling of the cushions starts on the ground at the reverse position. Then, the cross-braces and the bottom cable are placed. The next step is to inflate cushions and rotate the structure upside–down. Then, the structure is self-erected by means of tension of the bottom cable. The last two steps can be done also in reverse order. Finally, the bridge is placed over the obstacle.

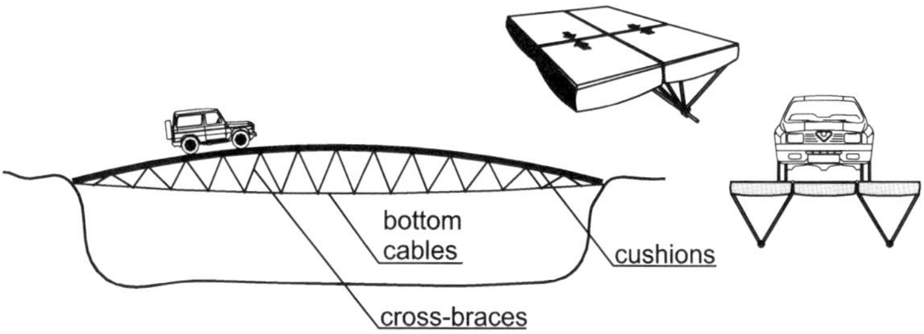

Fig. 22. Composition of deployable inflated bridge – type 1

Deployable inflated bridge – type 2

In general, composition of deployable bridge of typ 2 is the same as for the bridge of type 1. The main difference is in use cushions with variable thickness, instead of cross-braces, Fig. 23. Sequence of assembling is the same as for the bridge of type 1.

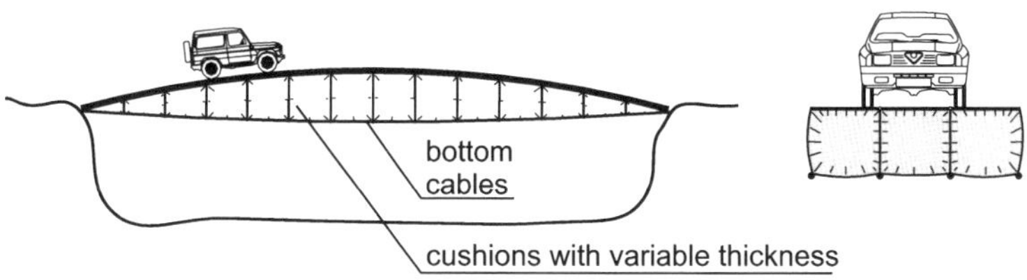

Fig. 23. Composition of deployable inflated bridge – type 2

Deployable inflated bridge – type 3

Deployable inflated bridge of type 3 is actually a bridge of type 2 with modified method of deployment. The idea is derived from the well-known children's toy "inflated tongue".

The bridge is pre-assembled, packed into a bundle and transported on a suitable vehicle. Assembling procedure starts with completely deflated and rolled set of cushions, Fig. 24.

With progress of inflation, the bridge begins to unroll. At this stage the structure reminds a "scorpion's abdomen", Fig. 25. The post-tensioning of the bottom chord starts together with inflation. This allows stabilizing the structure.

Completely inflated structure is placed over the obstacle and transportation vehicle is moving off the passage place, Fig. 26.

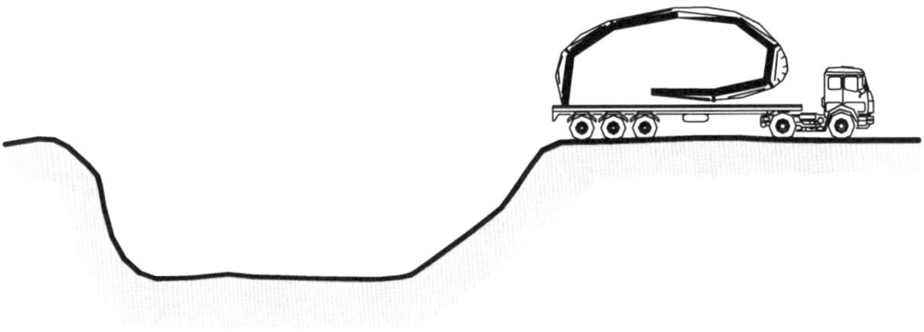

Fig. 24. Inflated bridge – type 3, bundled for transportation

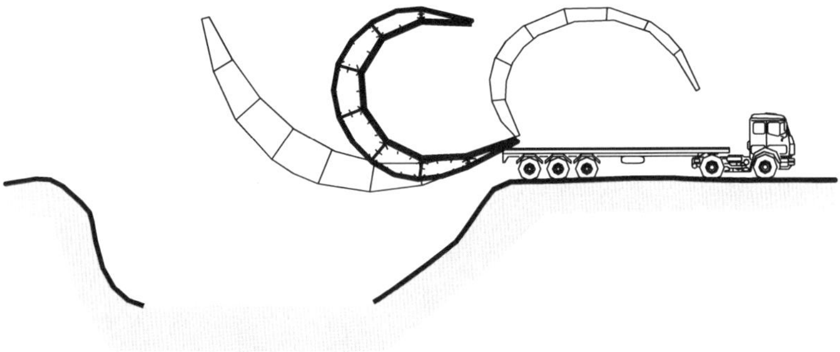

Fig. 25. Deployment of inflated bridge – type 3 ("scorpion's abdomen"position)

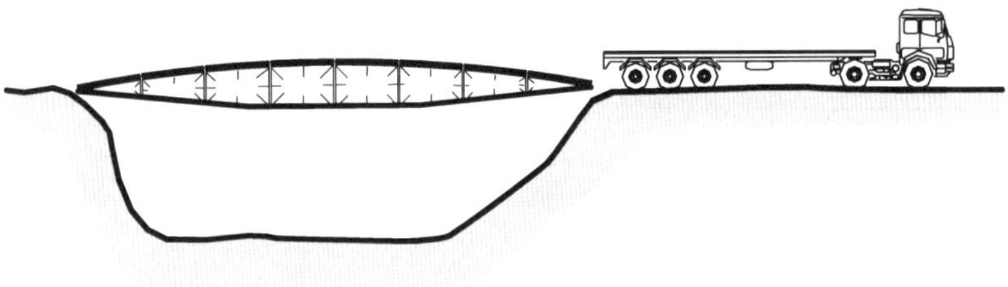

Fig. 26. Inflated bridge – type 3, ready to use

Described above, structural system for deployable bridges, exploits all advantages of air-inflated modular structures together with self-erection as a principle of shaping and erecting. This is a convenient solution for constructing bridges over precipices and rivers, in military and emergency applications.

7 Conclusive Remarks

Application of modular inflated elements to shell structures eliminates many disadvantages of air-supported structures as well as non-modular air-inflated structures. The span and (with some restrictions) shape of the structure can be determined just at the assembling. There is no need to use different structures for various purposes – many geometrical configurations can be reached. Modular inflated shell is a structure based on the idea of self-erection - a process that unifies post-tensioning and erection. Due to internal fit-out, the structure can be easy maintained. This type of structures can be used for many military and civil applications, where a fast assembled and adaptable solution is required.

References

1. Sumovski JA (1893) Tubular structure filled with gaseous fluid. US Patent 511 472
2. Geiger DH, Majowiecki M (1977) Acier-Stahl-Steel 9:281-288
3. Kawaguchi M (1993) A few nonlinear topics in structural design. In: Hangai Y, Mashita K, Mitsui K (eds) Nonlinear analysis and design for shell and spatial structures. Seiken–IASS–AIJ, Tokyo
4. Tarczewski R (1997) Bar and pneumatic self-erecting space structures. In: Ciesielski R, Ciszewski B, Gronostajski J, Hawrylak H, Kmita J, Kobielak S (eds) Challenges to Civil and Mechanical Engineering in 2000 and Beyond. Wroclaw University of Technology Press, Wroclaw
5. Tarczewski R (1996) Shaping of space structures by means of shortening of cable-type bottom chord. In: Obrebski JB (ed) Lightweight Structures in Civil Engineering. MicroPublisher, Warsaw
6. Tarczewski R (2000) Variable rigidity as a tool of shaping of self-erecting structures. In: Escrig F, Brebbia CA (eds) Mobile and Rapidly Assembled Structures III. WIT Press, Southampton Boston
7. Saar OS (2000) A new type of hardening structure. In: Escrig F, Brebbia CA (eds) Mobile and Rapidly Assembled Structures III. WIT Press, Southampton Boston
8. Tarczewski R (2002) Actively adaptable structures. In: Parke GAR, Disney P (eds) Space Structures 5. Thomas Telford, London
9. Bulson PS (1967) The behaviour of some experimental inflated structures. In: Feder D (ed) Proceedings of the 1st International Colloquium on Pneumatic Structures. University of Stuttgart–IASS, Stuttgart
10. Tarczewski R (2003) Deployable inflated bridge. In: Oñate E, Kröplin B (eds) Textile Composites and Inflatable Structures. CIMNE, Barcelona

Experiences in the Design Analysis and Construction of Low Pressure Inflatable Structures

Javier Marcipar[1], Eugenio Oñate[1] and Juan Miquel Canet[1]

International Center for Numerical Methods in Engineering
Universidad Politécnica de Cataluña
Gran Capitn, s/n, Campus Norte UPC
08034 Barcelona, Spain
marcipar@cimne.upc.es
http://www.cimne.upc.es

Key words: Inflatable structures, membrane simulation codes, temporal pavilions, inflatable structures for civil applications

1 Introduction

Inflatable structures have become increasingly popular in recent years for a wide range of applications in architecture, civil engineering, aeronautic (airships) and airspace situations [1-5].

Inflatable structures have unique features. Because of their foldability and air- or helium pneumatic stabilisation they cannot be compared to any classical struc-tural concepts.

The use of inflatable structures can be found in temporary and/or foldable structures to cover large spaces or to support other elements, in permanent roofs or shelters with a high degree of transparency, in mobile buildings as temporary housing in civil logistic missions (e.g. environmental disasters and rescue situa-tions), in the construction of tunnels and dams, in antennas for both ground and aerospace applications, as well as in extremely light airship structures among other uses.

Space Applications

Inflatable structures for space applications can be classified into two categories. The first one corresponds with satellite appendages and the second one with habi-tats.

E. Oñate and B. Kröplin (eds.), Textile Composites and Inflatable Structures, 241–257.

Inflatable structures in satellite appendages can be the primary structure of the appendage, or else be used as the deployment mechanism itself. In the first case the inflatable structure should become rigid after inflated. For deployment mecha-nisms the integrity of the inflatable structure is not requested after deployment and its rigidity is obtained by means of other structural members assembled during de-ployment.

Examples of typical satellite appendages where the inflated technology is ap-plicable include solar arrays and flat antenna arrays among others. A flat sunshield concept with inflated struts which become rigid after inflated has been recently developed by EADS CASA Space in the context of the INFLAST project of the Growth EC programme carried out in cooperation with the authors [6].

Inflatable habitats, on the other hand, are characterized by a big inflated vol-ume. These habitats can be used for space stations or Moon/Mars missions [7-8]. Other applications include the protection of rovers or other instruments in airspace vehicles or stations. Indeed, inflatable habitats request structures that can with-stand loads such as micrometeorites impacts and radiations.

Civil applications

Inflated structures for civil applications are competitive due to their minimum weight, high foldability, easy assembly and disassembly as well as easy and quick transportation with little manpower and maintenance.

For civil applications, different efforts have been made in the past years to de-velop inflated structures formed by assembly of high or low pressure tubes. This effort has increased the penetration of inflatable buildings in the market, but the growth on the use of this structural type is still very low.

Recent applications of inflated roof structures include air supported halls for sport or exhibition events with a low internal pressure. In the USA areas up to 30000 m2 have been covered with air supported membrane structures [i1]. Their disadvantage is the high energy consumption to keep the pressure permanent in the total inner volume. For structures with high pressure (higher than 1,0 kN/m2) the use of inflated elements has often been proposed, but they have rarely been built (Fuji-Pavillon, Japan). The reasons are that the necessary materials, structural design and manufacturing techniques have not been fully developed [i2]. Double layer membrane structures, usually known as inflated cushions, require rigid ele-ments at the edges or compression rings. They are used instead of glass elements due to their lower price, which is 1/3 of a glass covering. So far there are only two German companies dealing with these inflated cushions in the world market.

Generic technologies have already been tested on prototypes such as a hangar [i3], exhibition hall [i4, i5], various tube structures [i6] and an air-ship [i7]. A supporting structure for an exhibition hall was built out of high pressure in-flated chambers by FESTO in 1996. The structure has many new features. These include the use of double-layer-fabric for the walls as load bearing elements, flame-inhibiting elastomer coatings and a new translucent ethylene-vinyl-acetate coating as well as the active control of the structure against wind load. [i4,i5] References to structures filled with lighter-than-air gas i.e. helium, using the buoyancy forces to stabilize the structures, cannot be found in the literature. A structure of this type has been designed and built as helium filled hangar and as an air supported exhibition pavilion [i3, i6].

The knowledge needed to improve the inflatable construction system, the in-existence of tools for computer aided design and specifications makes difficult that professionals who design temporal and mobile structures choose the inflatable solution.

This article describes the experience of the authors in the design, analysis and construction of inflatable structures for civil applications using an assembly of low pressure elements such as tubes and double layer membranes with tension ropes.

The basis of this technology was developed by the authors in the INFLAST project [6] completed in October 2000 [Brite Euram Project Nr. BE96-3015 "New methodology for Design and Manufacturing of Inflated Structures" (INFLAST)]. This technology has been applied to the design and construction of a number of in-flated structures.

2 Inflatable Structures

An inflatable structure is composed by an exterior flexible membrane that contains a fluid inside (in general air or helium). The function of the interior fluid is to maintain the exterior membrane under strength. Final shape of the inflatable struc-ture and its structural resistance depends on the strength of the external membrane and the pattern design. The strength of the external membrane is directly related to the pressure of the air contained inside the structure and the volume between ex-ternal membranes.

The commercial scope of inflatable products ranges from those which are available on the market for example, inflatable boats, lifecrafts, escape slides, mat-tresses, toys and containers to others that are relatively new as airbags and special membrane structures

2.1 High Pressure Inflatable Structures

Some efforts have been made in the past years to develop inflated structures formed by assembly of high pressure tubes.

The obvious disadvantages of these structures are the design of the joints and their big vulnerability to air losses. In general, high pressure inflated structures are difficult to maintain and repair and have a high cost. An example of inflated pavil-ions formed by high pressure tubes is shown in Fig. 1.

Fig. 1. Example of inflated pavilion formed by assembly of high pressure tubes. "Fuji" Inflatable pavilion constructed by Mamoru Kawaguchi in 1974 for the "Expo Tokyo"

2.2 Low Pressure Inflatable Structures

Inflatable structures formed by an assembly of self-supported low pressure mem-brane elements are ideal to cover large space areas. They also adapt easily to any design shape and have minimal maintenance requirements, other than keeping a constant low internal pressure accounting for the air losses through the material pores and the seams.

Air supported system

One of the inflatable systems most commonly used is in the so called "air sup-ported system". In this structural case an inner space is generated from a com-pletely closed membrane perimeter and filled with air at low pressure.

There are areas of more than 30.000 m2 covered with this kind of structures. In any case, this system has several disadvantages:

- Excessive air volume is needed to support the cover.
- Complex access should be designed to ensure the stagnation of the inner space.
- Anchoring loads are really high in order to support the strength of the exterior membrane.
- Thermal resistance is very low. This aspect has a strong incidence in the climatization of the inner space.

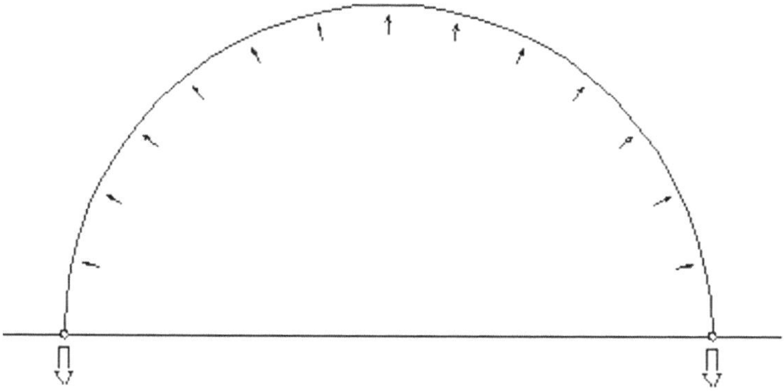

Fig. 2. Functional schema of the "Air supported system"

- Because of the contact or cold membrane with inner hot air, con-densation drops are formed and fall in a rainy way inside the cov-ered space.
- This system is impossible to move to another location because of the anchoring needed to support the membrane strength.
- This system has severe restrictions to adapt to any architectural design.

Air inflated system

Because of the disadvantages above listed, a new and more revolutionized system was designed called "air inflated system". This system is formed by tubes of dif-ferent textile materials filled with low pressure air. Textile materials are designed to support any external action, from fire resistance to climate aggression.

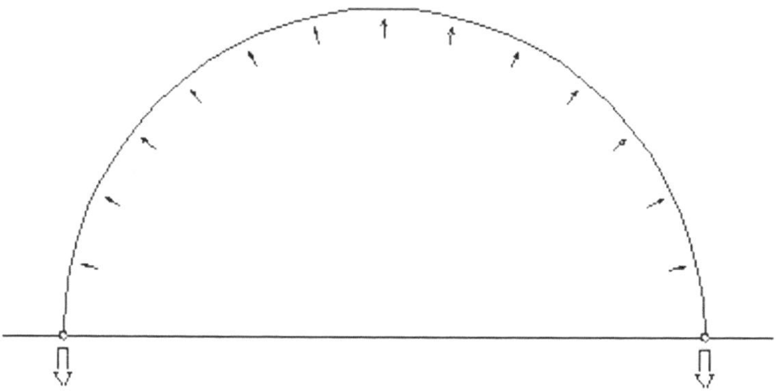

Fig. 3. Functional schema of the "Air inflated system"

Some innovations are needed to ensure the construction of this inflatable sys-tem such as new and powerful pattern design tools or simulation codes with the possibility to predict the behavior of the solution under different load cases.

Fig. 4. Inflatable pavilion (250m^2) formed by an assembly of low pressure tubes built for the mobile exhibition for the Gaudi Year of Construction Artisans, Barcelona 2002

Structures formed by inflatable tubes

The structure is formed by an assembly of low pressure tubes formed by different textile materials. The diameter of the tubes depends on the needs of the project and the dimensions of the space to be covered. The inflating pressure varies between 50 and 150 kg/m^2 accordingly to the structural stiffness required. This pressure can be obtained by simple fans with low noise production. Therefore, the use of air-compressors and expensive or noisy inflating equipment is not required.

The standard material chosen in the inflatable structures built by the authors is Polyamide 6.6 of high strength and 0.3 mm thickness. UV and fire resistance coat-ings which provide a M1 material type are used. Typical strengths in the transver-sal and longitudinal directions are 36 and 30 kg/cm^2, respectively. The limit strains are 24.5% along the transversal direction and 25% in the longitudinal direction. The weight of the material is 208 g/m^2. Obviously, higher strength materials and alternative coatings can be used to meet other structural and thermal requirements.

The double membrane layer helps towards the thermal protection of the inner space. A disadvantage, however, is the excessive amount of textile material, as the surface of the lateral walls of the tubes amounts for the 30% of the total surface of the membrane structure. This disadvantage can be overcome using the tension-rope system described next.

Low pressure membrane structures within tension ropes

These double membrane inflatable structures are formed by two membrane layers connected by tension ropes. The ropes are placed at strategic points in order to preserve the desired shape of the structure. The amount of membrane material in these structures is considerable less than in the tubular structures with the corre-sponding savings in the weight and the overall cost of the structure. An example of a structure of this type recently built by the company BuildAir (www.buildair.com) is shown in Fig. 5.

Fig. 5. Cylindrical pavilion formed by a double layer membrane with tension ropes

3 Structural Analysis and Pattern Design

Traditional methods for analysis of inflatable structure are based on membrane and tube theories [9,10]. A more detailed analysis of inflatable structures can be performed using membrane finite elements allowing for large displacement analy-sis combined with one-dimensional elements to model the tension ropes, if neces-sary. The membrane formulation used is an extension of the rotation-free shell fi-nite element formulation developed by the authors in recent years [11-15]. The following analysis conditions are typically considered.

- Coupled aerodynamic analysis in order to compute the evolution of the shape of the structure under different wind conditions. Aerodynamic load are also essential to design the anchoring system. The fluid-dynamic analysis is based in the stabilized finite element method developed by the authors [16]. The coupled fluid-structure interaction formulation used is an extension of that derived by the authors for ship hydrodynamic analysis [17].
- Thermal analysis in order to study the evolution of temperature in the textile material under prescribed external temperature changes. The thermal analysis can be complemented by the computation of the temperature evo-lution in the interior of the structure using the finite element method.

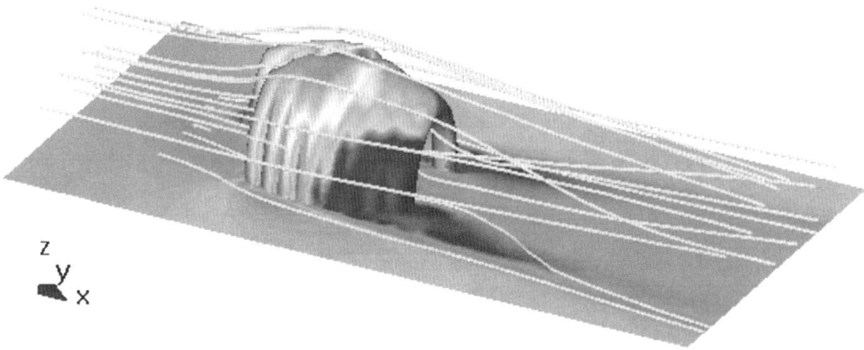

Fig. 6. Aerodynamic analysis results performed by CIMNE. Air particles around ECCOMAS 2000 inflated pavilion developed by BuildAir

The accurate pattern design is essential to ensure the correct shape of the struc-ture after the inflating process. The pattern design procedure developed by the au-thors is based on the isoparametric finite element concept. Patterns are assimilated to finite element patches and are defined using the pre-post processing system GiD developed at CIMNE [18]. A mapping into the isoparametric space allows repre-senting the pattern geometry in a 2D system, thus providing the necessary infor-mation for the cutting machine using a standard plotting system (Fig. 7). Fig. 8 shows a scale model in paper used to verify the pattern design of the ECCOMAS 2000 inflatable pavilion formed by low pressure tubes.

Fig. 7. Plotter system used to cut the membrane patches

Fig. 8. Paper model of an inflatable pavilion formed by low pressure tubes

4 Architectural Design

The use of structures formed by low pressure inflatable tubes is ideal to cover large temporal spaces. Innovative design allows offering clients unique products.

The exclusive and never seen design generates a great distinct impact for the users of inflatable system. The systems is ideal for applications that require tem-poral covers and pavilions with an exclusive design as a differentiation aspect. Ef-fectiveness and easy assembly are the main aspects to be considered. The design possibilities that the system allow are really enormous, offering to the client cov-ers with a great personalised and refined design value.

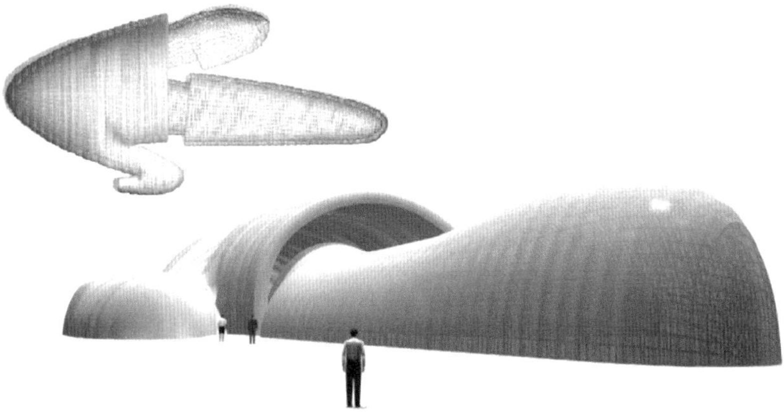

Fig. 9. Example of extremely complex design of an inflatable mobile theatre

5 Development Lines

New requirements have been identified and different solutions were developed in order to increase the possibilities of low pressure inflatable structures to success into the civil engineering and architecture market.

Some of the aspects developed by the authors in order to increase the use of in-flatable structures and to construct a new generation of inflatable covers for civil and architectural applications are:

- Regarding the manufacturing aspect, the development and optimi-sation of the procedures aiming to change from a non-industrial construction process to knowledge based industrial process using latest Information Technologies (IT) tools. An important objective was the reduction of production costs in order to make inflatable structures more competitive versus other temporary cover systems
- This requires multidisciplinary studies of the production chain, such as new lighter and cheaper materials and the optimisation of the production process.
- Much effort was dedicated to the pattern generation and optimisa-tion in order to reduce material lost and makes the material work in a better way. New software to generate and optimise patterns was developed considering

new material properties (anisotropy, multi layers, real shape of the fabrics, etc.)

- Alternative energy supplies and new ways to inflate the covers were developed to ensure the use of inflatable solutions in mar-ginal or disaster areas. It was important to optimise the use of gas (air or helium) inside the structure in order to minimise fluid lost.
- Material and mechanical behaviour of inflated structures follow different design principles from those typically used in standard concrete, steel or aluminium structures. Hence new conceptual and manufacturing concepts were developed in order to satisfy the in-creasing market demands while preserving necessary safety, com-fort and operational requirements. Structural control for wind pres-sure and temperature loads was mandatory in these cases.

The security of inflatable buildings was kept in mind in the implementation of every development.

6 Applications

6.1 ECCOMAS 2000 Pavilion

An inflatable structure formed by 11 low pressure tubes was designed and built as the Pavilion of the International Center for Numerical Methods in Engineering (CIMNE) at the ECCOMAS Congress held in Barcelona on 11-15 September 2000 [20]. The construction required 500 m^2 of Polyamide 6.6 material and some 1000 meters of sewing. Figs. 10-11 show a plant of the inflatable pavilion and a view of the exterior and interior during the exhibition.

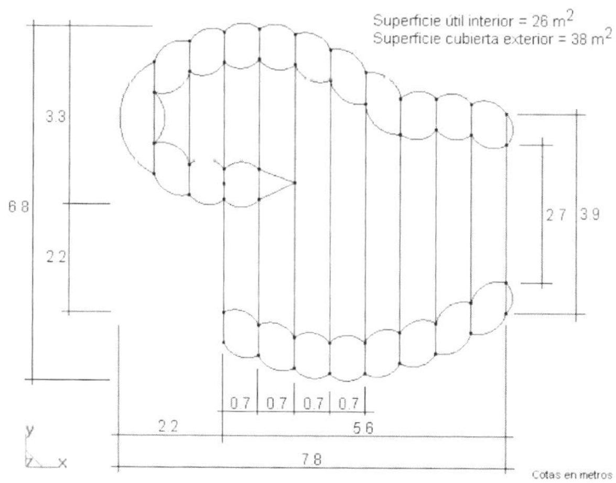

Fig. 10. Plan of the ECCOMAS 2000 Pavilion using 11 low pressure tubes. Distances in meters

Fig. 11. Exterior and interior views of the ECCOMAS 2000 inflatable pavilion

6.2 Gaudi Institute Exhibition Pavilion

An inflatable pavilion has been recently built by the company BuildAir in coop-eration with the authors for the Gaudi Institute of Barcelona for a mobile exhibi-tion on Construction Artisans.

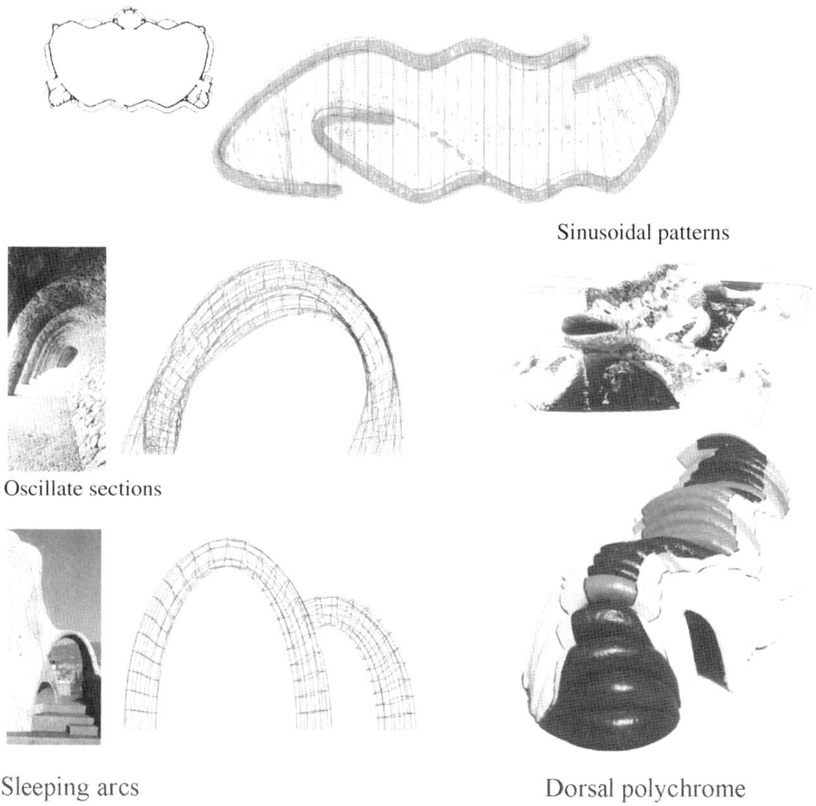

Fig. 12. Architectural concepts as basis of the Gaudi pavilion design

Architectural design and concepts

The construction technique was based on the low pressure tube technology de-scribed. A total of 58 tubes were used to cover a surface of 250m^2. A plan of the structure and different views of the inflated pavilion reaching height of 8.50 mts are shown in Figs. 13–15.

Public Engineering Constructions exhibition

This is another example of applications of inflatable pavilions to generate tempo-ral spaces for itinerant exhibitions. In this case, Buildair in cooperation with the authors had designed and constructed a space of 1.200 m^2 formed by an array of 6 pavilions of 11 meters by 16 meters.

This pavilion hold the exhibition "Las obras públicas en Catalunya" organized by Colegio de Ingenieros de Caminos Canales y Puertos, RENFE and CEDEX (Ministerio de Fomento) during November and December 2003.

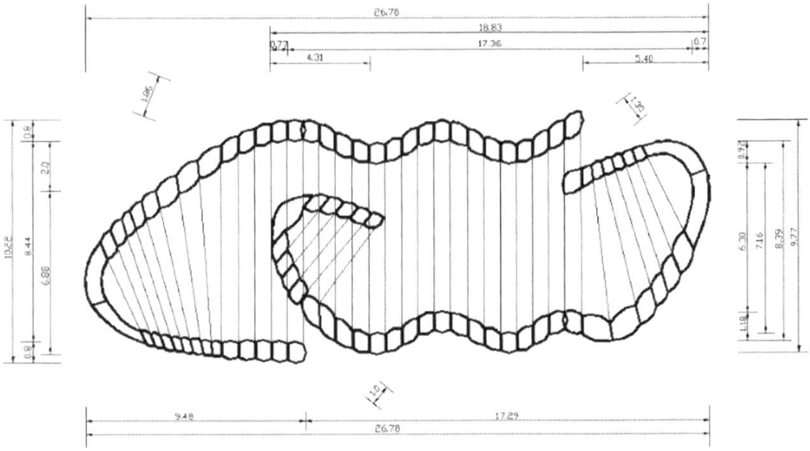

Fig. 13. Plan view of the inflatable pavilion for the Gaudi Institute Exhibition using 58 low pressure tubes

Fig. 14. Interior view of the inflated Gaudi Institute Exhibition pavilion

Fig. 15. Front view of the inflated Gaudi Institute Exhibition pavilion

Fig. 16. External view of inflatable pavilions, Francia Train Station. Barcelona, diciembre 2003

Fig. 17. Interior view of the pavilion

Fig. 18. Door detail, formed by an inflatable ball

References

1. Inflatable structures for engineering and architecture applications. BuildAir, www.buildair.com, 2002

2. Plant RH, Liapis S, Telionis DP (1996) Flood Protection using Inflatable Dams. Natural Disaster Reduction Conference. Washington December 3-5:264-265
3. Rehmet M, Bauder C, Schäfer, I Kröplin BH (1994) Solar Powered Airship Project. International Conference Remotely Piloted Vehicles, Bristol
4. Beukers A, Molder OV, Vermeeren CAJR (2001) Inflatable Structures in Space Engineering. Journal of the IASS
5. ILC Dover, World leader in innovative flexible solutions (2000). http://www.ilcdover.com
6. New Methodologies for Design and Manufacturing of Inflated Structures (INFLAST) (Brite-Euram Contract N BRPR-CT97-0448). Consortium: CIMNE: BAZAN, S.A., CASA, S.A., NOVURANIA, S.p.A., IRD a/s, Universitat Stuttgart, Airship Technologies, GmbH. Project finished on May 2000
7. Sadeh WZ, Criswell ME A generic inflatable structure for a lunar/martian base. Proceeding of the Ebgineering, Construction and Operations in Space IV 1146-1156
8. Nowak PS, Sadeh WZ, Morroni LA (1992), "Geometric modeling of inflatable structures for lunar base. Journal of Aerospace Engineering 5(3):311-322
9. Herzog T (1976) Pneumatische Kontruktionen: Bauten aus Membranan und Luft. Hatje-Verlag Stuttgart
10. Kawabata M, Ishii K (1994) Study of structural characteristics of Air inflated Beam Structures. Proceedings of the IASS-ASCE International Symposium, Atlanta 742–751
11. Rojek J, Pérez D, Cipriano J, Miquel J, Oñate E (1999) A finite elements methodol-ogy for analysis of inflatable structures. IASS Conference, Madrid
12. Oñate E, Zarate F (2000) Rotational free plate and shell elements. International Journal of Numerical Methods in Engineering 47:557-603
13. Marcipar J, Sarrablo V, Pérez D, Cipriano J, Miquel J, Oñate E (2000) Design and analysis of double membrane inflated covers presented. Presented in the MARAS 2000 Conference, Wessex
14. Taylor RL (2001) Finite element analysis of membrane structures (2001) Research Publication, CIMNE PI 103
15. Oñate E, Cendoya P, Miquel J, Rojek J (2002) Non Linear explicit dynamics analysis of shells structures using the BST rotation-free triangle. Submitted to Engineer-ing Computations
16. Oñate E (2000) A stabilized finite element meted for incompressible viscous flows us-ing a finite increment calculus formulation. Comput. Meth. Appl. Mech. Engng. 182(1-2):355-370
17. E. Oñate, J. Garca, "A finite element method for fluid-structure interaction with surface waves using a finite calculus formulation", Comput. Meth. Appl., Mech. Engng., 191, 635-660, 2001
18. GiD (2002) The personal pre/postprocesador. CIMNE, Barcelona, www.gid.cimne.upc.es.
19. European Congress on Computational Method in Appl. Science and Engineering (ECCOMAS) (2000), Barcelona 11-15 September, www.cimne.upc.es/ECCOMAS
20. Marcipar J, Oñate E, Sarrablo V (2002) Inflatable Structures. An integration of inno-vative materials, design, analysis and construction methods. Invited paper. Fifth World Congress on Computational Mechanics, Vienna

Recent Advances in the Rigidization of Gossamer Structures

B. Defoort[1], V. Peypoudat[1], M.C. Bernasconi[2], K. Chuda[3] and X. Coqueret[3]

[1] EADS SPACE Transportation
BP11 - 33165 Saint Médard en Jalles Cedex - France
brigitte.defoort@space.eads.net
[2] MCB Consultants
8953 Dietikon - Switzerland
[3] Laboratoire de Chimie Organique et Macromoléculaire
UMR CNRS 8009
USTL - 59655 Villeneuve d'Ascq Cedex – France

1 Introduction

The interest of using inflatable and rigidizable structures for space equipment (such as solar arrays, antenna reflectors ···) has been identified for many years, but this has not yet been implemented on operational space equipment in Europe, due to the lack of adapted materials and technologies. Recent improvements in these fields allow today the development of such projects and solar arrays have been identified as one of the most promising application. This paper focuses on inflatable and rigidizable lightly loaded structures. Typical driving requirements are high packaging efficiency, very low specific mass and large size. As with other flexible-wall structures, they exploit gas pressure for their deployment ("inflatable structures"). But, inflated structures unavoidably loose the gas that pressures them, and therefore require a pressure control apparatus and a gas supply to replenish the losses. This disadvantage is acceptable only for items that have to last for very short periods of time or for items where pressurization is a basic function (as in the case of habitats). It is commonly admitted that all inflated structures shall be rigidized in space as soon as their life time exceeds one week. As a result, the use of rigidizable materials that enable an inflated structure to become permanently rigid without relying on inflation is obviously a key technology in the field of Gossamer structures. For a given architecture, various kinds of rigidization techniques can be proposed: chemical rigidization [use of UV radiation (solar or with integrated light sources), thermal curing (using solar radiation and/or active heating)], physical rigidization [removal of volatile components in vacuum (solvent boil-off)], or mechanical rigidization [metal

E. Oñate and B. Kröplin (eds.), Textile Composites and Inflatable Structures, 259–283.

layer stretch/aluminum laminates]. One of the most promising rigidization techniques envisioned by EADS-ST is in-orbit UV curing of a composite structure. After a brief overview of Gossamer structures, exemplifying applications and potentialities of the technique by showing impressive realizations, we will detail a technology trade off related to the many rigidization processes that may apply to inflatable structures. Finally, we will focus on radiation initiated polymerization as a versatile tool to rigidize backbone structures.

2 An Overview of Gossamer Structures

2.1 Definition

Since the beginning of space flight, researchers and experimenters have been confronted with the problem of packaging into the restricted volumes available on the carrier vehicles items that they actually wanted to become much greater; and since those beginnings "inflatable" elements were among those proposed to master this challenge [1]. In recent years, NASA has introduced the "gossamer" expression to label those forms of spacecraft exceptionally low in mass and suitable for packaging into very small volumes, compared to conventional spacecraft: in general, it applies to inflatable and membrane structures for space use.

A more descriptive term, especially as it relates to the theme of the present book, is that of flexible-wall, expandable structures. The (initial) compliance of the walls allows the compact packaging and also enables the geometric efficiency of the materials that leads to the low mass, reinforced by the fact that one can design such structures for the space environment properly – not mainly to survive the launch phase. And, we refer to all those structures that are completely assembled at their manufacture site, then folded, stowed, packaged, or otherwise compacted for transport to their operational location, where they are deployed and installed for functional use. The installation sequence may include a rigidization procedure (mechanical, physical, or chemical – as discussed in the next Section), spinning the spacecraft (for a rotationally-stabilized object), pressurization (for continuously-inflated objects), etc. Expandable structures with flexible walls have been flown but only in a small number of cases. To date, the greatest majority of " large" structures deployed in space belong to the rigid-component, variable-geometry (RCVG) kind, that rely on actuation mechanisms to perform the transition from packaged to deployed state.

The range of technological approaches to the implementation of flexible-wall expandable structures is just as vast as that of the applications for which such structures can be used. One can organize the field using different discrimination criteria, deriving them from application-oriented considerations (e.g., the type of loads or geometrical requirements that drive a design), from characteristics of the structural elements' build-up (e.g., whether thin-walled membranes or thicker, more plate-like layouts), or from the methods used to stabilize shape of the object at installation. A first attempt classifies the structures according to four use and requirements criteria as follows:

- "Lightly-loaded", flexible-wall, expandable space structures - sized for the orbital environment (generally against buckling loads); typical requirements are: high packaging efficiency, very low specific mass, large size; the tension within an element's wall is of the order of 0.1 kN/m; a further subdivision distinguishes between:
  - support structures in general ("backbones"), in which a small-to-moderate integration between structure and system function occurs, and
  - precision structures, where the structural element and its shape have a direct system function impact
- "Heavy-duty" flexible-wall, expandable structures - sized to carry (internal) loads (generally, pressurization forces); typical requirements are: large enclosed volumes, compatibility with crew presence, moderate packaging efficiency and/or specific mass; the tension within an element's wall is of the order of 100 kN/m.
- "High-temperature" flexible-wall, expandable structures - sized to sustain significant temperature levels, as generated during planetary entry (although reduced thanks to the lower area loading such structures enable).

This article focuses mainly on lightly-loaded and heavy duty flexible-walls and not on high temperature elements which differ from the two first classes in technological terms even if some synergies exist and are used.Wilde and colleagues [2] give a summary of recent work in Western Europe on high temperature flexible walls.

2.2 Applications & Historical Background

Early Work: Inflatable Satellites

After the pioneering suggestions by Gatland and co-workers [1], the idea of inflatable spacecraft –in particular, to create optically observable orbital bodies– was developed by John O'Sullivan and his colleagues at Langley Research Center [3]. Soon, they prepared 12-ft (3.66-m), mechanically rigidized spheres, launched as Explorer IX and Explorer XIX for contributing to the measurement of the high atmosphere's density [4]. The Explorer spheres were sturdy enough to support themselves unpressurized under 1-g acceleration (Fig. 1). From this work evolved the concept and the technology for the passive communications satellites (Echo I & II, [5]), that eventually enabled the 40- m diameter PAGEOS (PAssive GEOdetic Satellite) [6]. Advanced concepts studied to achieve better mass efficiency than using spherical reflectors involved inflated lens/torus configurations and the wire grid sphere satellites, using photolyzable wall materials for the deployment and leaving eventually only a structure of stretched metal wires, making a radio reflector less sensitive to the solar pressure. Several such wire-grid spheres actually flew, e.g. the USAF OV1-8 satellite.

Fig. 1. An example of a balloon satellite: a mechanically-rigidized, 12-ft Explorer IX inflatable sphere under full-gravity testing. (NASA picture)

Precision Structures

Solar concentrators (for thermodynamic power generation) represent a further application that received extensive treatment using different gossamer technology approaches (with inflated membranes, with various foam-in-place techniques, with chemically-rigidized composites – both in form of membranes and of expandable-honeycomb structures), and with most designs adopting the lens-torus layout. Throughout the 1960s, they were studied in the US [7], but also in Germany where, around 1965, Bölkow investigated an inflatable foam-rigidized solar-thermal power collector [8]. Eventually, early in the 1970s, MBB built a 1-m inflatable and rigidized antenna reflector using glass- fibre-reinforced gelatin for the torus and the reflector shells, and a polymer-film radome [9].

In 1979, ESA began sponsoring a series of development contracts at Contraves (Zurich, Switzerland) that one of the authors (MCB) had the privilege to execute, lead, and participate in. For historical reasons, those development activities concentrated on microwave antenna reflectors, exemplified by the realization of the first rigidized offset reflector, but work was done in all classes of objects but for the "high-temperature" one.

Work on this technology –identified as Inflatable Space-Rigidized Structures (ISRS)– included a series of experimental activities using objects in the size range from 1- to 10-m aperture. First came three small models of a symmetric (center-fed) reflector to gauge issues such as folding and deployment, manufacture processes, and initially achievable accuracy. In successive phases, three 2.8-m reflectors (called LOAD-3 and designed for operation at 3.6 GHz) allowed the execution of following tests [10]:

- accuracy - improved manufacturing procedures adopted during that development phase allowed a reduction of the RMS error from 0.9 to 0.7 mm, while identifying the main sources of the remaining inaccuracies;
- packaging efficiency - were verified using the object that was successively subjected to electrical measurements; without degradation of surface quality as consequence of the folding and deployment exercises;
- controlled deployment in vacuum - a test within ESTEC's Dynamic Test Chamber demonstrated the quality of the residual air control procedures, the correctness of the deployment sequence, and the controlled deployment of the structure;
- electrical performance – measures were performed on the first complete object, after a full cycle of pressurization tests, folding, packaging, deployment, and cure;
- cure under (simulated) space conditions - a thermal-vacuum chamber solar simulation test demonstrated the correctness of the reflector's thermal design.

Further reflectors were manufactured and tested (under clean-room conditions):

- a 5.7-m diameter Test Article for the QUASAT radio telescope reflector, a center-fed layout [11], and
- a 10-m aperture offset-fed reflector (LOAD-10, Fig. 2), designed for operation at 1.6 GHz; after the folding and deployment cycle, the surface error had grown from 2.15 mm to 2.66 mm RMS, still yielding a gain of 42.6 dB and a sidelobe level of -33.8 dB [12].

The ISRS developments in Europe apparently found a resonance in Japan. Around the mid 1980s, a team formed around ISAS and began work on a modular, hybrid antenna reflector concept [13] – a variable-geometry truss backbone carrying ISRS reflector facets – as an unsuccessful candidate for the VSOP mission (the Japanese equivalent of QUASAT that eventually flew as HALCA).

Fig. 2. The LOAD-10 offset reflector

While the ESA initiative originally had but vague relations to previous US work, it contributed to the renewed interest there, when ESTEC personnel introduced the work done at Contraves to several JPL science projects teams. On the other hand, in 1980, L'Garde had proposed new approaches to continuously inflated antenna reflectors [14] and, after a number of development activities, in 1996 they finally achieved a test flight for a 15-m object, deployed from the Shuttle Orbiter [15]. Work on those inflatable reflectors continues [16].

Finally, under the USAF leadership, the inflatable solar concentrator was born again, this time to support the development of solar-thermal propulsion [17], a concept originally introduced by Ehricke [18]. While most designs foresee two offset parabolic reflectors, alternative configurations have investigated the use of flexible Fresnel lenses, also supported by gossamer elements. ESA has also sponsored studies for applying solar-thermal propulsion to upper stages for geocentric transportation [19] (Fig. 3).

Backbones

Concepts, type of applications, and study and development activities have been too numerous to attempt even a brief summary as done for the precision structures above. Many backbone structures (but not all by any means) involve skeletons, assembled from tubular components. Indeed, such a "one-dimensional" element forms the simplest backbone morphology. Following evolutionary considerations, one may discuss morphology and applications of backbones in the following order:

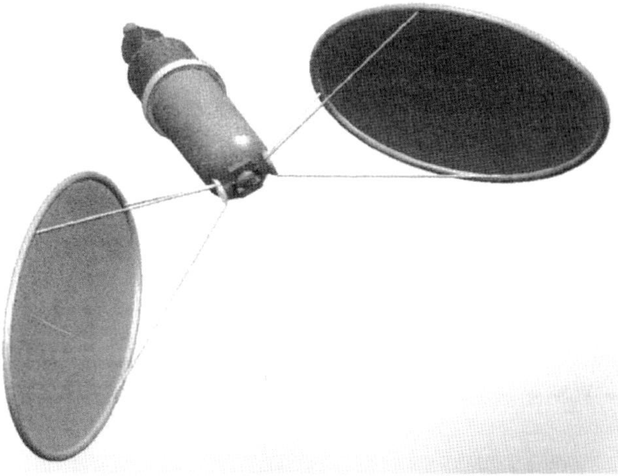

Fig. 3. European solar-thermal upper stage concept, with inflatable offset concentrators (EADS-ST image)

- *Planar Frames*: two-dimensional support for items such as, e.g., flat shields, solar sails [20], solar reflectors, and photovoltaic arrays [21] (Fig. 4), RF devices (reflectarrays, rectennae, lens,... [22]), or arrays of sensors.
- *Single-Tier Structures*: prismatic backbones (tripod, tetrapod, etc) for other functions, e.g. for light aerobraking [23], lens positioning, etc.
- *Two-Tier Structures*: Three-dimensional elements for telescopes tubes, cryogenic shield, hangars, and other unpressurized enclosures. The Contraves FIRST ISRS thermal shield concept belongs to this category: a complete 3.5-m skeleton [24] (Fig. 5), was manufactured and used for packaging, deployment, cure, and geometric tests.
- *Special Configurations*: Mast and booms, other (mostly) planar structures – for low-gain aerial structures (helix, Yagi), radiators.
- *Trussworks*: generic support structures, e.g. backbone structures both for Michelson [25] and Fizeau interferometers [26];
- *Polyhedral Skeletons & other, more complex forms*: Modified two-tier designs (e.g. for greenhouses), more complex lattice structures, spheres and spherical approximations.

Heavy-Duty Elements for Manned Flight

Gossamer structures hold the promise to provide significant capabilities in support of manned missions: throughout the 1960s, NASA and USAF studied and developed relatively small crew transfer tunnels and airlocks, orbital and surface shelters in support of exploration missions, full space stations,

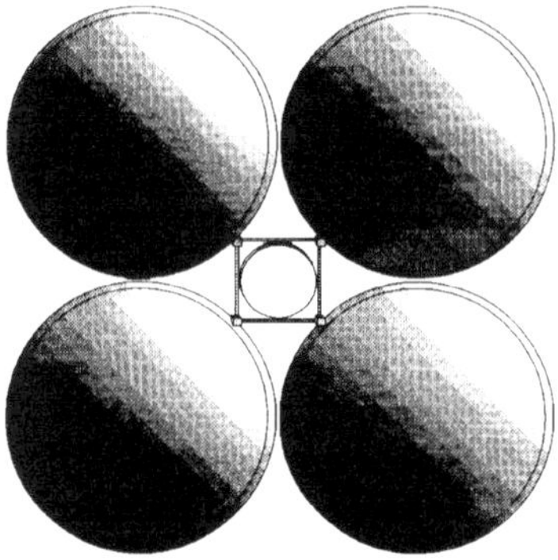

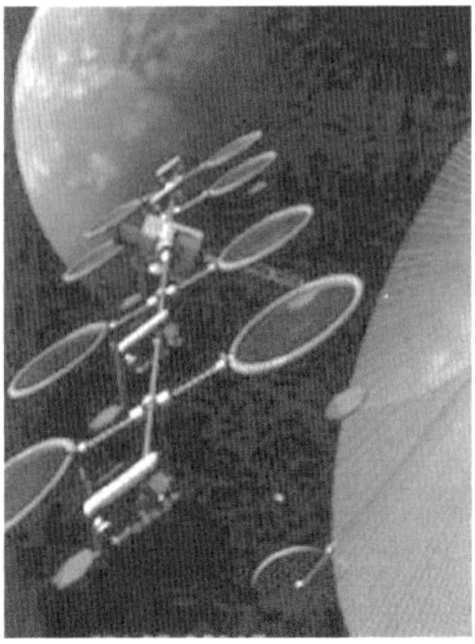

Fig. 4. Concept for a solar sailing spacecraft with four 2,500-m^2 saillets.(left); the "Sun Tower" solar power station builds on gossamer structures: supporting tori and flexible Fresnel-lens concentrators (right) (NASA picture)

Fig. 5. The 1/3-scale model of the ISRS skeleton for the FIRST's thermal shield (right) deployed out of an annular stowage volume around a simulated spacecraft central cylinder (left)

and pressurized hangar enclosures capable of holding entire spacecraft during scheduled maintenance/repair activities. The latest US entry in this class in the TransHab concept for a multi-storied habitat [27,28]. Activities along this direction have also been started in Europe [29,30].

3 Review of Rigidization Techniques

The use of rigidizable materials that enable an inflated structure to become rigid is a key technology in the field of Gossamer structures. The term "rigid" needs however to be clarified when discussing lightweight structures. For example, the 155 microns thick chemically rigidized material used for ISRS [31] is 39 times less rigid than a 100 microns thick steal foil in term of membrane stiffness and 280 000 times less rigid than a 10 cm thick foam plate in term of beam stiffness. Those ratio drop to 6.4 and 10500 respectively, once one considers the stiffness to surfacic weight ratio. This illustrates the fact that the weight and packed volume are the concepts that drive the development of thin flexible rigidizable walls.

Many technologies are identified for in orbit rigidization of Gossamer structures [32]. We firstly present a discussion of rigidization technologies, beginning with the identification and review of the different techniques and finally up to an evaluation of the existing technology. A set of evaluation criteria is defined and used to select the best candidates for a tubular solar array structure, to be suitable for Gossamer structures. The selection criteria include the material's ability to be folded, rigidization conditions (including power needs), thermal and mechanical properties, outgassing, durability in space environment, costs, rigidization reversibility... Discussions of specific materials for the different technologies are covered incidentally, to exemplify options and to assist the designer in his evaluation activity.

3.1 Rigidization Techniques and Associated Materials

Rigidization technologies can be classified depending on the nature of the phenomena that induces rigidization:

- Mechanical rigidization is obtained by stretching a polymer/aluminum laminate above its yield strain,
- Physical rigidization is obtained by phase transition (cooling a material below its glass transition temperature), using shape memory materials or by plasticizer or solvent evaporation,
- Chemically based rigidization is obtained either by thermally or UV induced polymerization. In orbit curing can be triggered or accelerated by gaseous catalysts carried by the inflation gas.

The different rigidization techniques are described below.

Mechanical Rigidization

This is class of structures deployed by inflation and rigidized by inducing through the pressure forces a stress higher than yield stress in a wall's metallic layer. Once the pressure is removed, the stressed aluminum maintains the structure's rigidity and shape. This concept is very attractive, and was used in flight on the Echo-2 satellites in the sixties [33], as well as on the Optical Calibration Sphere in 2000. The main advantages of this rigidization process are its reversibility, simplicity, predictability and rapidity. Furthermore it does not require additional power, has good space durability and no specific storage constraints. However, the anisotropy of the stresses and the need for an accurate control of the pressurization levels are issues that affect this technique for its application to cylindrical or toroidal objects. L'Garde proposes a solution to that problem which is based on fibers winding around a tubular structure [34]. Two main issues remain with regard to this technology: the first one is the different thermally induced dilatation of the constituting materials (polymer and aluminum) and the second is the compatibility of this technology with rolled-up storage.

Physically induced rigidization: cold rigidization, shape memory and solvent evaporation

The cold rigidization process relies on the exposure of originally flexible plastic layers – typically elastomers [35] – to the deep space thermal sink to cool them below their glass-transition temperature, rigidizing the structure essentially by freezing the matrix. This concept appears particularly indicated for shielding applications outside Earth's orbit, and was studied for shadowing shields of cryogenic stages for Mars flights. More recently, ILC Dover and L'Garde presented structures rigidized using this technique, respectively a hexapod structure [36] and the Space Solar Power Truss [37]. This technique is attractive mostly because of its reversibility, simplicity and low energy requirements compare to thermal curing. However, the need for temperature control and the coefficient of thermal expansion of the resins are serious drawbacks.

Recently, a number of studies have been conducted on shape memory composites, materials that mimic the behaviour of metallic shape-memory alloys [38,39,40]. The structure is completed on ground and consolidated at an elevated temperature, to set the material's geometric shape. The material will return to its original shape when heated above its glass transition temperature. For packaging, the structure is softened by heating it above Tg , taking care to keep it below its set temperature. After cooling, it is kept stowed. Prior to deployment, the stowed structure is again heated above Tg to make it flexible enough to be deployed by inflation. This is quite a complex process that limits the overall size of an object. The deployment in space requires a fair amount of power and control functions, as the heating must be rather uniform overall; also, presumably, the temperature should not drop below Tg.

Rigidization of a structure can also be obtained using evaporation of a solvent or a plasticizer in the material. The major issue of this solution is the large amount of solvent or plasticizer involved (e.g. between 13-50% for the Ciba polyimide tested during the Contraves ISRS program [41]). During the 1960s, a fairly large effort was dedicated to the study of rigidizable structures of this type utilizing fiber-reinforced gelatin prepregs. This approach has been improved more recently [42]. This solution presents the advantage to be a reversible process even if this is quite difficult on large structures. However it has many drawbacks related to packaging, outgassing, temperature variation sensitivity, weight...

Chemically induced rigidization : UV, thermal polymerization, gaseous catalyst and foaming in space

Thermal curing is a very classical path for aerospace composites. Various materials are available on shelf that can become rigid upon subsequent heating [43]. However, this technique has the major drawbacks to require a lot of power and energy to heat the structure. Numerous epoxy laminating resin formulation have been investigated in this class. The Contraves/ Ciba evaluation field alone included: conventional epoxy-based resin cured with an aromatic amine, epoxy resin cured with an amide, epoxy resin cured with an anhydride catalytically-cured cycloaliphatic epoxy and acryl-terminated epoxy resin [41]. The major drawbacks of this solution are the energy need and the additional weight due to the necessary thermal heaters [44], and/or the need for special coatings/MLI (Mulit Layer Insulation) that would increase the temperature of the structure [45].

Thermal curing can be triggered or accelerated using catalyst carried by the inflation gas. Pure catalytic cure systems offer the potential for "cure-on-demand". External catalysis involves the release of a gaseous catalyst within the inflated volume to activate and/or accelerate the reaction: such is the case of boron trifluoride with resin H developped by Contraves [31]. This solution has however the drawbacks that a secondary gas delivery system needs to be implemented on the Gossamer structure.

The use of solar UV for curing space-rigidized objects is extremely popular, in and beyond the literature. The first use of solar radiation to rigidify a structure was considered in the sixties for the Echo II balloon, and in the eighties by Contraves. An "on-command" cure capability is an aim for this class as well as Adherent Technologies proposes such a solution [46,47].

Thin walls based on foaming in space were also studied early in the sixties [48] and later in the nineties [49,50]. However this process faced non-uniformity and uncontrolled rigidization problems.

3.2 Technology Evaluation

Rigidization Techniques Selection Criteria

The elements to be taken into account for the evaluation of the rigidization methods and materials can be classified depending on the life phases of a Gossamer structure: non-rigidized, during rigidization and in the final rigid state. The main items are listed below.

- During manufacturing, and in the non rigid phase:
  - Cost and availability of the material
  - Shelf life, storage duration and constraint (raw materials and manufactured structure),
  - Weight,
  - Compaction ability,
  - Manufacture complexity (criteria linked to costs): foldability (ease to fold, damage risk), specific constraints related to the rigidization technique that directly impacts the design and manufacture of the structure (Multi Layer Insulation, specific coatings, heaters integration, specific deployment control system...), dimensional stability.
- During rigidization:
  - Rigidization reversibility and/or testability,
  - Process quality: reliability, rigidification on command, sensitivity, risk of uncontrolled rigidization, rigidization time, energy needs, in orbit specific constraints (rotation of the structure), outgassing.
- After rigidization:
  - Dimensional stability of the structure (Coefficient of Thermal Expansion),
  - Thermo-mechanical properties of the rigid material,
  - Specific properties regarding the application, - Aging in space environment (vacuum, UV, atomic oxygen, electrons, protons...),
  - Outgassing.
- Miscellaneous:
  - Adaptability to different design and architectural concepts,
  - Technology maturity (state of the art and user experience).

The accuracy and the relative importance of the selection criteria were discussed within this review. This analysis raises questions and considerations that need further discussion. Our technology evaluation was based on results presented in various up to date publications. However those deal in most cases with one specific material and as a result make it obviously difficult to evaluate technologies as a whole. Furthermore, the interest for inflatable structures was very strong in the sixties and is coming back on the scene nowadays; the age of some references should not hide the huge progress of polymer and materials related technologies since the seventies. Also, an important aspect relative to Gossamer technologies is the weight of the structure; but the rigidizable part in a Gossamer structure is in general 10 to 20 % of the total and, in the frame of rigidization technologies selection, one should not overrate this aspect. The rigidization reversibility is also often considered as a very important aspect. However, even if a specific technology is reversible, the applicability of such a concept is most of the time difficult on large structure. Finally it comes out that the most important criteria is the reliability of the rigidization technique and the required energy for rigidization.

Technology Evaluation

The evaluation criteria have been listed and weighed as a function of our preceding remarks. The criteria where we wished to put more weight are the reliability of the technology and the required energy for rigidization, the cost expressed in terms of material cost, but also in terms of manufacturing cost and the mechanical properties of the rigidized structures. As a result, the rigidization process itself accounted for 50% of the total. The materials properties after rigidization was evaluated as 14% of the total, taking into account that the structure (i.e. the weight) were dimensioned to fulfill mechanical specifications. Our evaluation led to the following results: 7 technologies – solvent evaporation, foam rigidization, thermal curing with the addition of a gaseous catalyst, solar UV curing, solar thermal curing and shape memory composites were ranked below 700 out of 1000. Four rigidization technologies were significantly better ranked than the others, according to our selection criteria and sensitivity: sub Tg rigidization, thermal curing with embedded heaters, UV curing with lamps and aluminum laminates.

3.3 Conclusion

This literature review allowed us to evaluate the performances, maturity, advantages and drawbacks of the different rigidization techniques, on the bases of an extended literature survey. The results highlight the diversity of the potential techniques and their very variable maturity. Selection criteria have been assessed in order to evaluate the different technologies. The reliability and maturity are especially important criteria. All the other selective criteria have been discussed.

As a result of the technology evaluation, Four technologies appear as especially interesting: stretching of aluminum/polymer laminates, thermal and UV curing with lamps of composite based material as well as sub Tg based composite rigidization. Among those four technologies, UV curing with internal light source comes out as the best solution.

As a result, the following part of this document is focused on UV based technologies for rigidization of Gossamer structure. The mature technology that is UV initiated polymerization will be detailed in the frame of its application to in space polymerization of lightly loaded composite structures.

4 Rigidization by UV-Visible Curing

UV-curing is now a well established technology finding a large number of industrial applications because of its distinct commercial, technical and environmental advantages. Initially developed for the fast drying of solvent-free formulations as printing inks, functional or protective coatings, adhesives and resins for microelectronics, polymerization induced by UV-visible light recently proved to be also an efficient method for curing various composite materials and gel coats [51].

In general, a formulated liquid or a molten powder is transformed almost instantly into a solid polymer simply by a short exposure to actinic light. A typical formulation for a clear coat consists of a mixture of functionalized oligomers mixed with low molecular weight monomers as thinners and with a photosensitive molecule or system, which is able to generate on demand the initiating species for the polymerization reaction. When powdery or fibrous fillers are added to the reactive blend, the penetration of light in the deeper layers of the material to be cured is a critical issue. Absorption, scattering and reflection phenomena decrease dramatically the amount of UV light that does penetrate into the material beyond a few tens micrometers of a coating with standard filler content (25–75 vol.-%).

The UV curing of composites and related filled materials is thus achieved by using a diffuse light source which has most of its emission in the long wavelength UV-visible light range. These diffuse lamps have the additional benefit of operating from a regular electricity supply. Stopping the exposure before completion of the polymerization process interrupts the cure and allows further working of the composite or gel coat if required, yielding extended processing flexibility.

There are essentially two types of UV-visible curable systems. They are based on free radical or on cationic mechanism. Both types of polymerization can be photochemically triggered by adequate initiators.

The majority of commercial light cure products are of the free radical type and use primarily acrylic (acrylate) components. Free radical systems are the most versatile in regard to product properties because many different types of monomers and oligomers are available for use to obtain the desired features. With a free radical system, polymerization stops almost as soon as the light is turned off. Free radical systems are also subject to oxygen inhibition, which means that oxygen in the air prevents the molecules at the surface from polymerizing, leaving an incompletely cured network.

Cationic systems contain epoxy and/or vinyl ether materials rather than acrylic components. Because only a restricted variety of monomers and oligomers are available for use in these systems, versatility in tailoring properties is limited. Unlike free radical systems, some cure does continue after the light source is removed, but it is sometimes minimal and often requires a thermal bump, or prolonged heating, to be effective. Cationic systems are not very sensitive to oxygen inhibition, but are easily poisoned by high humidity and nucleophilic contaminants.

4.1 Photo-Initiation

To obtain by this curing method rigid networks exhibiting a glass transition temperature T_g significantly above the operating temperature, monomers derived from bis-phenol A can be selected and adequately formulated [52]. The epoxy monomer DGEBA and its acrylated derivative EPDA (Chart 1) polymerize upon appropriate photo-initiation by a cationic or a free radical mechanism, respectively.

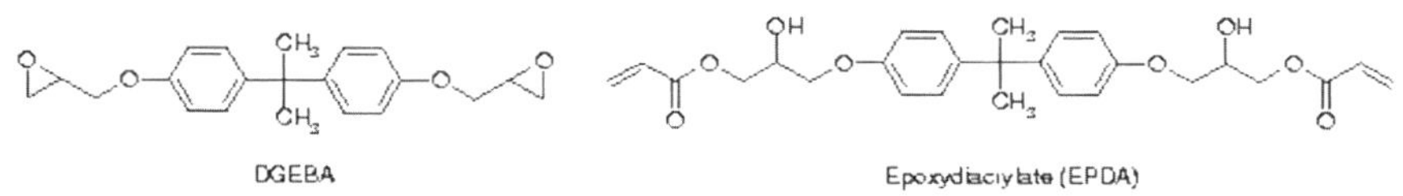

Chart 1. Examples of monomers derived from bis-phenol A

For an efficient exploitation of the incident UV-visible light, the absorption spectrum of the photo- initiating package has to be adjusted to the spectral characteristics of both the curable material and the light source [53]. In the case of a representative structure where the radiation curable material is sandwiched between protecting thermoplastic films, the light-filtering effect of the latter shall be overcome by minimizing the screen thickness, or alleviated by selecting an initiator exhibiting a long wavelength absorption, above the cut-off line of the enveloping film.

The transmission spectra of Fig. 6 clearly show the cut-off line at 320 nm by a 10 μm-thick PET film, whereas polyimide films do not allow the penetration of light for wavelengths shorter than 400 nm. Additionally, the dramatic reduction of transmitted light can be calculated at various operating wavelengths as a function of film thickness.

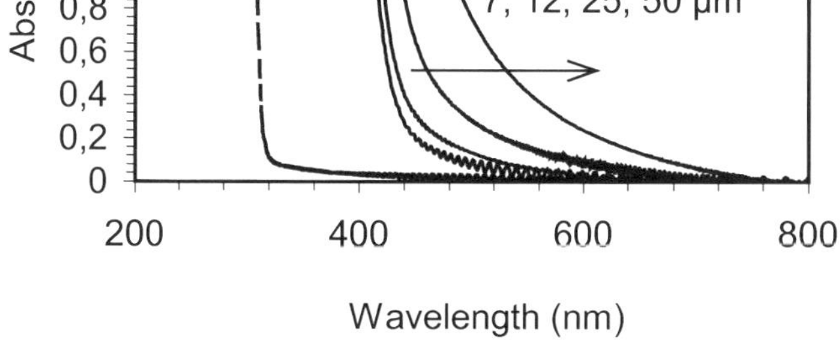

Fig. 6. UV-visible absorption spectra of protective thermoplastic films

Obviously, if PET absorption spectrum makes it possible to use a photoinitiator to cure the inner composite material with UV-A light, initiators sensitive to visible light are requested when using polyimide films.

Commercial phosphine oxide initiators, as mono-acylphosphine oxide (MAPO) and bis-acylphosphine oxide (BAPO) [54] exhibit long-wavelength absorptions that are not filtered by PET and that enable the curing of composites including transparent or white powdery fillers as well as glass fibres. Indeed the UV-curing of white-pigmented coatings is already state-of-the-art [55]. The bleaching of MAPO upon UV-visible irradiation filtered with PET (Fig. 7) allows the light to penetrate in the deeper layers of the material.

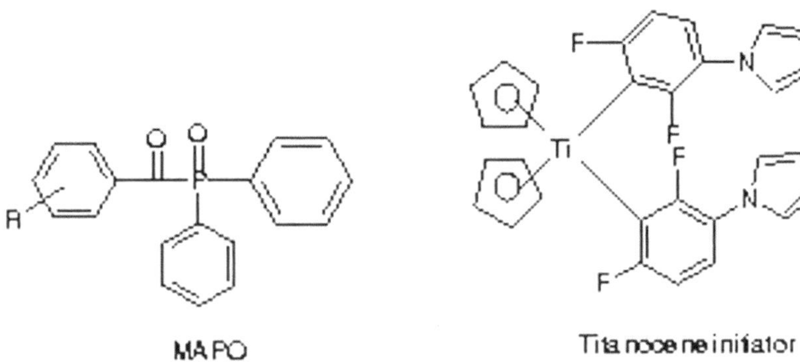

Chart 2. Acylphosphine oxide and titanocene photoinitiators

The progress of type I initiator photolysis can be modeled by simple absorption and decomposition laws owing to the monomolecular nature of the reaction. For a low initial absorbance at operating wavelengths, no significant gradient of energy absorption is expected and a first order description fits satisfactorily the observed rate of disappearance of the photosensitive compound.

The visible absorption spectrum of the titanocene initiator is convenient for initiating efficiently acrylate photopolymerization with 500 nm light, that is to say above the cut-off wavelength of the polyimide films mentioned above (Fig. 9). In the case of strong initial absorbance at operating wavelength, a gradient of light absorption in the curable material shall be taken into account, but as a consequence of gradual bleaching, in depth curing is achieved after predictable times of exposure. In the experiment corresponding to the spectrum of Fig. 9, zero-order kinetics is indeed observed over the main part of the photolytic process. This type of quantitative approach is particularly helpful to design and to control the rigidization process by radiation curing.

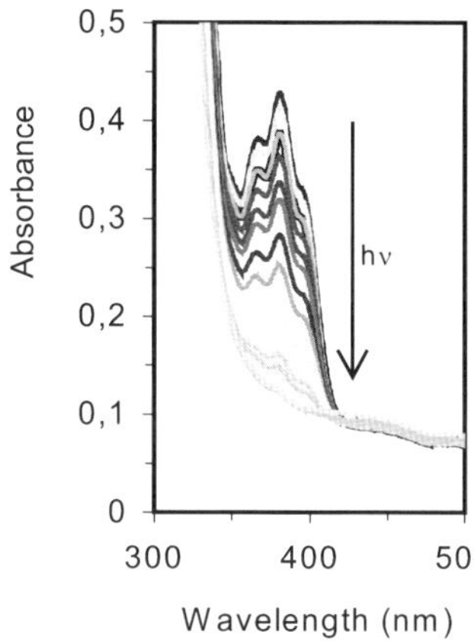

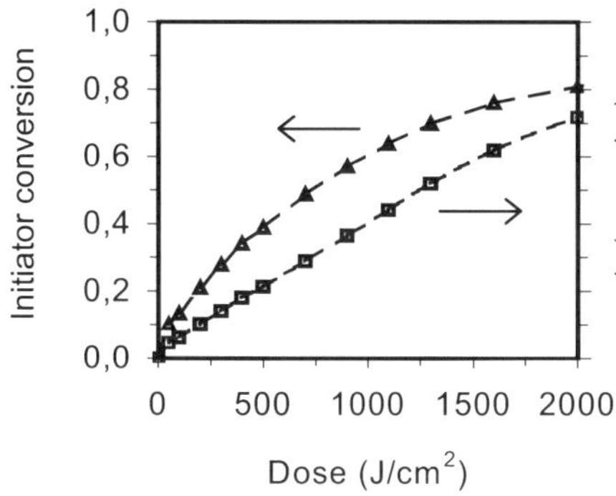

Fig. 7. (left) – Bleaching of a clear blend containing MAPO upon exposure to UV-visible light (PET filter)

Fig. 8. (right) – Progress of MAPO photolysis (same conditions as for Fig. 7)

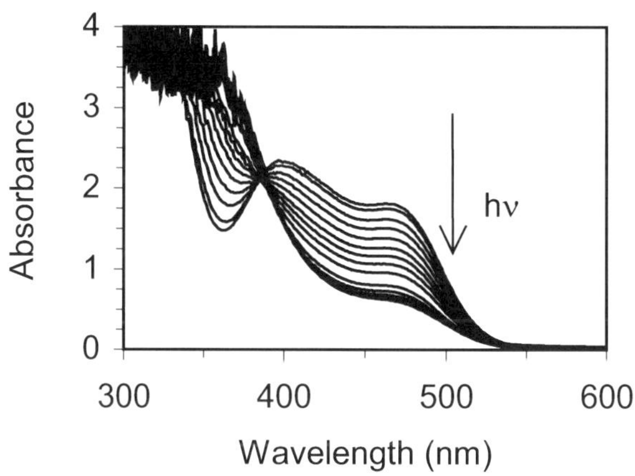

Fig. 9. Bleaching of a clear blend containing the titanocene initiator upon exposure to visible Xenon light filtered with a polyimide film

4.2 Optimizing Material Properties

The curing kinetics of acrylates as well as of other types of monomers can be monitored accurately by infrared spectroscopy [56]. Lab experiments for testing initiating system efficiency and for adjusting irradiation conditions can be conducted on samples of various thickness (from a few micrometers to some millimeters) using adapted sampling methods (transmission or ATR-IR)

[57,58]. The kinetic profile of EPDA photopolymerization at constant temperature can be varied by changing the nature of the photo-initiator and of the irradiation source. At 25°C, using various Type I initiators and appropriate filtered light, we have observed the curing within various timescales but with the same limiting conversion $\pi_{\max} = 0.37 \pm 0.02$ (Fig. 10). This demonstrates that vitrification exerts its control over the progress of polymerization in the different situations, provided that thermal control is effective.

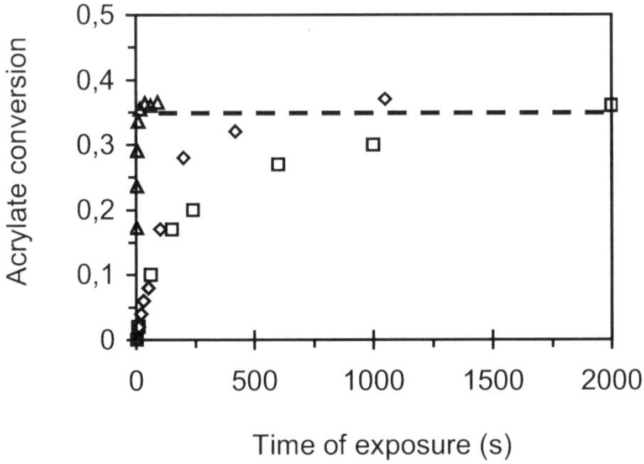

Fig. 10. Kinetic profile of EPDA photopolymerization under various conditions (initiator type and content, light source) at 25°C

The profiles recorded at various temperatures ranging between 10°C and 90°C are shown in Fig. 11, giving another evidence of the effect of mobility restrictions that are shifted to higher conversion levels as the curing temperature is raised.

The continuous progress of polymerization can be described by a phenomenological equation as shown below, with adjustable parameters c_{1-3}, which appeared more convenient than so-called autocatalytic models [59,60]

$$\pi(t) = c_1 t + c_2 - \frac{1}{c_3 t + c_2^{-1}}$$

After parameter adjustment by standard least square procedure, the model allows to predict satisfactorily the conversion degree of a sample submitted to given curing conditions (full lines in Fig. 11). The influence of initiator concentration, light intensity and temperature on the kinetic profiles can also be taken into account [61].

For slow photocuring reactions conducted under isothermal conditions, the curing temperature controls the final degree of conversion. A typical conversion-temperature relation for EPDA monomer is shown on Fig. 12.

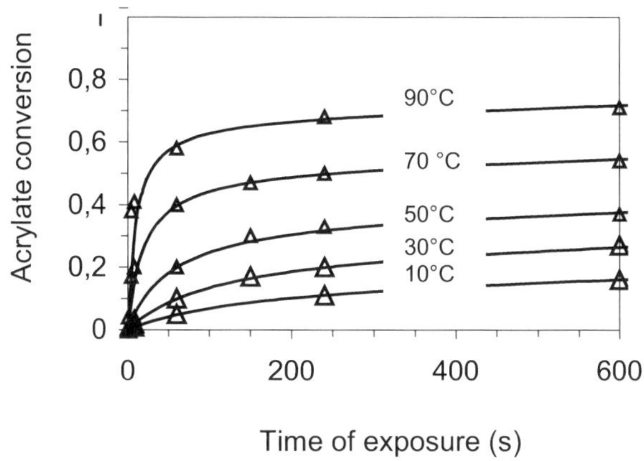

Fig. 11. Time-dependence of acrylate conversion as a function of curing temperature for EPDA photopolymerization with 1 wt-% type I photo-initiator under filtered light

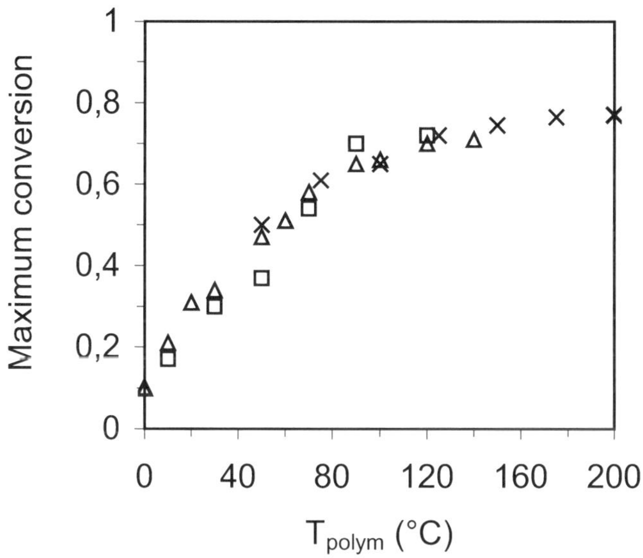

Fig. 12. Dependence of ultimate conversion as a function of curing temperature during isothermal EPDA photopolymerization

The determining influence of vitrification on the curing process is evidenced in Fig. 13, where the Tg of the network in the photocured samples is expressed as a function the curing temperature.

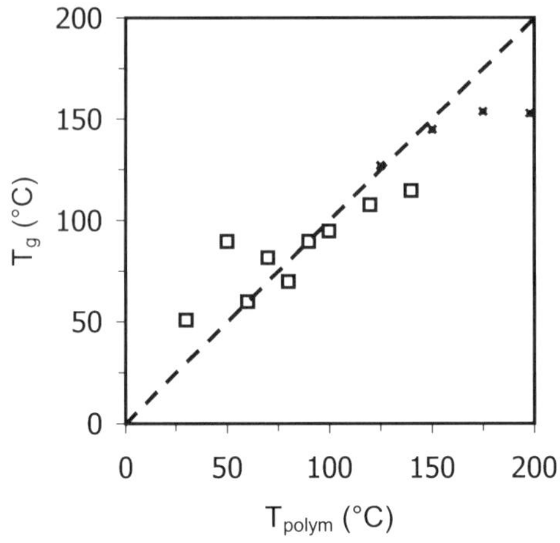

Fig. 13. Relation between the Tg of the cured EPDA network and the photocuring temperature

The adequate combination of thermal and irradiation effects on the cure kinetics of clear or filled resins therefore allows achieving the desired properties for the composite material of the rigidizable structure.

Alternative monomer chemistries and visible initiation packages have been assessed, exhibiting distinct advantages and limitations. A determining factor is the flexibility of the process in terms of light source spectrum and power as well as in terms of thermal conditions. Photosensitization of free radical and cationic initiators by visible light is a key issue. Several mechanisms including energy or electron transfer as well as redox reactions involving photolytic free radicals offer a broad range of combinations to be tested. Mechanical performance is rather easily achieved, but minimizing power and energy consumption of the light sources still appears as one of the most demanding features for the desired photocurable system.

4.3 Feasability Demonstration on Composite Structures

Large-scale UV rigidified tubes were recently manufactured based on light curable formulations at EADS-ST in collaboration with the LCOM.

The manufacturing process of the large scale demonstration boom consisted of the following steps: an inner polyimide bladder is assembled over a mandrel. Light curable prepreg fabric is then laid-up on the inner bladder, resulting in a 300 micron thick, structural composite layer. An aluminized, space qualified outer restrain is then placed over the composite layer and the whole tube is slid-off the mandrel. After boom end caps integration, the structure is compacted in order to occupy the least space possible. Finally, the boom is deployed by gas inflation and cured using internal lamps. The final boom is shown on Fig. 14. The typical sizes of the boom are respectively 2 meter long and 160 mm in diameter. The composite material was made of a light curable formulation for the matrix and of glass fibers. The rigidization of the boom was obtained after 8 hours of illumination. Further development shall include improvement of the resin formulation, compatibility of the process with carbon or organic fibers, and manufacturing processes development for larger structures.

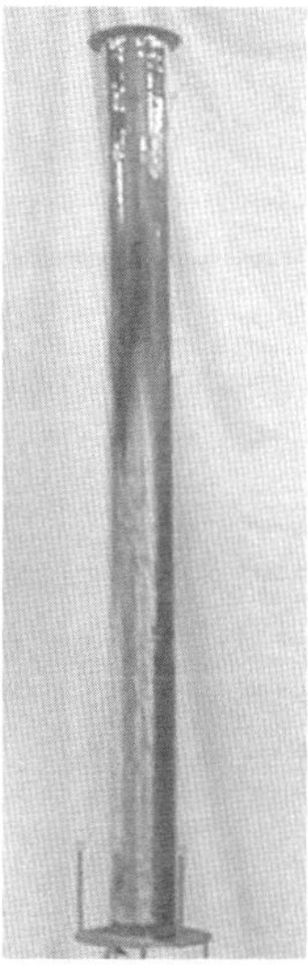

Fig. 14. 2 meter-long UV rigidized demonstrator manufactured, deployed and cured at EADS-ST

5 Conclusion

Despite the long-known potentialities of inflatable structure for space applications, this technology has not yet been implemented on operational space equipment in Europe, due to the lack of adapted materials and technologies. However, recent improvements in these fields now allows the development of such projects; and solar array has been identified as one of the most promising applications.

In-orbit rigidization of Gossamer structure is one of the key technology that shall be mastered. Radiation initiated polymerization is a validated on-ground technology in the field of thick and moderately filled composite materials. Due to its versatility and potentialities, it shall be able to fulfill the industrial need for a versatile, low power, low energy, controllable and predictable in-orbit rigidization technique. EADS-ST and the LCOM have a strong background in radiation curing of composite materials and in the formulation of radiation curable systems for specific high-requirement applications based on 10 years of close collaboration.

In the recent years, EADS-ST has heavily invested for developing Gossamer technology and is actively working in the frame of ESA funded activities on inflatable structures. EADS-ST aims at validating the technology in the course of a flight demonstration in the year 2007 (TRL 6) [62,63].

6 Acknowledgements

MCB dedicates this contribution to Karl BV Bentz (1935-2003), who was his supervisor at Contraves for many years and who introduced to him the idea of the "space curing bubble". Mr Bentz was a true gentleman who, by enduring interest, high technical competence, and by a quiet and respectful way, shaped much of my work and of my professional evolution.

XC and BD wish to thanks Dominique Lacour (EADS-ST), Emmanuelle Jesus (EADS-ST/LCOM), Christophe Christiaens (LCOM) for their contributions to the work presented in this paper.

References

1. Gatland KW, Kunesch AM and Dixon AE (1951) Minimum Satellite Vehicles. 2nd International Astronautical Congress, London (UK), September 7; also: JBIS 11:287-294.
2. Wilde D, Walther S, Pitchadze K, Alexsaschkin S, Vennemann D and Marraffa L (2001) Flight test and ISS application of the Inflatable Reentry and Descent Technology (IRDT). 52nd Astronautical Congress, Toulouse (France), October 1-5. aAlso: Acta Astronautica 51, 01:83–88, (2002).
3. O'Sullivan jr WJ (1959) Self Supporting Space Vehicle. US Patent 2,996,212, of 15/08/61.
4. Woerner CV and Keating GM (1962) Temperature Control of the Explorer IX Satellite. NASA TN-D-1369.
5. Slater RJ (1963) Design & Fabrication of Inflatable & Rigidizable Passive Communication Satellites (Echo I & Echo II). 1st Aerospace Expandable Structures Conference 576–604.
6. Teichman LA (1968) The Fabrication & Testing of PAGEOS I. NASA TN D-4596.
7. Rhodes MD (1969) Calorimetric Evaluation of Three 1.5-m Diameter Inflatable Rigidized Solar Concentrators. NASA TN D-5234.
8. Barthel G, Lanz S, and Mathes H (1965) Development of a Plastic Parabolic Mirror for Solar Power Supply Devices (in German). Bölkow rept 7.301.0-10/2.
9. LaRoche G, Roersch W, Mathes H *et al.* (1971) Development of an Expandable & Rigidizable Parabolic Antenna for Space Applications (in German). MBB rept RV1-08/16(624/05/66)/69Z.
10. Bernasconi MC, Pagana E and Reibaldi GG (1987) Large Inflatable, Space Rigidized Antenna Reflectors: Land Mobile Services Development. IAF 87 315 presented at the 38th International Astronautical Congress, Brighton (UK), October 1017.

11. van't Klooster K, Rits WJ, Pagana E, Mantica P and Bernasconi MC (1990) An Inflatable Parabolic Reflector Antenna Realization and Electrical Predictions. International Mirror Antenna Conference, Riga (Latvia), September 68. Also: ESA Journal 14[02], 211–216, (1990).
12. Savini D, Besso P, Tatalias P, van't Klooster K and Ritz W (1991) Electrical Performance of a 10-m Inflatable Reflector for Land Mobile Communications. ICAP 91 - 7th International Conference on Antennas and Propagation, York (UK), April 15-18, Proceedings 2, 853–856.
13. Natori M, Furuya H, Kato S *et al.* (1988) A Reflector Concept Using Inflatable Elements. Proceedings of the XVI International Symposium on Space Technology & Science, Sapporo (Japan), 459–467.
14. Thomas M and Friese GJ (1980) Pressurized Antennas for Space Radars. AIAA-80-1928 presented at the AIAA Sensor Systems for the 1980s Conference, December, 65–71.
15. Freeland R, Bard S, Veal G *et al.* (1996) Inflatable Antenna Technology with Preliminary Shuttle Experiment Results & Potential Applications. AMTA96, Seattle (Wash), September 30-October 3.
16. Chmielewski AB, Noca M and Ulvestad I (2000) ARISE Antenna. 2000 IEEE Aerospace Conference, Big Sky (MT), March 18–25, Proceedings 7, 439–445.
17. Malloy J and Miles B (1997) Benefits of the Integrated Solar Upper Stage (ISUS) to Commercial Space Systems. 1st Conference on Synergistic Power & Propulsion Systems Technology, STAIF-97 also: Mohamed S El-Genk, Ed. Space Technology & Applications International Forum. AIP Conference Proceedings 387[I], 445–451.
18. Ehricke KA (1956) The Solar-Powered Space Ship. ARS-0310-56, ARS Semi-Annual Meeting, Cleveland (OH), June 18-20.
19. Parkinson RC, Bernasconi MC, Bravais P and Corberand F Solar (2001) Thermal Orbit Transfer System (STOTS). 52nd International Astronautical Congress, Toulouse (France), October 1-5.
20. Bernasconi MC and Zurbuchen T (1994) Lobed Solar Sails for a Small Mission to the Asteroids. IAA L 070, IAA International Conference on Low Cost Planetary Missions, Laurel (MD), April 1-15. Also: Acta Astronautic, 35S, 645–655.
21. Bernasconi MC (2002) Inflatable Solar Arrays: Revolutionary Technology? - A European Perspective. Seminar at the Institute of Microtechnics, University of Neuchatel, Neuchatel (Switzerland), December 5.
22. Huang J (2002) The Development of Inflatable Array Antennas. XXVIIth General Assembly of the International Union of Radio Science (URSI), Maastricht (The Netherlands), August 17-24.
23. Gloyer P, Jacobovits A, Cohen D, Guerrero J and Wilmoth RG (2002) Aerobraking Technology for Earth Orbit Transfers. SSC02-VII-2, 16th Annual/USU Conference on Small Satellites.
24. Bernasconi MC and Köse S (1988) The Space-Rigidized Thermal Shield for the ESA Far Infrared Space Telescope (FIRST). 3rd European Symposium on Space Thermal Control and Life Support Systems, Noordwijk (NL), October 6. Also: ESA SP 288, 165–173.
25. Bernasconi MC and Rits WJ (1989) Inflatable Space Rigidized Support Structures for Large Spaceborne Optical Interferometer Systems. IAF338 89, 40th International Astronautical Congress, Torremolinos (Spain), October 7-13. Also: Acta Astronautica 22, 145–153.

26. Bely PY, Burrows CJ, Roddier FJ, Weigelt G and Bernasconi MC (1992) SISTERS: A Space Interferometer for the Search for Terrestrial Exo Planets by Rotation Shearing. ESA Colloquium on Targets for Space Based Interferometry, Beaulieu (France), October 13-16. Also: ESA SP-354, 99-. Also SPIE Proceedings 1947, 73–81, (1993).
27. Cadogan D, Stein J and Grahne M (1998) Inflatable Composite Habitat Structures for Lunar & Mars Exploration. IAA Paper IAA-98-IAA.13.2.04, 49th International Astronautical Congress, Melbourne (Australia), September 28-October 2. Also: Acta Astronautica 44[07], 399–406, (1999).
28. de la Fuente H, Raboin JL, Spexarth GR and Valle GD (2000) TransHab: NASA's Large-Scale Inflatable Spacecraft. AIAA-2000-1822, AIAA Space Inflatables Forum, Atlanta (Georgia), April 3-6.
29. Manoni G and Falvella MC (2002) Italian Experience and project in the Inflatable Space Structure Applications. 1st European Workshop on Inflatable Space Structures, ESTEC (The Netherlands), May 21-22.
30. Bischof B, Bolz J and Rieck U (2002) Inflatable Structure Compartment for Testing at Columbus ExPA's. 1st European Workshop on Inflatable Space Structures, ESTEC (The Netherlands), May 21-22.
31. Köse S (1987) FIRST Inflatable Thermal Shield / Final Report - Phase I. ESA CR(P)-2568 Contraves document SR/FIS/108(87)CZ.
32. Defoort B (2003) Review of rigidization technologies for flexible-wall expandable space structures. Textile Composites and Inflatable Structures, E. Oñate and B. Kröplin (Eds), CIMNE, Barcelona.
33. Cadogan DP and Scarborough SE (2001) Rigidizable materials for use in Gossamer Space inflatable structures. AIAA-2001-1417.
34. Lichodziejewski D, Veal G, Derbès B, L'Garde, Inc. (2002) Spiral wrapped aluminium laminate rigidization technology. AIAA 2002-1701.
35. Brewer W. *et al.*, Goodyear aerospace, Methods of evaluation of Inflatable Structures for space applications. AIAA 196X, 344-360.
36. Adetona O *et al.* (2002) Description of new inflatable/rigidizable hexapod structures testbed for shape and vibration control. AIAA 2002-1451.
37. Guidanean K and Lichodziejewski D, L'Garde Inc. (2002) An Inflatable Rigidizable Truss Structure based on New Sub Tg Polyuretane Composites. AIAA 2002–1593.
38. Cadogan DP *et al.* (2002) Shape memory composite development for use in Gossamer space inflatable structures. AIAA 2002–1372.
39. Abrahamson E *et al.* (2002) Shape memory polymers for elastic memory composites. AIAA 2002–1562.
40. Lake M *et al.* (2002) Development of coilable longerons using elastic memory composite materials. AIAA 2002–1453.
41. Bernasconi MC, Seiz W and Reibaldi GG (1984) Inflatable, Space-Rigidized Structures: Recent Development of the Material Technology. IAF-84-384. Also: Aerotecnica Missili & Spazio 64, 02:71–85 (1985).
42. Derbès B, L'Garde Inc. (1999) Case studies in inflatable rigidizable structural concepts for space power. AIAA-99-1089.
43. Cadogan DP (2001) Rigidization mechanisms and materials. Gossamer spacecraft: Membrane and inflatable structures technology for space applications, Christopher H.M. Jenkins and P. Zarchan (eds.), AIAA, Reston. Va 191:257–279.

44. Carey J, Cadogan D *et al.* (2000) Inflatable sunshield in space versus next generation space telescope (NGST) sunshield – A mass properties comparison. AIAA-2000-1569.
45. McElroy PM, Cadogan D *et al.* (2000) Analytical evaluation of thermal control coatings for an inflatable composite tube. AIAA-2000-1640.
46. Allred R, Hoyt A *et al.* (2002) Inflatable spacecraft using Rigidization On Command concept. AIAA-2000-1637.
47. Allred R, Hoyt A, Cadogan D *et al.* (2002) UV rigidizable carbon-reinforced isogrid inflatable booms. AIAA-2002-1202.
48. Schwarz S (1963) Delayed action, foam-in-place polyurethane for use in aerospace environment. USAF ASD-TDR-62-416.
49. Cassapakis CG and Thomas M (1995) Inflatable structures technology development overview. AIAA 95-3738.
50. Schnell R *et al.* (2002) Deployment, foam rigidization and structural characterization of inflatable thin film booms. AIAA-2002-1376.
51. Holman R (1984) UV and EB curing formulation for printing inks, coatings and paints. SITA Technology, London.
52. Allen NS and Edge M (1993) Radiation Curing in Polymer Science and Technology. J.P. Fouassier, J.F. Rabek (eds.), Vol. 1:225, Elsevier Applied Science, London.
53. Fouassier JP (1995) Photoinitiation, photopolymerization and photocuring: Fundamentals and applications. Hanser, Munich.
54. Dietliker K (1993) Radiation Curing in Polymer Science and Technology. J.P. Fouassier, J.F. Rabek (eds.), Vol. 2:155, Elsevier Applied Science, London.
55. Pietschmann N (1994) FARBE & LACK 100, no. 11:923–929.
56. Defoort B, Coqueret X *et al.* (2001) Electron-beam polymerization of acrylate compositions 6: influence of processing parameters on the curing kinetics of an epoxy acrylate blend. Macromol. Chem. Phys. 202(16):3149–3156.
57. Rabek JF (1993) Radiation Curing in Polymer Science and Technology. J.P. Fouassier, J.F. Rabek (eds.), Vol. 1:329, Elsevier Applied Science, London.
58. Patacz C, Defoort B and Coqueret X (2000) Electron-beam initiated polymerization of acrylate compositions 1: FTIR monitoring of incremental irradiation. Radiat. Physics and Chem. 59:239-337.
59. Defoort B, Coqueret X, Larnac G and Dupiller JM (2000) Modelling of the EB curing of fiber-reinforced composite materials for aerospace applications. 23rd Adhesion Society Meeting, Myrtle Beach, USA, 392–394.
60. Defoort B, Coqueret X, Larnac G and Dupiller JM (2000) Electron Beam curing of acrylate resins for composites: modeling reaction kinetics. 45th International SAMPE Symposium, May 21-25, Long Beach-CA, USA, 2223–2234.
61. Chuda K, Smolinski W, Defoort B, Rudz W, Gawdzik B, Rayss J and Coqueret X (2004) Polimery, 49:504-513.
62. Bonnefond T, Peypoudat V and Loukiantchikov A (2003) Flight testing of an inflatable ans rigidizable structure demonstrator. Structural Membranes 2003, June30th–July2nd, Barcelona (SP), and Textile Composites and Inflatable Structures, E. Oñate and B. Kröplin (eds.), CIMNE, Barcelona.
63. Bonnefond T, Peypoudat V and Loukiantchikov A (2004) Flight testing of an inflatable ans rigidizable structure demonstrator. To be presented in the 5th Gossamer Spacecraft Conference, Palm Springs, USA, April 19-21.

Form-Optimizing Processes in Biological Structures. Self-generating structures in nature based on pneumatics

Edgar Stach[1]

University of Tennessee, College of Architecture and Design
1715 Volunteer Boulevard, Knoxville, TN 37996, USA
stach@utk.edu

"If architects designed a building like a body, it would have a system of bones and muscles and tendons and a brain that knows how to respond. If a building could change its posture, tighten its muscles and brace itself against the wind, its structural mass could literally be cut in half."

Guy Nordenson, Ove Arup and Partners

Summary. *This case study is an investigation of self-generating forms in nature based on pneumatic structures and their use in architectural theory. It focuses on the concept of self organization as a defining principle in nature and in particular, on the mathematical, geometrical and physical properties of bubble clusters and shows examples from nature, biology and engineering. Part of the research resulted in a series of digital models and renderings of different bubble clusters and there polyhedral configuration. Advanced structural design methods are already using systems based upon self-generated models rooted in biological and genetic forms. Engineers are able to input a series of variables into a computer program which in turn, derive a structure using a genetic algorithm resulting in the most efficient use of materials, etc. Numerous examples of such procedures already exist in nature today, in particular, biology. Blueprints for these forms are stored in the genetic code of the DNA[1] of all life forms. Until recent advances in computer technology, the ability to put such genetic algorithms to use has not been possible*

[1] The DNA (deoxyribonucleic acid) is the carrier of our hereditary characteristics and that it is based on two strands twisted about one another forming a double helix. The strands consist of alternating carbohydrate and phosphate molecules. On each carbohydrate sits one of the four nitrogenous molecules **A**denine, **C**ytosine, **G**uanine and **T**hymine. A DNA strand can thus be compared with a long sentence (sequence) of code words, where each word consists of three letters that can be combined in many different ways, e.g. CAG, ACT. Each code word can be read by components inside the cell and translated into one of the twenty amino acids that build proteins. The three-dimensional structure, and hence the function, of the proteins is determined by the order in which the different amino acids are linked together according to the genetic code. (www.nobel.se/chemistry).

E. Oñate and B. Kröplin (eds.), Textile Composites and Inflatable Structures, 285–303.

Key words: Form-optimization, lightweight membrane-construction, radiolarian, genetic algorithm polyhedra

1 Introduction

The study of form-optimizing processes in biological structures has a long history starting with Frei Otto, Werner Nachtigall and followed by many researchers [2][3][4][10]. These researchers have outlined in a number of forms the mathematical relationships that control the overall geometry of polyhedral in biological structures [12]. The research centers on an investigation how optimizing processes in biological structures are possible starting points to generate optimized architectural forms and structures. For this particular study the bubble cluster based on the pneus was selected. The pneu is a system of construction comprising of a non-rigid envelope having a certain tensile strength, and an internal filling, which is in most cases pressurized. This system of construction can be translated into the architectural world in the form of pneumatic structures. This structural system, which can be found in many lightweight structures today, is based on the principals of those pneumatic structures found in nature.

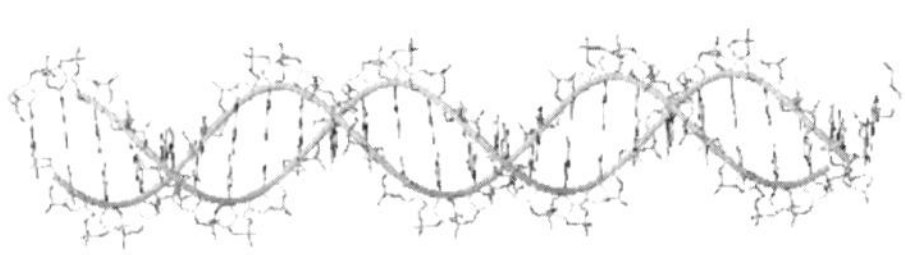

Fig. 1. DNA double helix

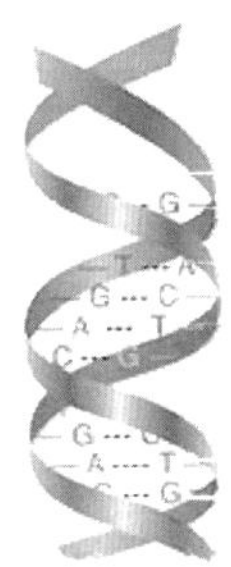

Fig. 2. DNA, the genetic code

2 Pneumatic Structures in Nature

One example of a pneumatic structure in nature is the soap bubble. In soap bubbles, growth is achieved through a system of division and inflation. This increased internal pressure encased in a reinforced membrane subject to tensile stress causes the bubble to grow in a process known as isomorphism or self-generation.

Free-floating bubbles collect and form dense clusters known as foam. If three bubbles are placed on a glass surface and a fourth is added, the fourth bubble will relocate to the top of the three bubbles to form the simplest three-dimensional cluster consisting of four bubbles. If further bubbles are added they will automatically form a foam structure. If the bubbles are of equal size the liquid edges of the foam are straight and of equal length and the angles of incidence at the nodal points are equal. The total structure forms a net of equal mesh size called the "basic net" (Fig. 13).

Fig. 3. Rendered computer model: Tetrahedron bubble structure

Fig. 4. Rendered computer model: Octahedron bubble structure

Fig. 5. Cube bubble structure

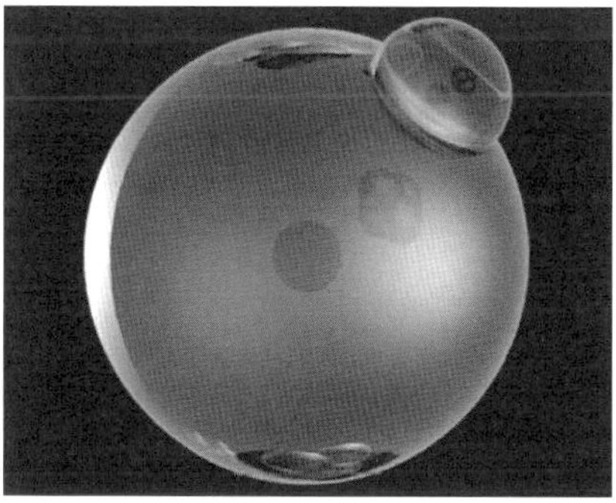

Fig. 6.

Fig. 7.

3 2-D Bubble Clusters

Net structures are formed through the solidification of a 2-D bubble cluster. Bubble clusters occur when bubbles are freely dispersed within a cell without touching each other. In the next phase, the bubbles are introduced to each other through points of contact and form patterns by agglomeration (Figs. 5–13). These patterns are based on geometric forms such as cubes, tetrahedrons and octahedron (Figs. 3–5). As solidification takes place, the membrane of the bubble dries out and the fiber net hardens (Fig. 12). The bubble membrane then dissipates and the net structure is left (Fig. 13).

Fig. 8.

Fig. 9.

Fig. 10.

Fig. 11.

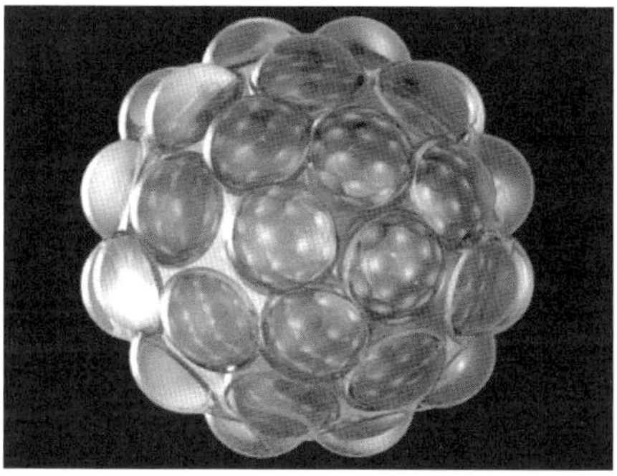

Fig. 12.

Fig. 13.

4 Mathematics/Geometry

Closest packing

One of the fundamental geometric principles that drives the repetitive, self-generating forms in nature is the notion of *closest packing* [1] of spheres. It is this, which defines the curvature of an insect's compound eye or creates the formwork to mold a radiolarian's skeletal structure.

As spheres are packed closely together, certain laws of physics cause geometric shapes to occur, such as hexagons. These polygons create repetitive surfaces among and around the spheres. In some cases, these surfaces find themselves useful for a number of functions, such as in an insect's eye. In other instances, these surfaces interlock together to create volumes, polygons create polyhedrons. These volumes may be used to serve a purpose.

Often times, it is not the surface or volume that is put to use in these systems. Quite likely, it is the edges along which these spheres meet that are of use to the organism. Once again, the radiolarian and diatom gather silicate deposits along the edges where the spheres meet around the outer surface. It is along these edges that a skeletal system is formed.

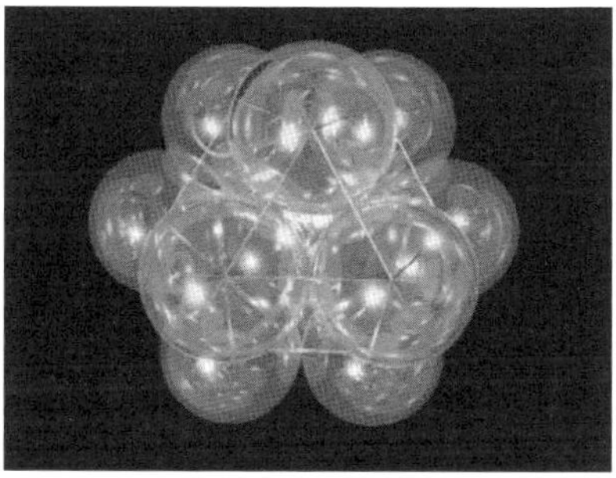

Fig. 14. Closest packing of spheres

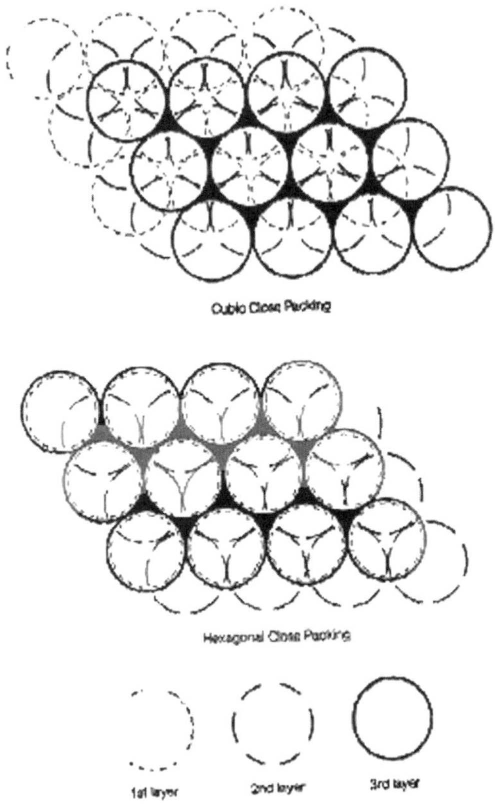

Fig. 15. Closest packing diagram

Configuring and integrating form systems

One of the greatest advantages to geometric systems based on the closest packing model is the great variety of configurations from which to choose. Repetitive, self-generating form can be derived in the shape of hexagons, pentagons, and even triangles (Figs. 3–5). These can be arranged independently or between various types [1].

After a particular form is created, it, too, can be arranged with other similar forms to create even more shapes or, in terms of architecture, spaces. Some examples of this can be seen here (Fig. 16). Another important quality of these systems is the ability to obtain similar forms with varying degrees of complexity in terms of number of members, scale, etc. As you can see, very comparable forms can be achieved in different ways. The structural complexity of a geodesic dome is probably too complicated for that of a radiolarian's skeletal system. Yet, these two structures share obvious formal qualities with one another. At the same time, the mathematic and geometric basis from which both are derived are practically identical.

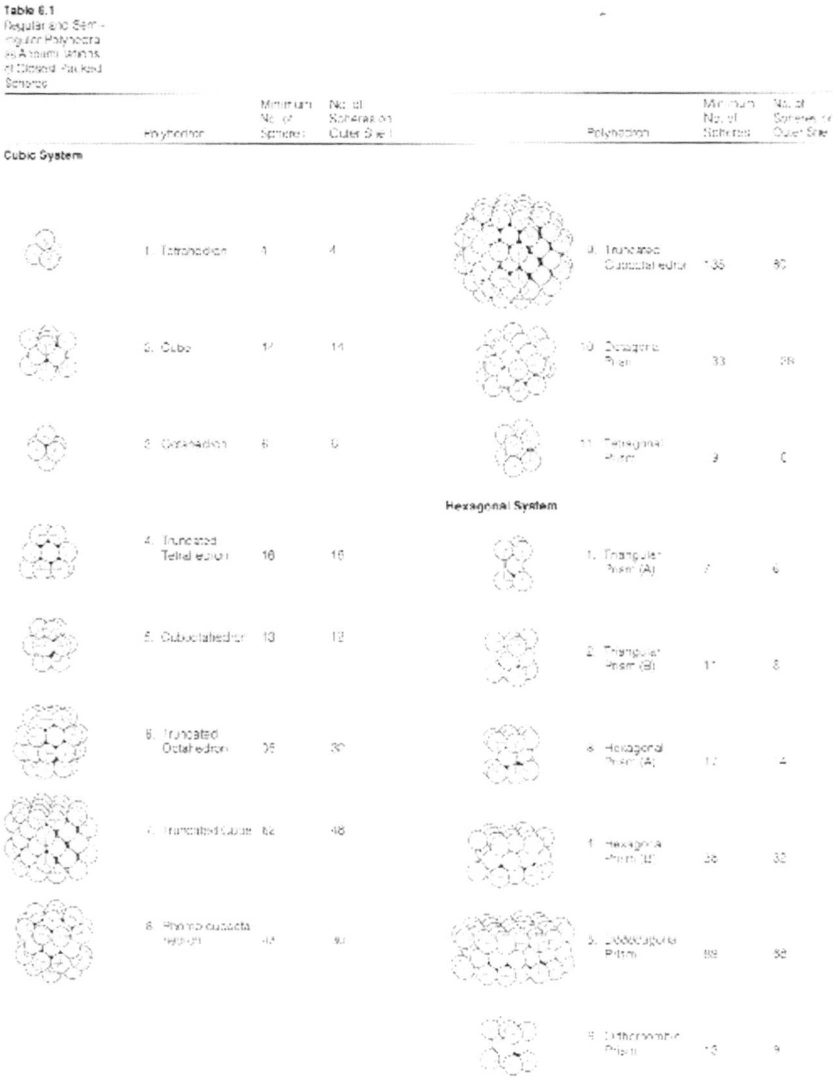

Fig. 16. Closest packing chart, Spheres as Morphological Units [4 Pearce pp57]

5 Structural Optimization in Engineering

Genetic algorithms

In engineering fields, accomplishing an objective with a minimum of effort, either in terms of material, time or other expense, is a basic activity (Figs. 18, 19 and 20). For this reason it is easy to understand the interest designers have in different optimization techniques. Mathematical, as well as, model based tools have traditionally been

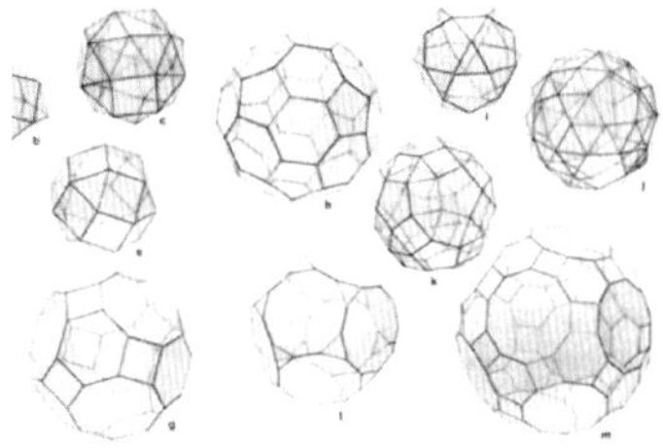

Fig. 17. Various geometric configurations

Here are a few varieties of geometric configurations that achieve similar overall results, spheres, while blending various geometric shapes together in a number of ways.

employed for such optimization. In recent times, mathematical methods executed on computers have become predominant. Unfortunately, computer derived solutions often obscure the range of possible solutions from the designer by only exhibiting a final, 'best' solution. Naturally, optimization methods can only respond to the objective parameters which are coded into the problem, and as a result, non-coded parameters, such as aesthetics, or context are left out of the optimization process, and ultimately left out of the final design solution.

Structural optimization in engineering takes natural constructions as an example. Similar to nature itself, computer-generated genetic algorithms[2] can be calculated using stated goals to achieve global optimization - the search strategy is, like in nature, goal-oriented. An evolutionary algorithm maintains a population of structures (usually randomly generated initially), that evolves according to rules of selection, recombination, mutation and survival, referred to as genetic operators. A shared 'environment' determines the fitness or performance of each individual in the population. The fittest individuals are more likely to be selected for reproduction (retention or duplication), while recombination and mutation modifies those individuals, yielding potentially superior ones. Using algorithms, mechanical selection, mutation and recombination improves generationally with a fixed parameter size and quality.

[2] A genetic algorithm generates each individual from some encoded form known as a 'chromosome' and it is these which are combined or mutated to breed new individuals. The basis for the optimization is a vast array of possible solutions (population), where every solution (individual) is defined through a particular parameter (chromosome). The individuals within a generation are in competition with one another (selection), in other words, the value (fitness) of the individual is what allows the survival of the parameter (gene) until the next generation. The results of this computer-supported process are automatically generated and optimized. Evolutionary computation is useful for optimization when other techniques such as gradient descent or direct, analytical discovery are not possible. Combinatory and real-valued function optimization in which the optimization surface or fitness landscape is 'rugged', possessing many locally optimal solutions, are well suited for evolutionary algorithms.

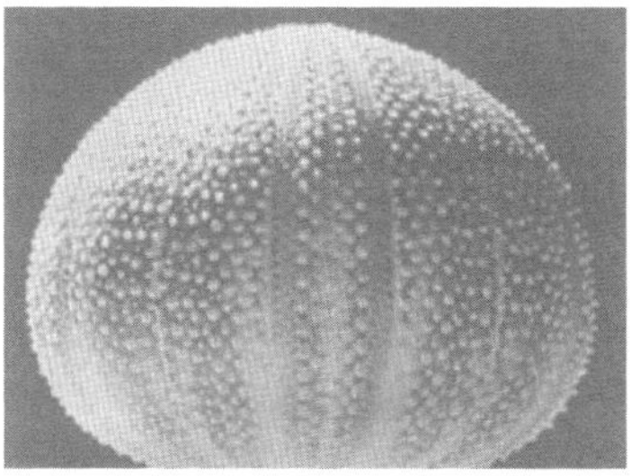

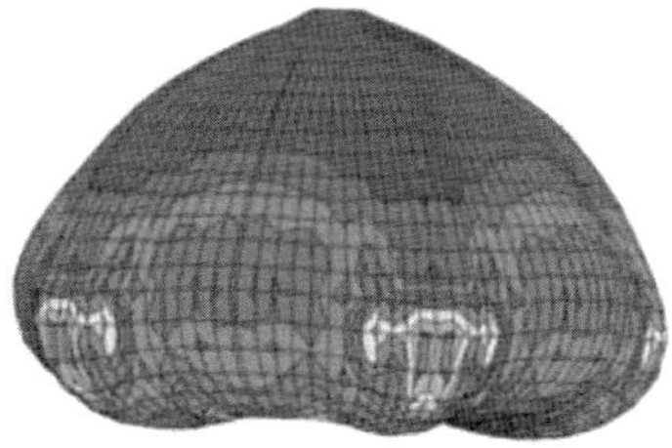

Fig. 18. Structure optimization in the shell structure of a sea urchin.

Fig. 19. Finite element analysis[3] of sea urchin shell, color coded stress analysis [15 *Process und Form*, K. Teichmann].

Computer-compressed evolution

Design space and finite elements

Computer-compressed evolution follows the same construction principle that nature employs to promote for example the shell growth of a sea urchin (Figs. 18/19) or the silica structure of radiolarian (Figs. 23/25). Building material can be removed wherever there are no stresses, but additional material must be used where the stresses are greater. This is the simple principle that evolution has used for millions of years to produces weight optimized "components". Using computer programs based on computer-generated genetic algorithms like the SKO method[4], scientists are now able to simulate this evolution and compress it into a short time span [9].

In order to simulate lightweight engineering strategy according to nature's guidelines, scientists using the SKO method must first define a virtual design space, which represents the outermost parameters of the component being developed. To subdivide this design space into many small individual parts, the finite elements, a grid is applied. If now a virtually external load applied, the computer calculates the resulting force exerted on every one of the finite elements. The FE model shows exactly where there is no load stress on a component and in turn shows where it is possible to make savings with regard to the materials used. On the other hand, for areas that bear heavy stress the simulation program indicates the need to reinforce the construction material. Like nature the computer let repeat this "finite element cycle" several times. As a result, they can refine a component repeatedly until the optimal form –one that evenly distributes the stresses within a component– is found.

[3] The finite-element-method is a procedure used to solve structural-mechanical calculations with precedence given to the three-dimensionality of the system. As a result, the construction is broken into discreet elements - Finite Elements (FE – such as columns, beams, plates, shells, etc. characterized by the individual connections (discreet points) where they are combined with one another.

[4] The DaimlerChrysler Reaserch Center Ulm and Uni Karlsruhe, Prof. Claus Mattheck, in Germany developed the SKO method (Soft Kill Option). The method is based on the idea that is it only possible to achieve a combination of the lightweight and maximum strength in a design when the stresses are constant over the structure's entire surface area, ensuring that no area is under- or overstressed.

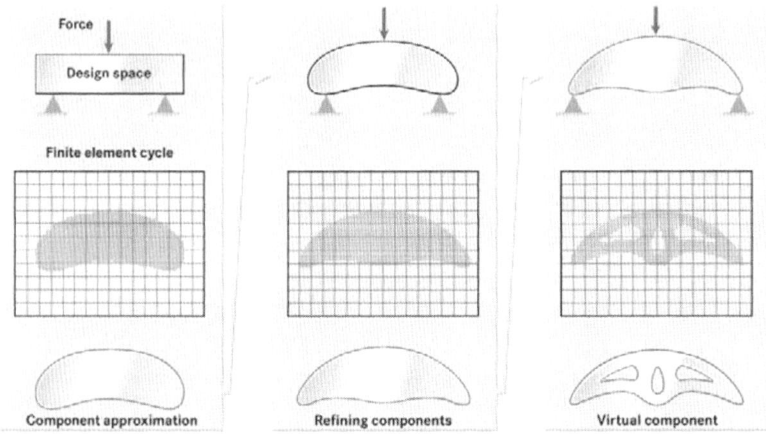

Fig. 20. SKO method (Soft Kill Option). [9 HIGHTECH REPORT 1/2003, pp60-63]

6 Biological Models

Radiolarians

A number of self-generated, biological models based on the bubble cluster theory exist. One of the best examples of this is the Radiolarian. Radiolarians are single-celled, marine organisms. These microscopic creatures extract silica from their environment to create a skeleton. Highly articulated geometric patterns define the usually spherically shaped structures. The resulting form resembles that of a dome.

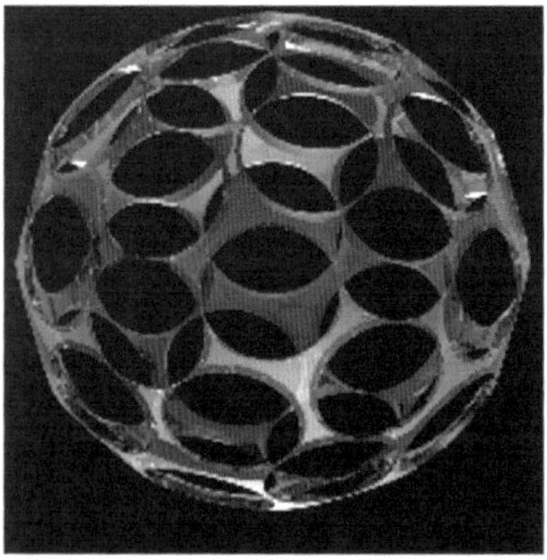

Fig. 21. Computer generated spherical cluster skeleton based on the bubble clusters theory (Figs. 6–13)

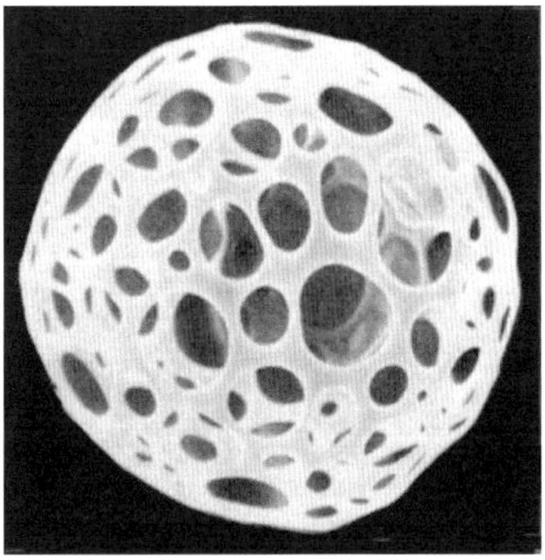

Fig. 22. Fossilized skeleton Radiolarian [6]

The process carried out to produce such as resulting structure is a relatively simple one. A great number of vesicles, tiny sacs of fluid, are created. These bubbles are essentially tiny versions of the larger soon-to-be cell (Figs. 6–13). They are the common modules for the cell. As the vesicles begin to pack closely together in a radiating pattern, the resulting form is a spherical mass, the cell. The unique geometry of the cell's surface, as produced by the closely packed bubbles, is the formwork for a skeleton [4].

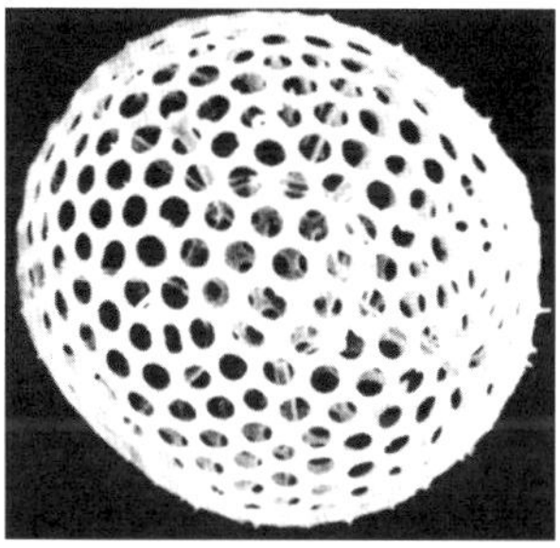

Fig. 23.

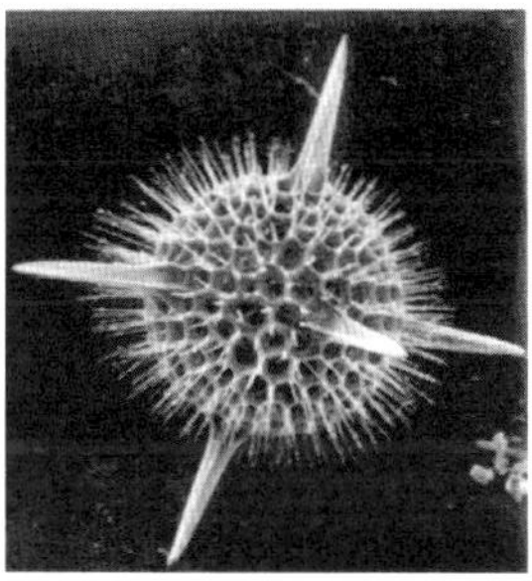

Fig. 24.

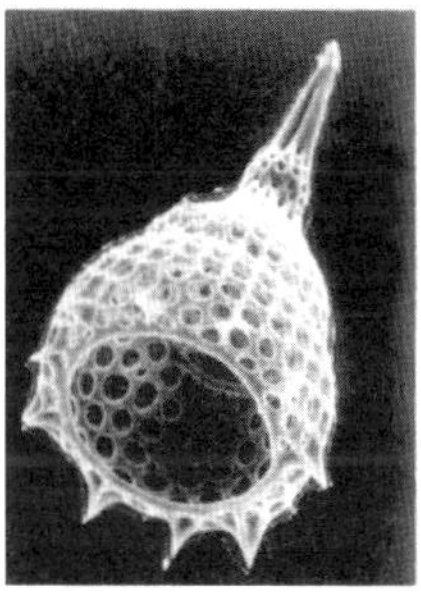

Fig. 25.

Radiolarian and fossilized skeleton [5][6]

Within the crevasses formed at the edges where the bubbles contact one another, silica particles settle and join together. The geometry created by the bubble connections allows the silica to form a series of interconnected members of similar size and shape. The result is a complex-looking skeletal structure made up of a great number of simple, relatively similar members.

To create such a seemingly complex system for so simple an organism, it is necessary to build upon highly efficient mathematic and geometric principles. These principles are embedded in the genetic code of the radiolarian, and similar life forms.

Diatoms

A number of diatoms share similar characteristics with their single-cell cousin, the radiolarian. Some of these include: radial form and a ridged skeletal structure. The skeletal structure of a diatom is not necessarily spherical, but it is radial. Tiny bubbles radiate in a closely packed form to create the cell shape. This bubble form is decrypted as a "foam-raft". Once this occurs, silicate deposits meet along the bubble edges to form the skeleton. This material forms what can be considered actually a "glass" skeleton.

Variations of diatoms are numerous. Not all diatoms are perfectly round in configuration. For instance, some are actually triangular in shape. These diatoms still adhere to the closest packing principles, just a different version than most. Other types of diatoms are oblong in shape. Instead of a repetition of similarly shaped spheres, or bubbles, these adaptations pack together larger, tubular forms to create their skeletons.

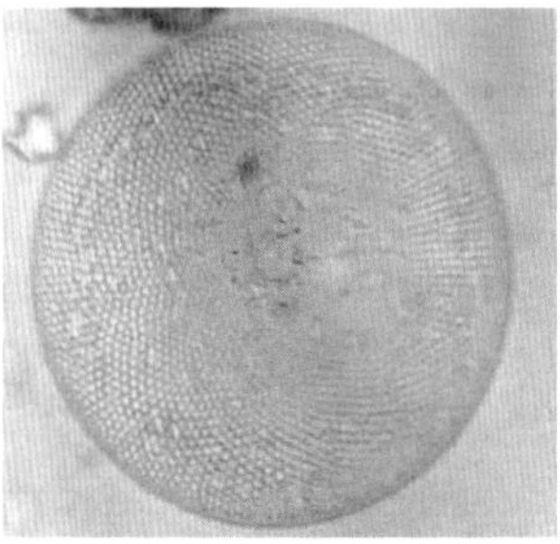

Fig. 26.

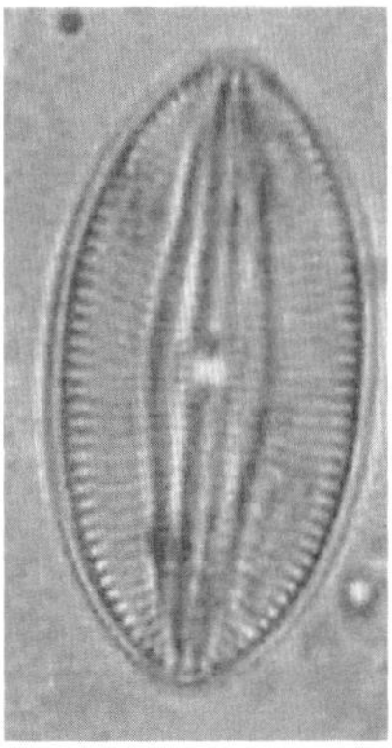

Fig. 27.

Diatom cells

Diatoms, and radiolarians alike, are chief examples of highly efficient structural systems with simple enough geometries for organisms with limited genetic information storage capabilities.

Cork cells/honeycomb

Cork cells and honeycomb structures are prime examples of minimalist, self-repeating structures. Both rely on simple members to shape themselves along with a certain rigidity and efficiency of space within its structure.

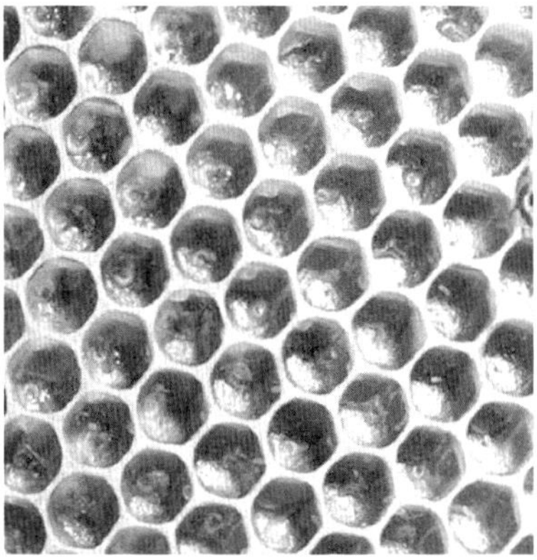

Fig. 28. Honeycomb

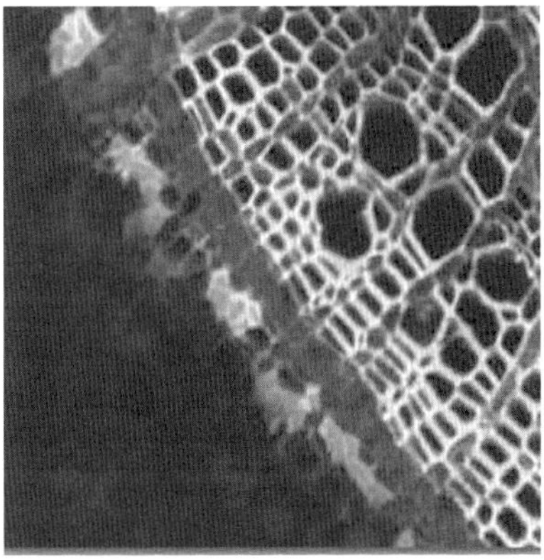

Fig. 29. Cork cells

The structure of cork cells is the purely biological representation of a honeycomb structure. The cells are roughly hexagonal and do adhere to the closest packing rule of geometry [4, Pearce pp16]. Cork cells, however, are not entirely uniform. Honeycomb, while not biological in nature, is created by biological creatures and is rooted in genetic information stored within these beings. Its purpose is quite functional, in that, it stores honey for food. The most efficient way to contain the honey in relatively small compartments is in a hexagonally shaped container. This simple shape is quite important for a number of reasons.

First of all, a great number of honeycomb cells are necessary to store the food. A hexagon is made up of six sides of identical length. This makes the construction process easier on the bees as only one dimension of material is necessary. Second, due to equal member size as well as a number of possible orientations, the hexagons fit together very easily. This is important because many bees work at the same time from different places and, therefore, need to be able to connect the structure with little effort.

Insect compound eye

One of the most recognizable examples of closely packed systems is that of the compound eye (Figs. 30-31) found on most insects.

Many characteristics of insects, most often the ability to fly, facilitate the need for a relatively large eye with maximum surface area and range of view. To fulfill these needs in such a small creature, a simple method of order is necessary to construct such a complex system. Naturally, a highly repetitive organization was used. A tightly packed network of smaller, hexagonally shaped eyes is wrapped around the curvature of the larger, whole eye, which bulges from the insect head.

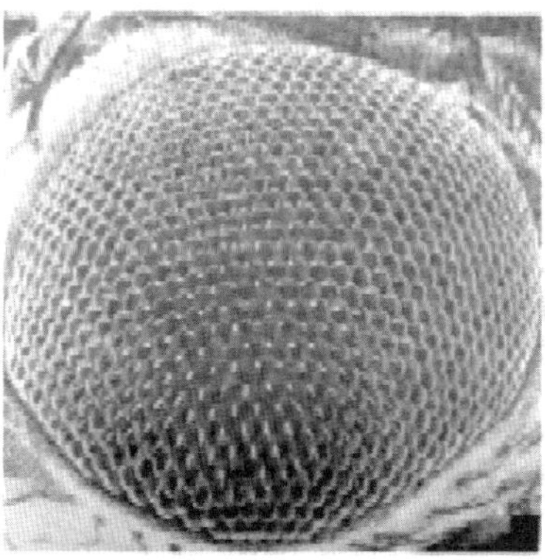

Fig. 30. Compound eye

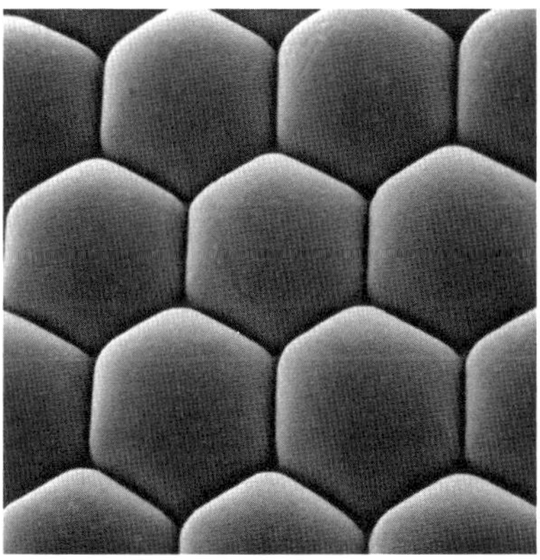

Fig. 31. Detail of insect compound eye

The flexibility of the hexagonally shaped eye units allows this to easily be achieved. Permitting the blanket of eyes to be wrapped in any direction on the insect's head gives it a view toward most any angle possible, which is necessary for such erratic flight patterns. Compound eye configurations can vary from species to species, but the hexagonal sheet configuration is the most common.

Fullerene[5]

The smallest known and just recently discovered structures based on the principles of self generation, closest packaging and polyhedral are the fullerene[6]. Fullerenes (Figs. 32–33) were first discovered in 1985 when the soccer ball shaped C60 (Buckminsterfullerene) was synthesized. The novel phase of carbon was named after the engineer and architect Richard Buckminster Fuller (1895-1985) as the molecules share the architecture of his geodesic domes. Fullerenes are ranging in size from 20 to over 500 carbon atoms.

The carbon atoms in C60 are arranged in a geometric shape consisting of 12 pentagons and 20 hexagons. Other spherical fullerenes (collectively known as buckyballs) were subsequently synthesized with a different number of hexagonal faces. The smallest possible fullerene is the dodecahedral C20, a shape consisting of 12 pentagonal faces and no hexagonal faces [7]. Larger, fullerenes have been found to exist in nature [8]. Nanotubes, nanohorns and buckybowls are other examples of fullerenes [14].

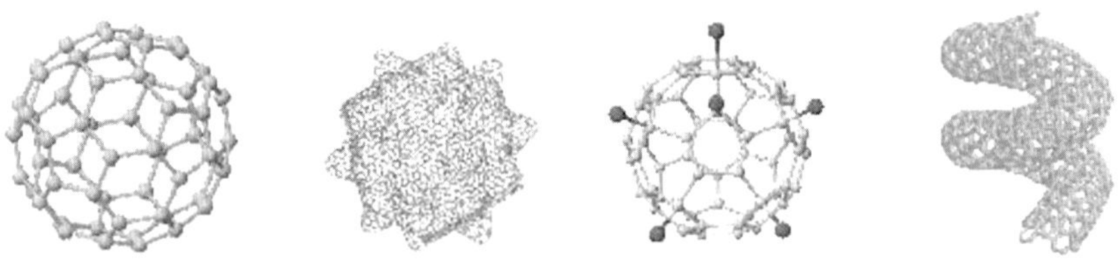

Figs. 32–33.

Different models of fullerenes. Fullerene Research Centre, [14 University of Sussex www.susx.ac.uk] Left to right: C60, C60-, Exohedral Fullerene Compounds C60, Cl6, Nanotubes

7 Architectural Applications

commonsense nature - producing maximum effect with minimum resources

Structural configurations

Just as volumetric and formal configurations are quite various, so, too, are the structural configurations of closest packed organizations. Structures with the qualities of domes are not the only forms that can be used and turned into a building. Variations in the manner in which the polygons go together can create long- spanning tubular structures and volumes. The blending of these various methods of employing the system can be seen here.

[5] Definition: A class of cage-like carbon compounds composed of fused, pentagonal and/or hexagonal sp^2 carbon rings.

[6] In 1996 Prof. Sir Harold Kroto was jointly awarded the Nobel Lauriate for chemistry for discovering the fullerenes. Fullerenes are large carbon-cage molecules. By far the most common one is C_{60} –also called a "buckyball", other relatively common ones are C_{70}, C_{76}, and C_{84}.

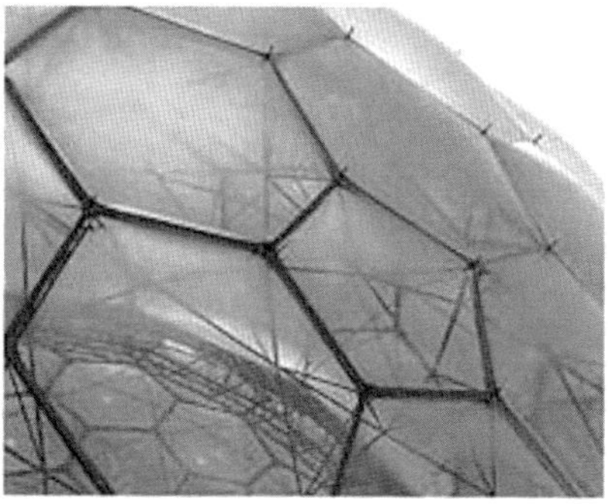

Fig. 34. Polyhedral structure Eden Project, Cornwall, GB, Nicholas Grimshaw Architects

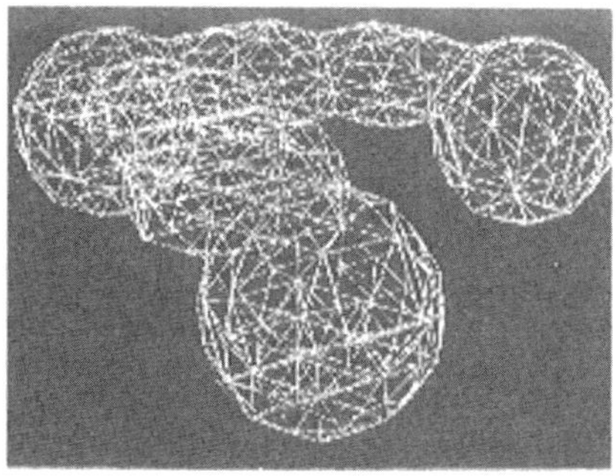

Fig. 35. Structural system by B. Fuller

The basic way to derive these structural schemes is to place relatively thin, load carrying members along the edges where the polygons meet. When configured correctly, these are the points at which forces naturally occur. Based upon the geometries of that form, it is likely that a highly efficient system is the result. One of the greatest benefits of this type of structural design is the relatively small number of member sizes. When multiple polygon geometries are involved, other members and connection types are necessary, but all can be accounted for in the initial design.

Fuller's geodesicpatent

Applying the science of self-generation to architecture has yet to be fully realized. Attributes of biological examples, like the radiolarian, do currently exist in structures like Buckminster Fuller's geodesic dome. Fuller recognized the closest packing of spheres phenomena in nature and used it as a model for his dome based on those mathematical principles. One of the most revolutionary innovations Fuller developed with his geodesic system was the repeated use of highly efficient, similar members. This allowed for the self- generating/repeating method of construction already in use by nature.

Fig. 36. Geodesic sphere

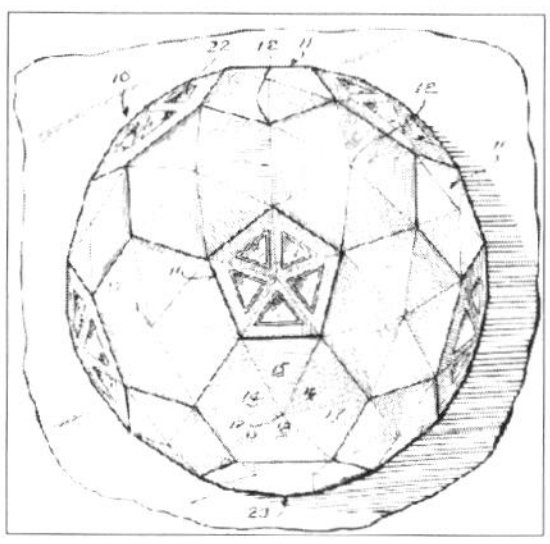

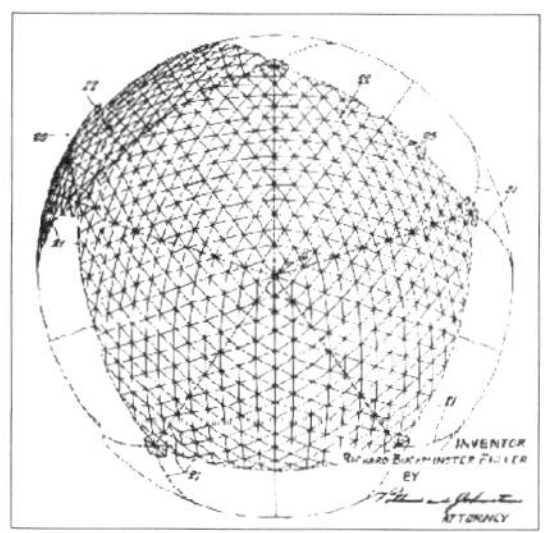

Figs. 37-38. Fuller's geodesic patent drawing

To take Fuller's discoveries one-step further using today's technology, it is possible to generate structural optimized structures in a computer by using for example the SKO method to create not just a representation of a biological form, but tu reproduce the evolutionary steps taken to make a structural system most efficient.

Eden Project, Cornwall, GB, Nicholas Grimshaw Architects

A recent example for polyhedral structures in architecture is the Eden project completed in 2001 by Nicholas Grimshaw. The Eden project is a botanical garden and education centre within a former china clay quarry. The construction of enormous greenhouses (biomes) created a sheltered micro-climate and enable large numbers of the world's tropical and Mediterranean species to be represented within the plantings. The building foundations follow the complex contours of the pit and support the lightweight tubular steel geodesic domes which are interlinked with arches. The largest of these domes is 100m across and 45m high internally.

Figs. 39-40. *Eden Project, Cornwall, GB*, Nicholas Grimshaw Architects, Anthony Hunt Associates engineers, structure: MERO Membrane ETFE-Pillows diameter 9 m, Dimensions: 15 590m^2, 100m × 220 m height 55 [www.anthonyhuntassociates.co.uk]

The Lightweight ETFE foil pillows form the cladding system between the dome members with panels up to 11m diameter providing maximum light and UV transmission. These hexagon shaped bubbles were used as they can settle perfectly on to any shaped surface.

The Biomes' steelwork is extremely light and is anchored into the foundations with 12-metre long steel ground anchors. The design comprised a two-layer steel curved space frame, the hex-tri-hex, with an outer layer of hexagons (the largest 11m across), plus the occasional pentagon, and an inner layer of hexagons and triangles (resembling huge stars) all bolted together like a giant Meccano[7] kit. Each component was individually numbered, fitting into its own spot in the structure and nowhere else.

[7] Meccano is a metal construction set consisting of nuts, bolts, strips, girders, brackets, wheels, axles, motors, gears and pulleys, Patented in 1901 by Frank Hornby of Liverpool, England.

Figs. 41-43. The transparent foil 'windows', made of 3 layers of ETFE (ethylenetetrafluoroethylene-copolymer), form inflated 2- metre-deep pillows. ETFE has a lifespan of over 25 years, transmits UV light, is non-stick, self-cleaning and weighs less than 1the equivalent area of glass and has a great stress redundancy. [www.anthonyhuntassociates.co.uk]

National Swimming Centre Beijing Olympics 2008, Peking(CN) PTW, Sydney & China State Construction Engineering Corporation, Peking & Ove Arup Pty Ltd., Beijing, London

The winning project for the international design competition' National Swimming Centre Beijing Olympics 2008' by the Australian architectural firm of PTW is an other example for the efficient combination of polyhedral structures with ETFE cushions. The design, called 'Watercube' is a simple and concise square form that ultimately uses the water bubble theory, the natural formation of soap bubbles, to create the structure and building cladding. The structure system, a space frame, is based on polyhedral cells in different dimensions, the most effective sub-division of three dimensional structures. It is also based on the way that structure in nature tiles spaces. The building's skin, made from ETFE, has been designed to react specifically to lighting and projection.

Figs. 44-46. The Rendering, model and structure model shows the three dimensional arrangement of the polyhedral structure, a common natural pattern seen in water bobbles and organic cells. The building envelope acts as an very efficient green house where90area. [Ove Arup Pty Ltd, www.ptw.com.au]

Construction and Internal structure

The internal structure within the depth of the roof and walls is highly repetitive. There are three different nodes and four different members. These elements can be cast, rolled or fabricated in different ways [13, Ove Arup Pty Ltd].

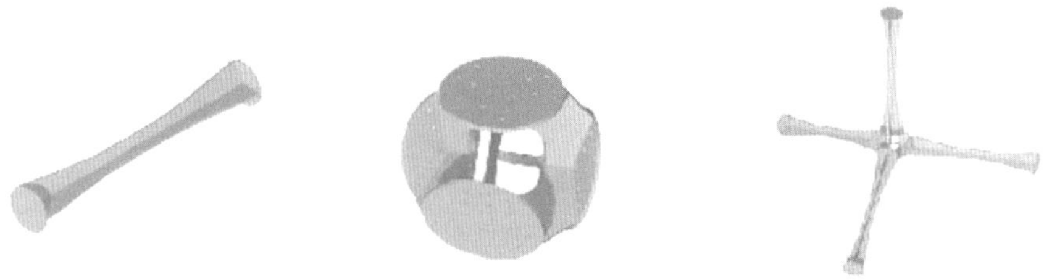

Figs. 47-49. The members are fabricated from three plates with circular end plates. The nodes are simple steel fabrications with circular plates for the end plates on the members and corresponding holes to receive the bolts. These two elements are simply bolted together to form an assembly. The assemblies are bolted together to form the space frame. [Ove Arup Pty Ltd]

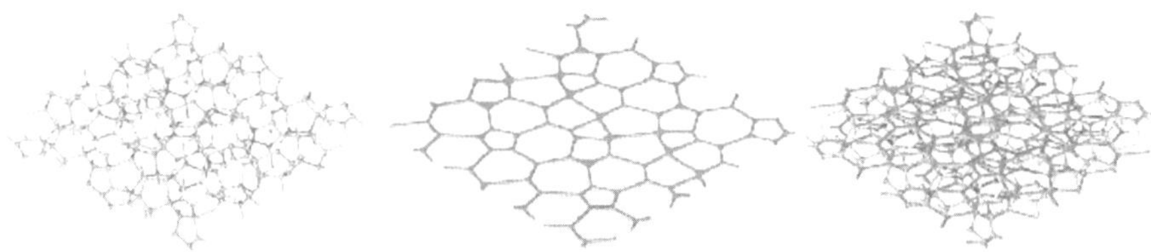

Figs. 50-52. From left to right: polyhedral space frame, flat face structure, combination of both. The face structure comprises a flat web of rectangular box sections either welded or bolted together on site. The face structure is added to the top and bottom of the space frame to compete the structure. [Ove Arup Plt ltd]

8 Intelligent Structures and Materials

Self-organization as the defining principle of nature

Polyhedral structures based on the bubble principal are perfect study models for self-generating structures in nature because of there relatively simple physical and morphological principles and geometries. Self- organization it the defining principle of nature. It defines things as simple as a raindrop or as complex as living cell - simply a result of physical laws or directives that are implicit in the material itself. It is a process by which atoms, molecules, molecular structures and constructive elements create ordered and functional entities.

Engineers are using this concept already successful for optimization processes in a white range of applications starting in mechanical-, medicine-, air and space engineering. Architects are only one step away adopting the same technique for designing in a macro scale buildings and structures. Material scientists[8] are already designing and producing new materials or smart materials in a Micro scale using the self organizing principles. In the future, the material engineers will develop constructions out of self-structuring materials that consciously use the principles of

[8] Saarbrückener Institut für Neuer Materialien [INM], Frauenhofer Institute für Fertigungsmechanik und Angewandte Forschung, Bremen

self-organization, creating not only materials with brand new properties but also inspiring architects to define their constructions in a more intelligent way.

At its best, intelligent structures and materials will influence the entire philosophy of construction. Engineers will no longer ensure safety through quantity of material and cost. Simple structural analysis will no longer suffice; instead, self-organizing structures will define the new construction principles.

Acknowledgements

Research Assistant and Renderings: Christian Nobel, University of Tennessee, College of Architecture and Design John Falko, University of Tennessee, College of Architecture and Design

University of Tennessee, College of Architecture and Design Research Grand.

References

1. Makishima Shoji (2001) Pattern Dynamics: A Theory of Self-Organization. Kodansha Scientific, Ltd., Japan.
2. Otto Frei (1984) Diatoms I: Shells in Nature and Technics. Institut fur Leichte Flachentragwerke, West Germany.
3. Otto Frei (1990) Radiolaria: Shells in Nature and Technics II. Institut fur Leichte Flachentragwerke, Germany.
4. Pearce Peter (1978) Structure in Nature Is a Strategy for Design. MIT Press, Massachusetts.
5. The Rad Page (2001) http://radpage.univ-lyon1.fr/rad_en.html#Introduction, June [September 2001].
6. Radiolarians (2001) http://oceanlink.island.net/oinfo/radiolarians/ /radiolarian.html, [September 2001].
7. Prinzbach H, Weiler A, Landenberger P, Wahl F, Wšrth J, Scott LT , Gelmont M, Olevano D and Issendorff BV (2000) Gas-phase production and photoelectron spectroscopy of the smallest fullerene, C20. Nature 407:60–63.
8. Becker L, Poreda RJ and Bunch TE (2000) Fullerenes: An extraterrestrial carbon carrier phase for noble gases. Proc. Natl. Acad. Sci. USA, 97(7):2979–2983.
9. HIGHTECH REPORT 1/2003, 60–63.
10. D'Arcy Wentworth Thompson (1992) On Growth and Form. Dover Publications, Inc.
11. Tomaso Aste and Denis Weaire (2000) The pursuit of perfect packing. Bristol, PA: Institute of Physics Pub.
12. François Gabriel J (1997) Beyond the cube: the architecture of space, frames and polyhedra. Willey & Sons, Inc.
13. http://www.arup.com.au/beijing.php
14. University of Sussex, www.susx.ac.uk
15. Teichmann K and Wilke J (1996) Prozess und Form. Ernst und Sohn.

Making Blobs with a Textile Mould

Arno C.D. Pronk[1] and Rogier Houtman[2]

[1] Department of Architecture, Building and Planning
Technical University of Eindhoven
P.O. box 513, NL-5600 MB Eindhoven, NL
a.d.c.pronk@bwk.tue.nl
http://www.blob.tue.nl

[2] Department of Civil Engineering
Laboratory of Building Engineering
Delft University of Technology
P.O. box 5048, NL-2600 GA Delft NL
Tentech Design & Engineering
P.O. box 619, NL-2600 AP Delft NL
rogier@tentech.nl
http:/www.tentech.nl

Summary. *In the last decade complex buildings i.e. with unregular curved surfaces have been designed. . The subject of this paper is the construction of those complex buildings. One of the main characteristics of a membrane structure is its geometrical complexity, which can be seen in multiple curved surfaces and complicated connection elements. Modern sophisticated computer technologies can be used to produce easily these complex three-dimensional shapes out of flat strips of fabric. Due to a lack of suitable production methods the expression of the natural stress flow in supporting and connecting (rigid) structural elements is difficult. This paper assumes that it is possible to achieve the architectural desired free forms by manipulation of structural membranes. To prove that it is possible to achieve the architectural desired free forms different cases are described in which this technique is used. The first case describes the design of an indoor Ski run. The second and third case describes the building of a lightweight stage covering and an art pavilion. In all the three cases physical models have been used in the design phase. The structural design of the membrane mould has been engineered with the program easy. The rigidized structures have been analyzed using different FEM programs for each case. The transformation of a form-active structure (membrane) into a surface-active structure has been researched to make domes ore dome-like structures.*

Key words: Blobs, textile mould, free geometry architecture, tensile structures, pneumatic structures, formfinding, structural optimisation

E. Oñate and B. Kröplin (eds.), Textile Composites and Inflatable Structures, 305–322.

1 Blobs

In 1994 K. Michael Hays [10] writes that in reaction to fragmentation and contradiction there is a new movement in architecture, which propagates a combination not only of forms, but also between different media like film, video, computers, graphics mathematics and biology. He recognizes that architecture is influenced by the development of an increasing complexity of information and communication is changed into information and media. This has lead to a development that is being referred to as blob architecture (Figs. 1,3). The characteristics of blobs are: smoothness, irregularity and a double curved skin.

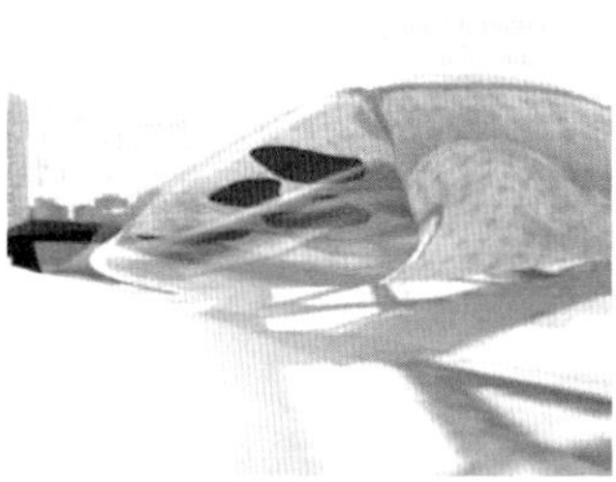

Fig. 1. by Michael Bittermann

Fig. 2.

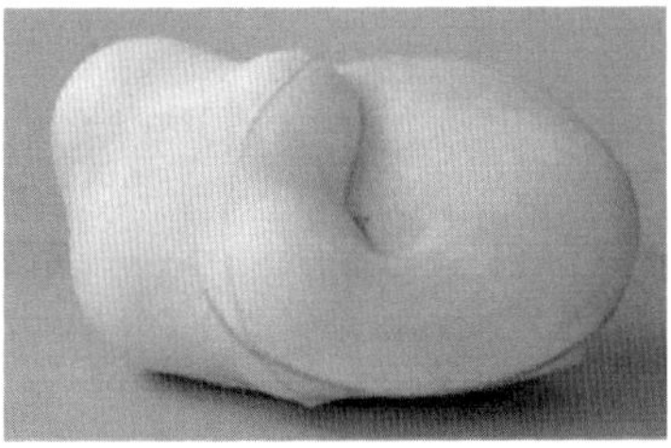

Fig. 3.

Modeling by means of nylon stockings and balloons

2 Blobs with a Textile Mould

The similarity between form active structures, like tent- and pneumatic structures on the one hand and blobs on the other hand is so striking that it is obvious to try to make blobs with techniques, that are being used for constructing tent- and pneumatic structures.

In the past numerous possibilities have been examined. Frei Otto for example has demonstrated the possibilities of influencing the form of pneumatic structures by stretching nets and cables over them. Another possibility of manipulating a tensile form is the combination of cloth and a pneumatic structure into a blob design. An example is the floating theatre at the Expo 1970 in Osaka designed by Yutaka Murata. One of the latest examples of transforming the shape of a pneumatic structure is the tensile structure of the Swiss pavilion (Figs. 4,5) at the Expo 2002. The edges of the structure are transformed by using bending stiff elements. The connection with nature is obvious if we realize that a human body can be seen as a membrane (the skin) stretched over bones (wire-frame) and muscles (pneumatic structure).

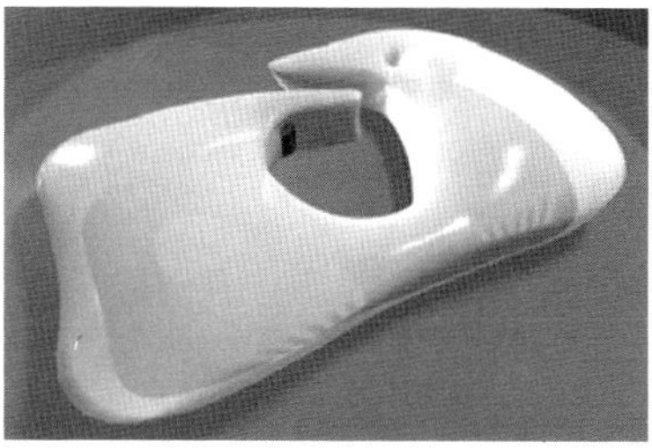

Fig. 4.

Fig. 5.

Nouvelle DestiNation Bundespavillon, Swiss Expo 02 (Eckert Eckert Architekten)

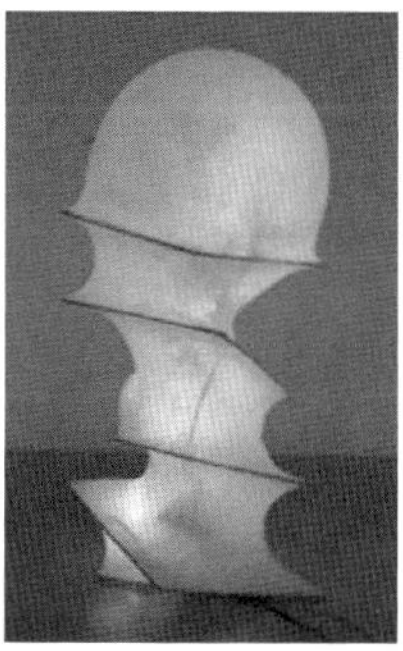

Fig. 6. Rigidized inflatable structure (A. Pronk)

3 Form-Active/Surface Active

In the open-air theatre in Soest a pneumatic structure was used as a mould. This mould was then rigidized, which resulted into a bent stiff beam that was combined with cloth. The result was a tensile structure. This technique was then studied. The purpose was to use this technique to realize complete buildings.

Heinz Isler has already demonstrated that it is possible to rigidize a pneumatic mould to construct buildings. The same principle is used in aerospace engineering for realizing antennas and space habitats. (In Soest the same principle is used to construct architectural shapes.) The surface of the building was not the result of the mechanics but the result of an architectural design process. At the Technical University of Delft and Eindhoven a group has been formed that has taken on the challenge of finding a way to realize blobs by means of transforming and rigidizing pneumatic structures. As a first study a model has been build that consists of balloons and a wire-frame that is placed in a nylon stocking (Fig. 2). It is possible to make many different forms with this technique. After modeling the shape a polymer resin is applied (Fig. 3). This physical model can be analyzed by means of a finite element computer program that looks at the active behavior of the surface of the structure. The input for the program is generated by a 3d scan (Fig. 17).

4 Stage Covering for an Open-Air Theatre

4.1 Introduction

This semi-permanent membrane structure covers the stage of the open-air theatre in Soest (the Netherlands) Fiber reinforced plastics are used for the production of a structural optimized and therefore lightweight and complex arch shaped structure. By using an inflatable mould the arch could be produced more economically (30% cost reduction). In the production the vacuum injection method is utilized for stiffening flexible fibers.

The owner of the Soest open-air theatre asked for a protection against bad weather for the stage. Therefore we suggested covering it with a lightweight membrane structure. A suspended membrane floats above the stage, so that visual relations with the natural environment are still preserved (Fig. 7). Outside the theatre-season the structure could partially be dismantled in this way the environment that is protected by national government is not visually disrupted. Two guyed columns are part of a dismantling system and could be used for hoisting the temporary membrane. The form of the spatial membrane is, beside the indirect support of the columns, the result of an arch. Because of this arch the protective area of the covering is increased and additional curvature in the membrane is improved (Fig. 8). In this way the membrane structure is a combination of two highpoint surfaces and an arched surface, the stage covering works like a tensegrity structure. The columns and arch transmit compressive loads. Both the Tensile loads and the stabilization of the whole structure are transmitted and organized by the prestressed membrane and cable structure.

Fig. 7. The stage covering for the Soest open-air theatre in the Netherlands (H. Werkman)

Fig. 8. An arched beam ensures an increase of the protective area and the curvature of the membrane structure

Due to its position in the audience's view and its proportions, the arch contributes in certain extent to the architecture of the structure. Therefore, special attention is given to the elaboration of the structural arch. The arch' dimensions exceed several times the thickness of the membrane and the cables. To avoid an abrupt change between the 'thick' arch and 'thin' membrane, a tapered arch section is desirable. The result is a conical arch. Because the mass of the arch would influence the membrane shape, a lightweight construction is necessary. This conical arch, which is characterized by geometrical complexity due to multiple curvature, and the necessity of a lightweight structure, asked for the use of an unconventional construction material and production technology.

4.2 Materialisation

Conventional construction materials like steel and aluminium and accompanying production technologies are not suitable for making lightweight multiple curved arches. The material properties and production methods of fibre reinforced plastics (FRP) matches the arch requirements. Some advantages of fibre reinforced plastics are: rigid and lightweight construction possibilities, fatigue resistance, chemical and corrosion resistance, freedom in design and form and the possibility to integrate parts. Important disadvantages are the cost prices of material, mould, production (labour) and engineering. In the case of complex shapes, for example a conical arch, approximately 50% of the production costs consist out of model costs. Therefore an effective way of cost reduction is to decrease the mould price.

4.3 Geometrical Complexity and Production Technology

Through the utilisation of a pneumatic mould the cost price of the arch is reduced with 30% (Fig. 9). In the production of the mould the same computer applications (EASY, FEM-based software) and production technologies are used as those used for the development of the membrane structure. After modelling and formfinding in EASY cutting patterns are generated and used for the production of the mould. The internal over-pressure ensures the rigidity of the inflatable mould. The general dimensions, like the distance between the supports, are controlled by an auxiliary structure (Fig. 10).

Fig. 9. Pneumatic mould supported by the auxiliary structure (Tentech/Buitink Zeilmakerij)

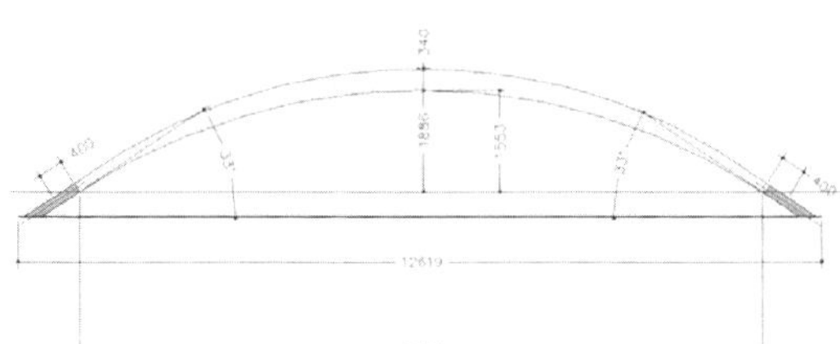

Fig. 10. General dimensions of the mould

4.4 Rigidizing Inflatable Structures

A rigidizable inflatable structure can be described as a structure that is flexible when inflated and becomes rigid after exposure to an external influence. Therefore it is not necessary o maintain the overpressure. There are several rigidizing techniques developed and more are still under development. They can be divided into three categories: thermosetting composite systems, thermoplastic composite systems and aluminum/polymer laminate system. Advanced rigidizing systems used for space applications are designed for specific structures and may be very expensive. In civil engineering, vacuum injection, which is a thermosetting composite system is a feasible way of rigidizing membranes.

4.5 Structural Optimisation

The structural engineering of the membrane structure is done with the use of the software package EASY, which is based upon the finite element method (FEM). To examine the shape of the structure and its reaction to external forces, the structure is first modelled with an arch set up as a spatial truss with a defined stiffness. Then this model is used for the structural analysis of the complete structure, consisting of a membrane and supporting cables and columns. Also the deformations due to extreme loading (wind and snow loads) are examined.

In order to be able to produce the synthetic arch the stiffness has to be determined. The pre-stress in the membrane and boundary cables causes an axial compression in the arch. Hence the curve of the arch will increase. The arch consists of synthetic fabrics rigidized by injecting resin into this fabric. By varying the use of material (e.g. thickness of the layers, layers of different materials) a range of E-moduli can be obtained. Also the moment of inertia is a variable. Therefore a variety in stiffness and bending resistance is possible. As said before, in order to find the desired shape the initial arch is designed having less curvature than ultimately was needed. - an E-module of 210 GPa (210.000 N/mm^2, comparable to steel) is used. The initial moment of inertia (Iy) was set at 855·10^4 mm^4, resulting in stiffness 1.8·10^{12} Nmm^2. First the deformations of the arch under pre-stress are calculated. The pre-stress in the membrane and the boundary cables cause the arch to deform and result in an increase of curvature.

EASY-BEAM is used to determine this initial curve. Then the stiffness is used to calculate the composition of the synthetic arch, (a specific E-module with needed I_y.)??

To be able to be more material-efficient a second step is taken. By adjusting the E-module from 210 GPa to 60 GPa a new stiffness is found ($EI_y = 5.14 \cdot 10^{12}$ Nmm^2). The initial curve of the arch is also adjusted to its new stiffness. Deformations of the curve under pre-stress are calculated, as are the deformations under extreme loading. These deformations turned out to be more than desired.

A third step had to be taken. The stiffness had to be increased considerably. This is obtained by a change in the moment of inertia (I_y). In the first step of the design the diameter of the arch was determined at 200 mm. By enlarging this diameter to 360 mm a factor 20 of increase in Iy is achieved (also a change in layer composition was introduced). Because of architectural consideration and in order to economize the use of material even more Iy is varied within the arch. This is translated in a tapered cross-section, with a decrease in diameter towards the ends of the arch.

Fig. 11. Bending for power, a pole vaulter using a beam's bending stiffness and deformation

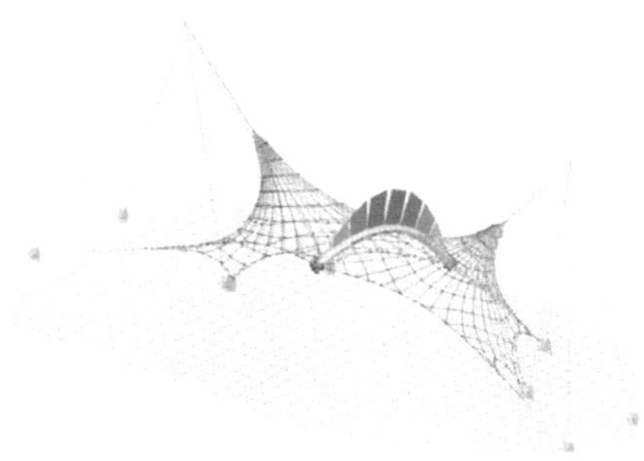

Fig. 12. Bending forces in beam-elements of conical arch, calculated in EASY-BEAM

In this third model Iy varies between $16170 \cdot 10^4$ mm^4 in the middle section to $8170 \cdot 10^4$ mm^4 at the ends. The deformations under pre-stress and extreme loading are checked and are within the design boundaries.

These insights resulted in a tapered glass- and carbon fiber beam, with its diameter varying between 150 to 360 mm.

4.6 Vacuum Injection

For the production of the arch the vacuum injection method is used to impregnate the resin in the woven fibres (Figs. 13–15). Around the pneumatic mould alternately layers of fibres and resin are placed. To create a closed system the whole package is wrapped with some airthight and protective foils. In the closed system a pressure differential is applied that impregnates the fabric with resin. The pressure differential of the technique is obtained by means of a vacuum. The injection has to take place within the cure time of a resin. The following formula (1) expresses the filling time (t_{fill}) as a function of the porosity (φ) and permeability (K) of the reinforcement, viscosity (ν) of the resin, flow distance (l), and applied pressure (ΔP).

$$t_{\text{fill}} = \frac{\varphi \cdot \nu \cdot l^2}{2 \cdot \kappa \cdot \Delta P}$$

The objective is to design a channel layout that ensures full wetting of the fabric at each location. Three distinctive injection strategies for a three-dimensional object can be followed, viz.. edge injection: downward, upward and sideways. Downward injection is sometimes disadvantageous because bubbles will be entrapped more easily and there is the increased risk of dry spots due to race tracking by the runner channels. The choice between the other two injection strategies depends on the geometrical shape of the product. Factors that are of influence are the number of inlet ports and the total injection time that, when minimal, are both at an optimum. In this case upward injection is used.

Fig. 13.

Fig. 14.

Fig. 15.

Production of the conical arch, by using the vacuum injection method. Both glass and carbon fibres are applied (photos: Rep-air Composites)

5 Indoor Ski Run

After the case of the open-air theatre students carried out several studies, Henno Hanselaar has carried out a very interesting one that shows the possibilities of the structures. He designed an indoor ski run with blob appearances and analyzed the mechanical behavior of this structure. The design has been made with the aid of the computer program Maya 4.0. This program is designed to make virtual animations, which are used for example in video games. It is also easy to design blobarchitecture and kinetic buildings. A three-dimensional site was drawn with the help of geodetic information from the local government.

Two lines were drawn on the ground of the slope that acted as the edges of the shell structure. Profile lines were drawn between the ends of these lines (they will function as rails) and on arbitrary distances between the ends of these lines. With the "Birail 3+"-function Maya generates a surface between the drawn profile lines. The surface can easily be transformed by changing the profile lines. The "Rebuild"-function generates an even smoother surface. When the final shape is obtained, the drawing can be exported as an Iges file type.

5.1 Surface-Active Analysis of the Mechanical Behavior

This file type can be imported in the computer program DIANA. It is the FEM package that is used to make a structural analysis of the rigidized shell. For pre and postprocessing DIANA makes use of the FEMGVX program. In the main menu of FEMGVX there are two options. The first is Femgen. This can be used for generating a 3d model and modifying the properties. The second option is Femview. With Femview the calculation results can be viewed. The building of the model has been done, as described above, in Maya.

5.2 Conclusions from the Structural Calculations

After all the results of the structural analysis have been processed the next phase was the evaluation and the possible material adjustments. If it appears that certain values are not satisfying a different solution has to be found Most striking is the

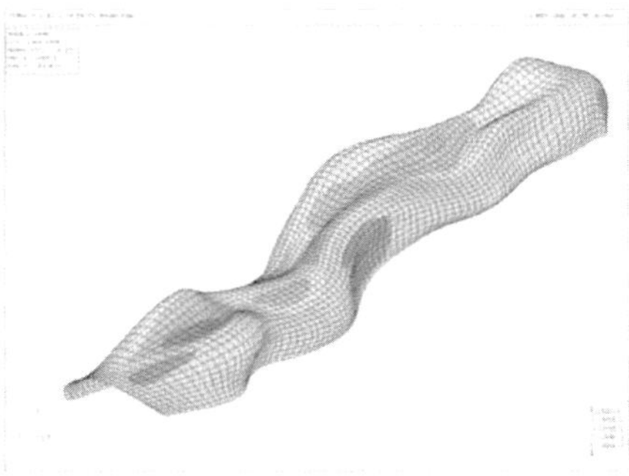

Fig. 16. Model in DIANA

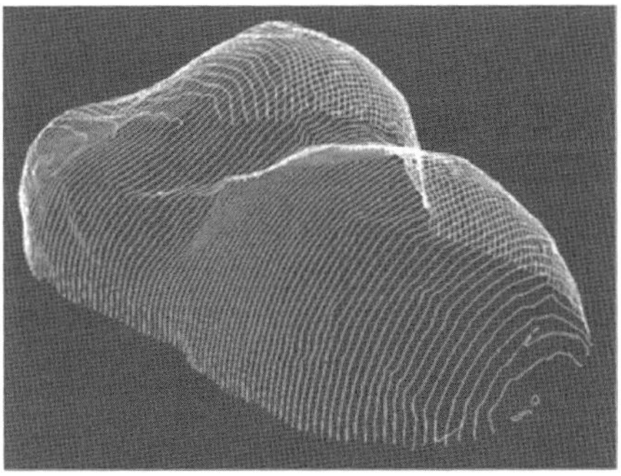

Fig. 17. 3D scan of Blob oject

large deflection of the system. But due to all the irregular bent surfaces there is no reference for the deflection. In this case a deflection of for example 400mm or 800mm cannot be seen. The structural system partly functions as a shell. At places where there is a transition from one curvature to another the outer forces are transferred by means of a bending moment. This is of course a bad situation for a thin walled structure. There are different solutions for this problem. From the solutions that were thought of the option of varying the wall thickness was chosen.

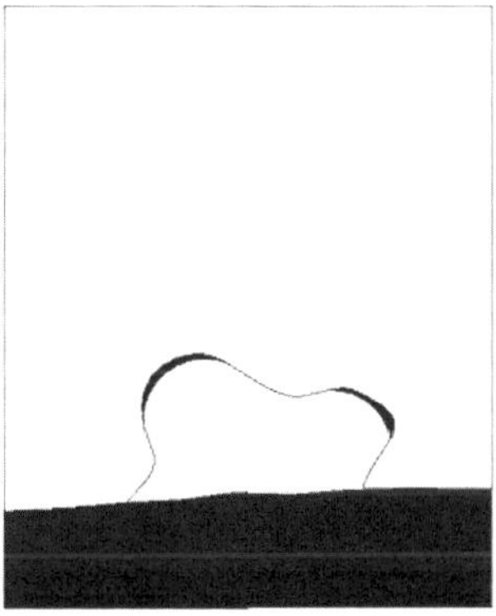

Fig. 18. Wall locally strengthened around problem area

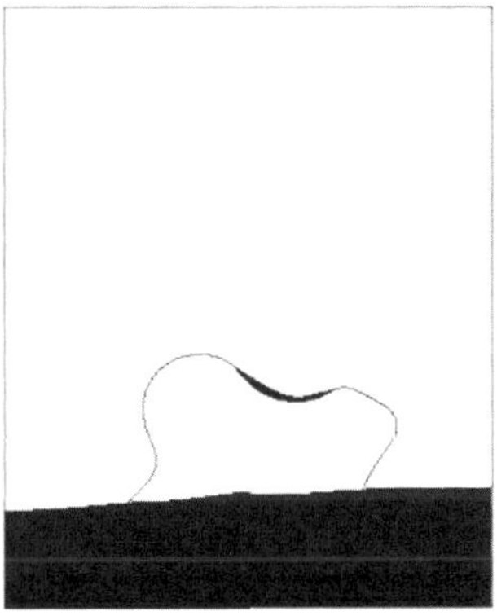

Fig. 19. Wall locally strengthened in problem area

5.3 The Form-Active Analysis of the Structure

The form of the indoor ski run was analyzed by means of describing the form by sections. To achieve the designed form there are a number of possibilities for the inflated structural elements that are put under the skin. At first the cross sections in width direction are shown. Next the different inflated structural elements are explained.

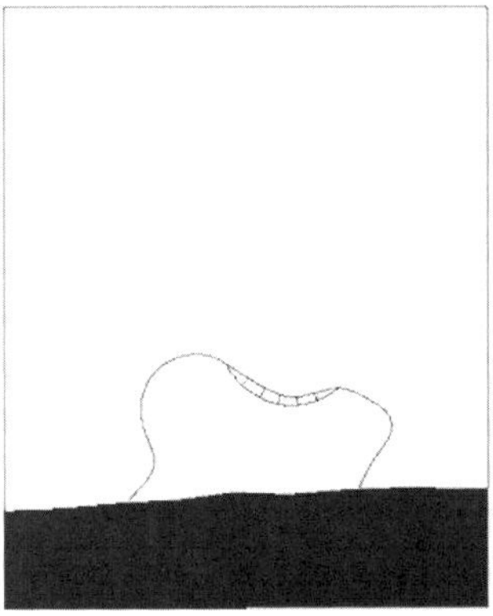

Fig. 20. Support construction placed under the roof

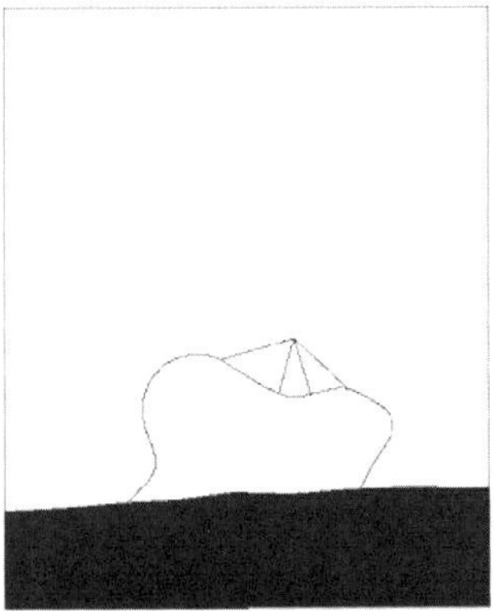

Fig. 21. Support construction placed on top of the roof

The width cross sections are more or less sinclastic. At several spots there is an anticlastic curvature. This indicates that there will be no structural inflated element underneath. The tension in the skin in longitudinal direction will have to apply the anticlastic curvature (Fig. 23). The longitudinal sections also show a global sinclastic shape and locally anticlastic curvature. This has to alternate with the width anticlastic curvature (Fig. 22).

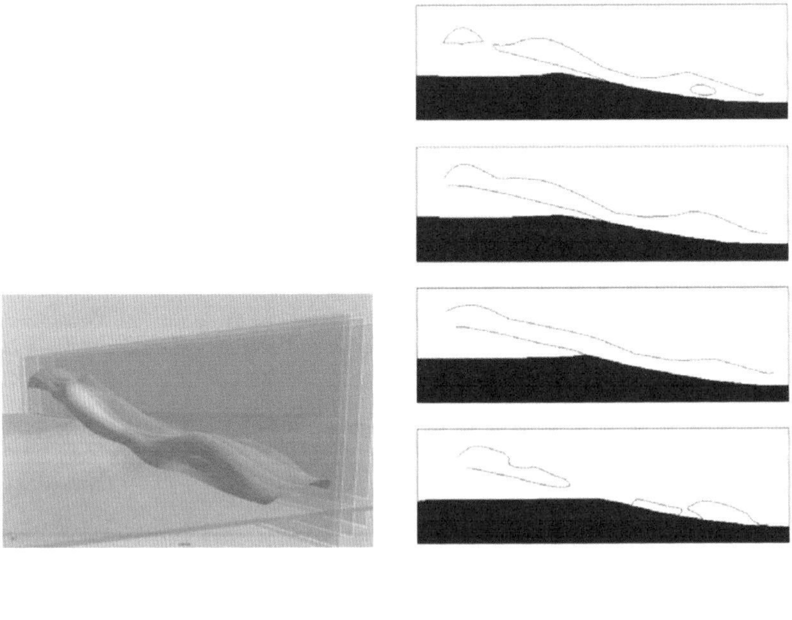

Fig. 22. Longitudinal cross sections

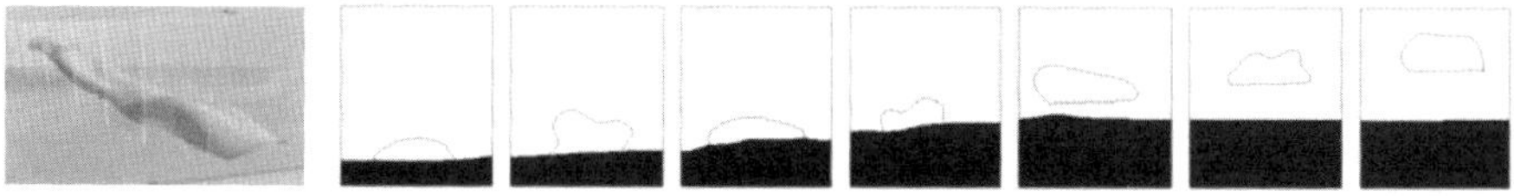

Fig. 23. Width cross sections

5.4 Manipulating Inflated Elements

The possibilities to fundamentally change the form of an inflated structure are limited. The form can merely be influenced. The curvatures are always existent and that is also the intention. There are simple examples from daily life where inflated structures are used and deformed. For example a car tire has a flat outer surface. In the rubber of the tire steel rings are embedded to prevent the tire from having a round cross-section (Fig. 24). An airbed is also an example where the form is influenced. There are partitions that hold the upper and lower side together when the airbed is inflated. These partitions make the cavities that exist in an airbed. In this way different sorts of shapes can be made with the aid of pneumatic structures. A number of possibilities to influence pneumatic forms will shortly be described.

Pressure Surfaces

By pressing two sides of a pneu with the aid of 2 so-called pressure surfaces, the pneu gets an elliptic shape. At the contact surface with the pressure surfaces, the pneu will follow the form of the pressure surfaces. The same effect appears when two pneus are pushed against each other. The contact surface is in this case flat (Fig. 25).

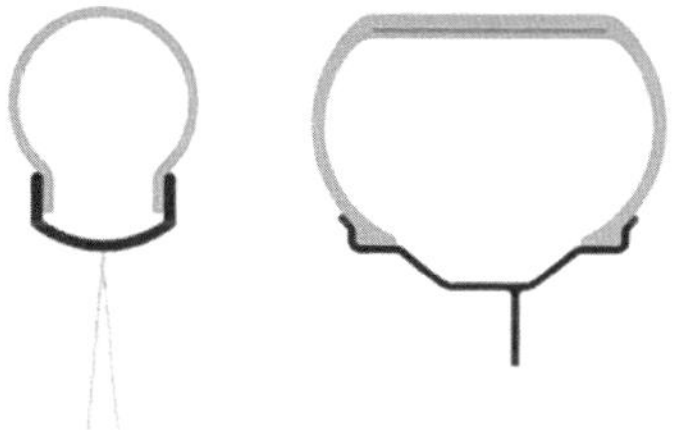

Fig. 24. Cross-sections of a bicycle tire and a car tire

Fig. 25. Pneu with pressure surfaces

Tension Cable

By tightening a tension cable around a pneu, the pneu can be laced up. At the position of the tension cable a sharp line appears (Fig. 26).

Partition

By applying a partition it is possible to get the deformation to follow a certain line. This is the line of the partition. For example in the air mattress the partitions are there to hold the upper and lower side of the air mattress together (Fig. 27).

Fig. 26. Pneu with pressure surfaces

Fig. 27. Airbed with partitions

Combination of Internal Pneus and Outer Skins

Fabric material is highly suitable to make negative curves. By combining an outer skin with internal pneus, positive as well as negative curves can be made. The following cross-sections can be made (Figs. 29–31).

Fig. 28. Combination of two pneus (torus)

Fig. 29. A small pneu is used as a distance holder

Combination of Two Pneus

In this case a pneu is wrapped around a torus (donut shape). The torus holds the pneu in its place during the inflation. In this way a reasonable flat form like M&M's can be achieved (Fig. 28).

5.5 Variety of Pneu Combinations

In Maya a number of combinations of different pneu forms were researched that are necessary to realize the form of the indoor ski run. Three alternatives have been made. In the alternative that was chosen two trunks can be recognized. During construction the free ends, at the top and at the bottom of the ski run, will be stiffened and thereby holding the tent cloth. It is possible that the form of the pneus deviates from the original design form. This can be seen in Maya where the dark gray areas are the deviating pneus (Fig. 33). The deviation is between a few centimeters and a decimeter. To construct cross-section 6 a number of pneus have to be placed on top of each other (Fig. 32).

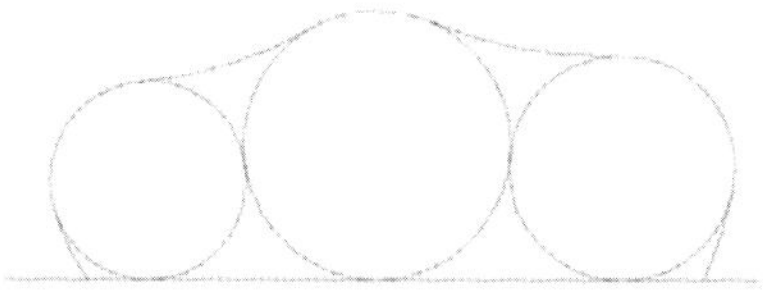

Fig. 30. To create a declining facade the tent cloth has to be fixed under the pneu

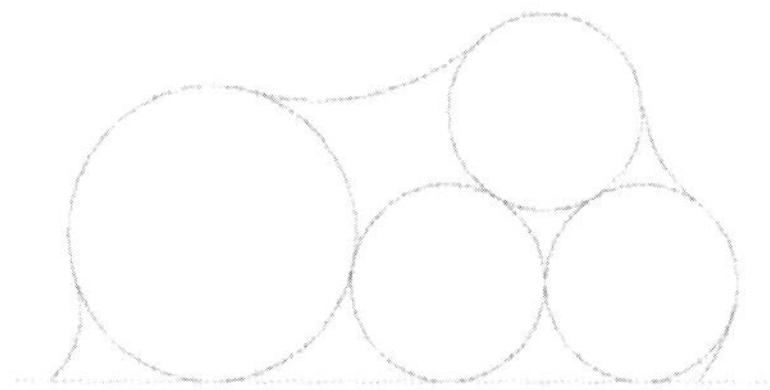

Fig. 31. To create a declining facade the tent cloth has to be fixed under the pneu

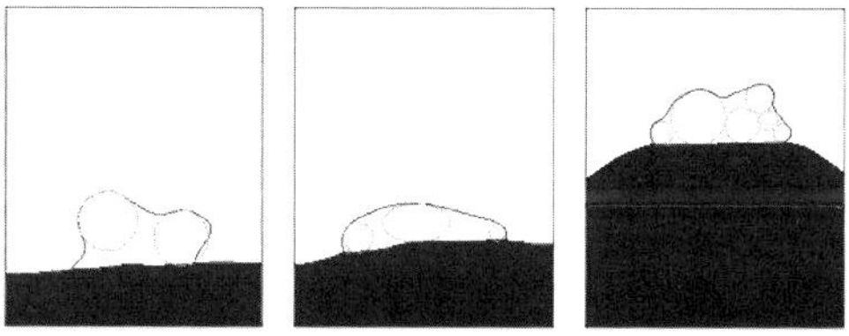

Fig. 32. Cross-sections 2, 3 en 6 with pneus

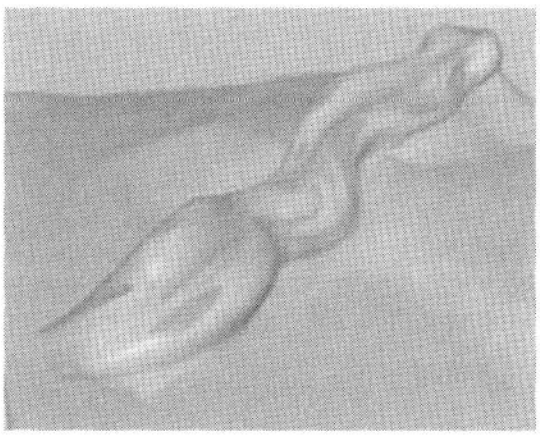

Fig. 33. Alternative 3

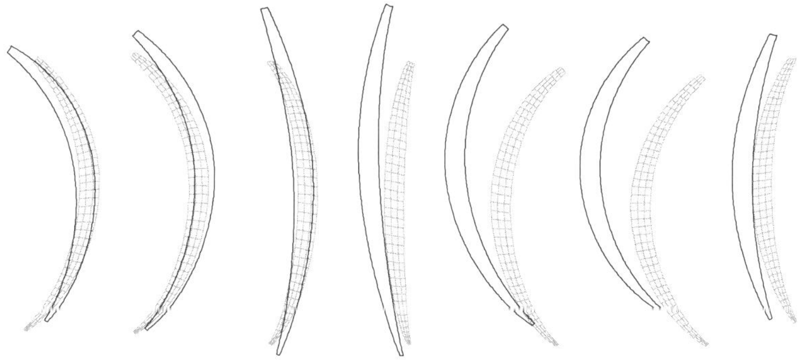

Fig. 34. Top view of 3D and 2D

5.6 Scale Model

To verify whether the proposed technology is a sensible way of working, a part of the ski run is made as a scale model. To be sure that the result of the model was close to the designed form, cutting patterns were made from the inflated structural elements. The different cutting patterns were stitched together and made airtight. Patterns from a banana, an M&M, 2 types of trunks and a part of the outer skin are made. All the different elements are grouped on the ground plan made out of cardboard (Fig. 35). The skin is put over the inflated elements and rigidized (Fig. 36).

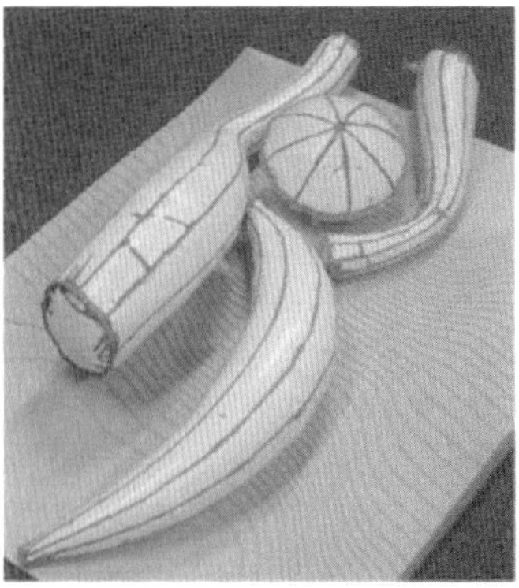

Fig. 35. Inflated structural elements

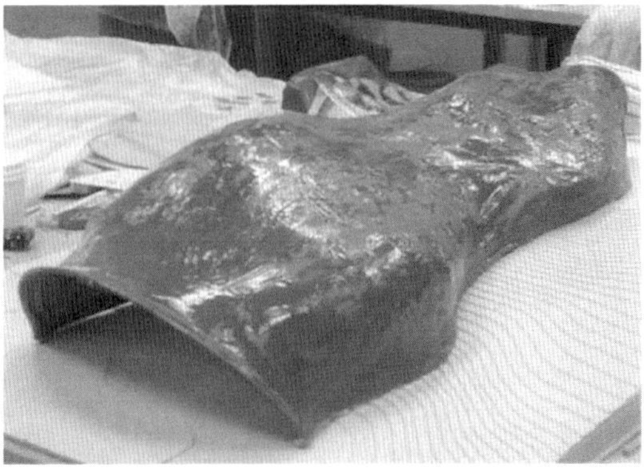

Fig. 36. Rigidized skin of Ski Run

It is hard to create geodesic lines on inflated elements because of the sinclastic curvature. There are numerous solutions for the geodesic line algorithm. Therefore the straight model out of Maya is used. An inflated element is made out of a mesh. This element is divided into separate strips. The strips follow the curvature of the element. The Ski run has quite strong curved inflated elements that results in strongly curved 3D patterns.

Because the 3D strips are not according to geodesic lines, the resulting 2D strips are very curved. When it would have been a real-sized structure, it would not have been very economical to use these kind of cutting patterns. In this case it only concerned a scale model and waste of material was not that large. For flattening the 3D strips the EASY Flatten module from the TECHNET Company was used.

5.7 Concluding Remarks

The study of a blob structure as a cover of a Ski run has resulted in a very convincing structure(in what way?). In a controlled way a design of the outer skin is made supported by inflated elements. After rigidizing the skin (by means of concrete) it is very well possible to solve occurring problems in the structural system. The section analysis of the inflated elements suggests a possible way of supporting the outer skin. Missing link here is the numerical analysis of the total system, outer skin combined with the inflated elements. This is still what has to be done. Then also will be investigated what the effect will be on the shape during the application of the rigidizing material. This study must be seen as a first step on the way of a new design and production method of Blob blowing structures.

6 The Art Pavillion

This case describes the concept and engineering design for an art Pavilion in Eindhoven Holland. The surface of the pavilion is made out of Glassfiber-reinforced Polyester. The mould of the structure is made of a PVC coated Polyester membrane.

The manipulation of form-active structures makes it possible to use membranes as a mould to make a surface-active structure. The engineering is divided in two parts. The first part describes the way the artist (Jurgen Bey) made his model and how we used this model as a starting point for the engineering.

The pavilion will be used for art exhibitions, filmprojections on the screen, as street lighting and as a piece of art. The general theme of the activities in and around the pavilion is the influence of technical innovations in art. The pavilion has two positions, vertical and horizontal. To put the pavilion in upright position, aircushions are used to raise the pavilion from a horizontal position to a vertical position (Fig. 37).

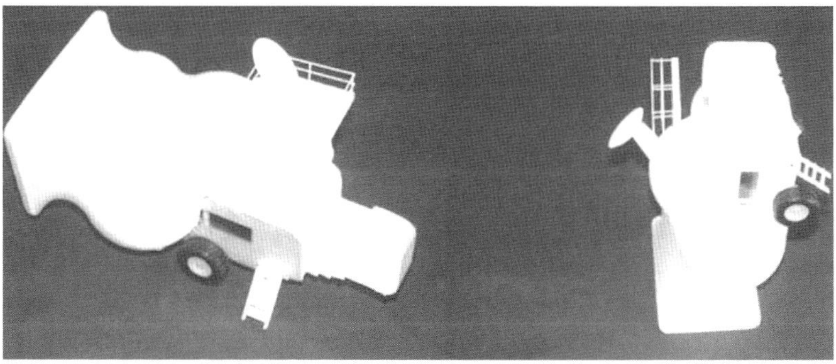

Fig. 37.

The artist was made familiar with form-active modelling. The design the artist had to conceive should be able to be realised as an inflated inner fabric that supports a tensioned outer fabric. This is the so-called Blowing Structure Method. The first design sketches were in 2D, the final design only existed out of a 3D physical model, made according to the Blowing Structure method.

From the model that is shown above (Fig. 37) a 3D scan was made. It was represented as an IGES file which made it possible to generate cross sections. This digital model with cross sections was used to analyse the form the inner and outer skin. The first form analyses were done in the program Rhino. Here the size and position of the inflated elements were derived from the 3D scan. The inflated elements were imported into the program EASY of Technet GMBH. To be able to form find a fabric over the inflated elements, a new module for the program was developed. This program is called Conformer Alpha and is based upon a sliding support system and bary-centric coordinates. This makes it possible to investigate the interaction between the inner inflated elements and the outer tensioned surface. Both inner and outer surface will need to be translated into cutting patterns for production.

Fig. 38 shows the order of the inflated elements. Fig. 39 shows the cutting patterns of the outer membrane. Fig. 40 shows a rendering of the tensioned outer skin and the inflated inner elements. Fig. 41 shows a view of the Conformer Alpha program with a part of the inner and outer skin.

The second part of the engineering was the behaviour of the surface-active structure and the materialisation. The 3D scan made from the model shown in part one

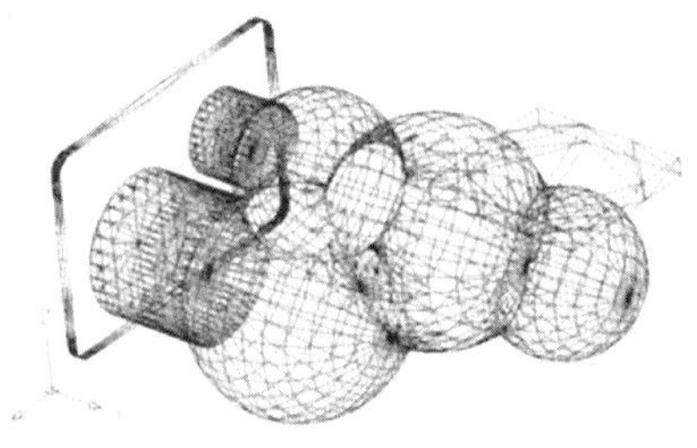

Fig. 38.

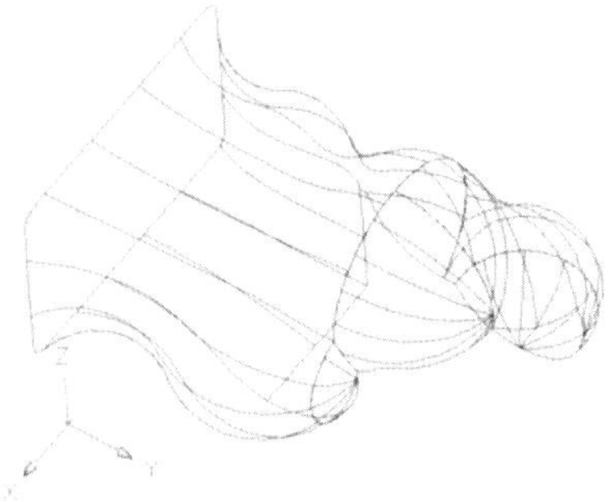

Fig. 39.

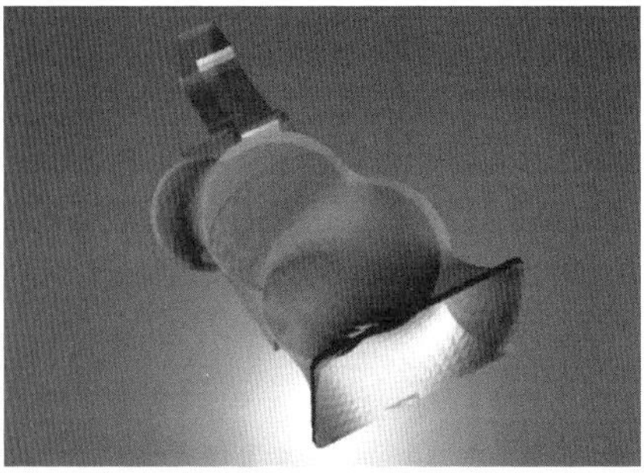

Fig. 40.

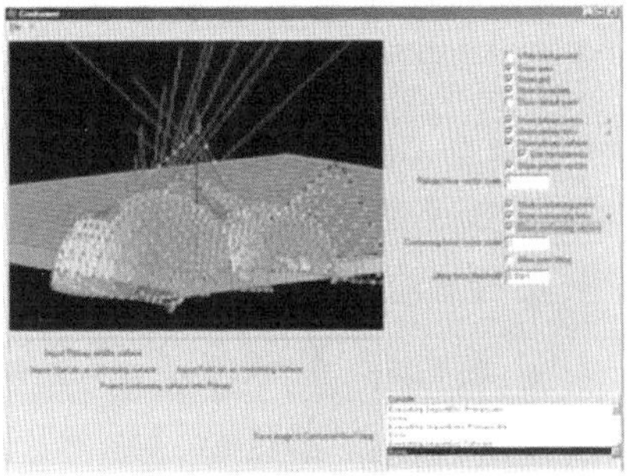

Fig. 41.

was also used for the structural analysis. This was done with the Finite Element program Marc-Mentat. Marc Mentat is able to import the 3d scan in STL format. The STL file consists out of triangular elements. In Marc it is possible to calculate with triangular elements but in order to calculate with a thick shell element, quadruple elements were needed. We didn't succeed in transforming the triangular elements into quadruple elements automatically. Therefore we had to redraw the mesh of the model by hand in Autocad. The benefit of the program Marc is that if the model works it is quite easy to calculate different types of materialisation for the skin of the pavilion. Another benefit is that the calculation time of the model is short in comparison to other Finite Element programs. The results are shown in the Figs. 42–45.

The pictures show the way the skin of the pavilion is divided in translucent and non translucent parts. Two factors determined this outcome. The first and most important was the structural behaviour of the pavilion. The second factor was the artistic input of the artist (Jurgen Bey). The yellow and orange parts are translucent, the other parts are non-translucent sandwiches. Parallel to this process different kind of materialisation experiments were made. We did al kind of materialisation experiments. For example in one experiment the behaviour of ropes in a polyester composite structure has been studied (Fig. 46).

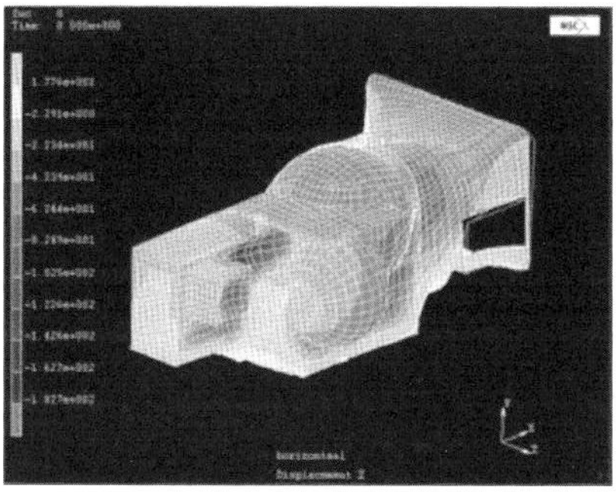

Fig. 42

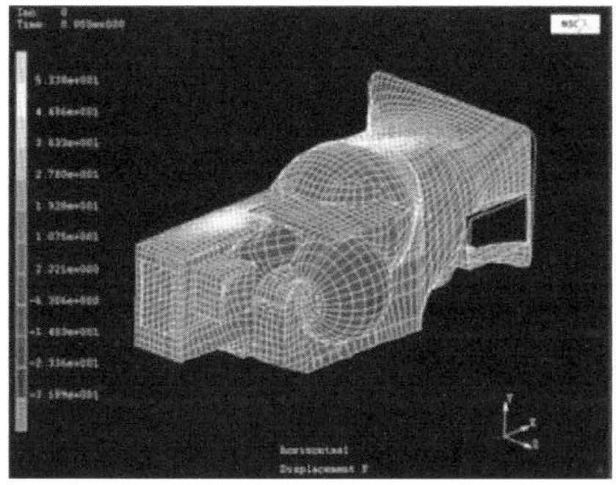

Fig. 43

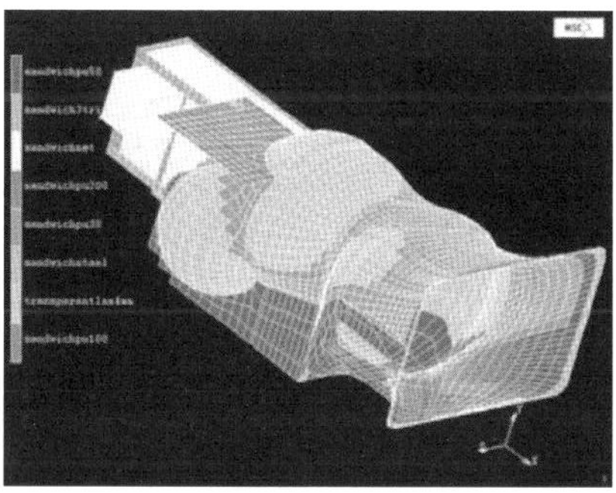

Fig. 44

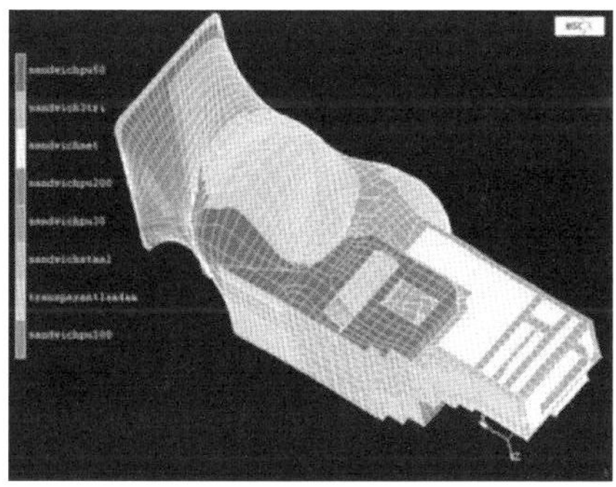

Fig. 45

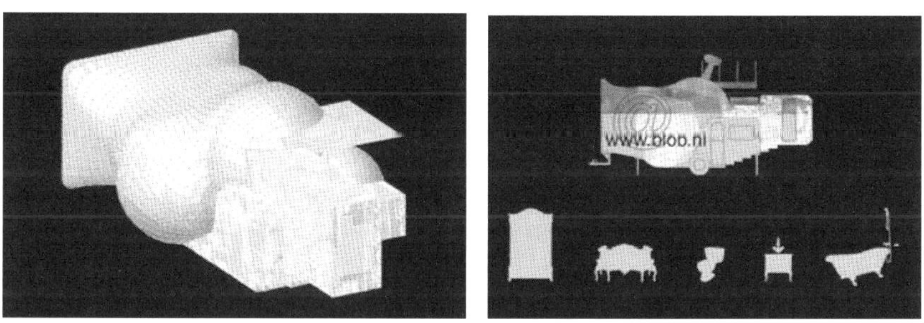

Fig. 46. Artist impression of the pavilion by Jurgen Bey

7 Conclusion

Modern sophisticated computer technologies offer new possibilities in designing free-geometrical architectural forms and geometrical complex structural elements. Pneumatic and tensile structures, in combination with bent stiff elements, play a great role in the development of non-rectangular shapes, due to their minimal dead load and great freedom of shapes. Examples are the moulds, which have been used, in the described cases of this paper.

References

1. Houtman R (1996) Van idee tot tentconstructie. Handleiding voor het ontwerpen en uitvoeren van gespannen membranen, TU Delft.
2. Werkman HA (2003) Een podiumoverkapping voor het openluchttheater in Soest. TU Delft.
3. Schaur E et al (1979) IL 21 Form – Kraft – Masse 1, "Grundlagen", Karl Krämer Verlag.
4. Pronk ADC, Veldman SL (2002) Making blobs with air-cushions. Proceedings of the International Symposium on Lightweight structures in Civil Engineering Warsaw.
5. Hoebergen A, Herpt E, van Labordus M (1999) The manufacture of large parts using the vacuum injection technique. Proceedings of the 21st International SAMPE Europe Conference, Paris, France, Apr. 13-15, SAMPE Europe.
6. Pronk ADC, Houtman R (2003) A fluid pavilion by rididizing a membrane. Textile Composites and inflatable structures, CIMNE, Barcelona.
7. Beukers A, van Hinte E (1998) Lightness: The inevitable renaissance of minimum energy structures. 010 Publishers. Rotterdam.
8. Chilton J (2000) Heinz Isler, The enigineer's contribution to contemporary Architecture. August, ISBN 0 72772878 4.
9. Engel H. (1997) Structure Systems. Stuttgart.
10. Hays KM (1994) I'm a victim of this song/good spirit come over me.
11. Veldman SL, Vermeeren CAJR (2001) Inflatable structures in aerospace engineering – An overview. Proceedings of the European conference on spacecraft structures, materials and mechanical testing, Noordwijk, the Netherlands 29 November–1 December 2000, (ESA SP-468, March 2001).